Nonparametric Statistical Methods for Complete and Censored Data

Nonparametric Statistical Methods for Complete and Censored Data

M.M. Desu
D. Raghavarao

CHAPMAN & HALL/CRC

A CRC Press Company
Boca Raton   London   New York   Washington, D.C.

Library of Congress Cataloging-in-Publication Data

Desu, M. M.
   Nonparametric statistical methods for complete and censored data / M.M. Desu and D. Raghavarao
     p. cm.
   Includes bibliographical references and index.
   ISBN 1-58488-319-7
   1. Nonparametric statistics. I. Raghavarao, Damaraju. II. Title.

QA278.8.D47 2003
519.5—dc22   2003060194

To: Aruna, Subbarao, Anu, Sheila, and Alyssa

M.M.D.

To: Lakshmi, Venkatrayudu, and Sharada

D.R.

Preface

Nonparametric statistical methods are extremely useful for researchers in biostatistics, pharmaceutical statistics, business, psychology, and social sciences. These methods are precursors for the tools used in analyzing right-censored data. Few books deal extensively with nonparametric statistical methods and pave the way to the analysis of censored data.

This book fills this gap and discusses most of the commonly used nonparametric methods for complete data and then extends those methods to censored data settings. This book can be used as a textbook for a one-semester junior-senior or first-year graduate course. It will also be a useful reference book for researchers who are analyzing censored data or complete data with nonparametric methods.

This is not a theorem–proof format book. While most of the available books are either cookbook type or highly mathematical, this book attempts to introduce the concepts intuitively with minimal mathematical statistics background. Most of the methods discussed are in relation to a univariate response variable. Methods for the analysis of complete data with binary, categorical, and continuous variables are given initially in each setting and then extended to right-censored data on a continuous response. The main text is free of difficult mathematical details, which enables the reader to follow the discussion easily and master the details. The omitted mathematical derivations and other details are given in Appendix A at the end of each chapter. These details can be mastered by individuals with one or two semesters of mathematical statistics training. To facilitate the understanding of the methods, computer programs are given in Appendix B to each chapter. These programs are written in the SAS language so they can be run on the SAS system. The coding for the programs can be found on the CRC Press website, www.crcpress.com, under electronic products/downloads/updates.

In addition to nonparametric methods for analyzing complete and censored data, this book provides excellent discussions on

1. optimal linear rank statistics

2. clinical equivalence

3. analysis of block designs

4. precedence tests

We want to thank Professor Richard N. Schmidt for his continued encouragement and his enormous help in the preparation of the manuscript. We thank our families for their encouragement and continued support.

M. M. Desu
D. Raghavarao

Contents

CHAPTER 1

Procedures for a single sample

1.1 Introduction

In this chapter we consider procedures for analyzing a random sample on the response variable X. Two cases are of interest: (1) X is binary and (2) X is continuous. First we discuss some statistical problems concerning a sample with binary data. Then we discuss procedures for dealing with data on a continuous response variable. We discuss methods for complete data, then methods for censored data situations.

1.2 Binary response

A researcher is interested in studying the effectiveness of a new drug under development. For this purpose, suppose 14 patients were recruited and treated. The researcher will be interested in further investigations of the drug if the drug is effective in more than 20% of patients. The researcher may like to know how many of the 14 treated patients should find the drug effective in order that further study is warranted. Furthermore, if 4 of the 14 treated patients found the drug effective, the researcher may like to set up a confidence interval for the probability of effectiveness of the drug. Similarly, a marketing company developed a new commercial and showed it to 30 respondents. Five people liked the commercial. The company wants to set up a confidence interval for the probability of liking this commercial. If 4 out of 30 examinees answered a question incorrectly, does this constitute evidence that 10% of the examinees answered the question incorrectly? Problems of this type also occur in other branches of research and we will discuss these issues in this section.

Consider a random experiment with only two possible outcomes. Traditionally, the outcomes are called *success* and *failure* and the experiment is usually referred to as a *Bernoulli trial*. The probability model for this Bernoulli trial is

$$P(success) = \theta, \quad and \quad P(failure) = 1 - \theta,$$

where $0 < \theta < 1$. In order to learn about θ, the *success probability*, one usually repeats such a Bernoulli trial a fixed number of times, say n, where the repetitions are independent. The entire experiment is called a *binomial experiment with n trials*. In relation to each trial we define a random variable. Suppose that X_i is the random variable denoting the outcome of the ith trial $(i = 1, 2, \ldots, n)$. The variable X_i takes the value 1, when the outcome is a "success," and the value 0, otherwise. Thus the probability model for X_i is defined by the probability function

$$f(x; \theta) = P(X_i = x) = \theta^x (1 - \theta)^{1-x}, \quad x = 0, 1, \tag{1.1}$$

where $0 < \theta < 1$. The data are the set of observations on the random variables $X_1, X_2, \ldots, X_n$, where these variables are i.i.d. (independent and identically distributed) random variables with the common distribution defined by the probability function $f(x; \theta)$ given in (1.1). This common distribution is called the Bernoulli distribution with the parameter θ and the data is called a random sample, of size n, from a Bernoulli distribution.

The two statistical problems of interest are: (1) the estimation of θ (point estimation and interval estimation, and (2) testing a hypothesis about the value of θ. The researcher also may be interested in determining n, the sample size to meet the objectives of the study.

1.2.1 Estimation of success probability

The point estimate can be obtained from the maximum likelihood method. It is known that the maximum likelihood estimate of θ is the proportion of successes, i.e.,

$$\hat{\theta} = \Sigma_i X_i / n = S_n / n = \bar{X}_n. \tag{1.2}$$

It should be noted that the statistic S_n denotes the number of successes.

Binomial distribution

Let X be the number of successes in a binomial experiment with n trials and probability (of success) θ. Then the probability function of X is

$$f(x; n, \theta) = P(X = x) = \binom{n}{x} \theta^x (1 - \theta)^{n-x}, \tag{1.3}$$

for $x = 0, 1, \ldots, n$. Here $0 < \theta < 1$.

We denote such a variable X by $Bin(n, \theta)$ and X is said to have the binomial distribution. Sometimes the parameter n is called the *index* and the parameter θ is called the *probability*. The probability function (1.3) reduces to the probability function (1.1) of the Bernoulli distribution when $n = 1$. In later sections we need to use the cumulative distribution function (cdf) of the binomial distribution and so we note some results about the cdf. For real x, the cdf F is

$$F(x; n, \theta) = P(X \leq x).$$

Clearly

$$F(x; n, \theta) = \begin{cases} 0, & \text{for } x < 0, \\ 1, & \text{for } x \geq n. \end{cases}$$

However, for $0 \le x < n$, we have

$$F(x; n, \theta) = \sum_{i=0}^{j} \binom{n}{i} \theta^i (1 - \theta)^{n-i}, \tag{1.4}$$

where j is the integral part of x. This sum can be related to an incomplete beta function, which is an integral.

Incomplete beta function

The *incomplete beta function* $I(x; a, b)$ is defined for positive constants a and b and for $0 \le x \le 1$ as

$$I(x; a, b) = \frac{\Gamma(a + b)}{\Gamma(a)\Gamma(b)} \int_0^x u^{a-1} (1 - u)^{b-1} du,$$

where $\Gamma(.)$ is the usual gamma function. It may be noted that $I(1; a, b) = 1$, and $I(0; a, b) = 0$. It can be shown that, for $0 \le x < n$,

$$F(x; n, \theta) = (n - j) \binom{n}{j} \int_0^{1-\theta} u^{n-j-1} (1-u)^j du = I(1-\theta; n-j, j+1), \tag{1.5}$$

where j is the integral part of x. The proof concerning the integral representation appears in Appendix A1. From the integral representation (1.5), it is easy to see that the cdf of the binomial distribution is a decreasing function of θ. We also note that

$$E[Bin(n, \theta)] = n\theta, \quad and \quad var[Bin(n, \theta)] = n\theta(1 - \theta). \tag{1.6}$$

We recall that the statistic S_n follows the binomial distribution with parameters n and θ. Hence from (1.6), it follows that

$$E(\hat{\theta}) = \frac{1}{n} E[Bin(n, \theta)] = \theta. \tag{1.7}$$

So $\hat{\theta}$ is an unbiased estimator of θ. Further,

$$var(\hat{\theta}) = \frac{1}{n^2} var[Bin(n, \theta)] = \theta(1 - \theta)/n. \tag{1.8}$$

For the construction of a confidence interval we need an estimate of this variance. An unbiased estimator of this variance is

$$v^2 = [\hat{\theta}(1 - \hat{\theta})]/(n - 1).$$

In large samples, the distribution of

$$Z = (\hat{\theta} - \theta)/v$$

can be approximated by the standard normal distribution. Using this result, a $100(1 - \alpha)\%$ confidence interval for θ is (θ_l, θ_u), where

$$\theta_l = \hat{\theta} - z_{1-\alpha/2} \cdot v, \quad and \quad \theta_u = \hat{\theta} + z_{1-\alpha/2} \cdot v, \tag{1.9}$$

with z_p the $100p$ percentile of the standard normal distribution.

A detailed discussion about the confidence intervals is given in Subsection 1.2.7.

1.2.2 Testing one-sided hypotheses about θ

First we consider the problem of testing the simple null hypothesis

$$H_0 : \theta = \theta_0, \tag{1.10}$$

against the simple one-sided alternative hypothesis

$$H_A : \theta = \theta_1 (> \theta_0). \tag{1.11}$$

The Neyman-Pearson lemma can be used to get the most powerful test. This test is to

$$reject\ H_0\ if\ S_n \geq C_+, \tag{1.12}$$

where the constant C_+ is chosen so that

$$P(type\ I\ error) \leq \alpha.$$

In other words, C_+ is the smallest integer such that

$$P(S_n \geq C_+ \mid \theta_0) = P(Bin(n, \theta_0) \geq C_+) \leq \alpha. \tag{1.13}$$

In some applications, it is appropriate to use the composite version of (1.11), which is

$$H_+ : \theta > \theta_0. \tag{1.14}$$

For this problem we also use the test (1.12), since the critical value C_+ depends only on θ_0, not on θ_1.

The most general problem is concerned with testing the composite null hypothesis

$$H_0^* : \theta \leq \theta_0 \tag{1.15}$$

against the composite (one-sided) alternative hypothesis H_+ of (1.14). It turns out that the test (1.12) is also the most powerful test for this general testing problem. This assertion follows from Theorem 8.3.2 of Casella and Berger (1990).

Now let us consider testing the null hypothesis (1.10) against the other one-sided alternative hypothesis,

$$H_- : \theta < \theta_0. \tag{1.16}$$

Table 1.1 *Tests for one-sided alternatives*

Null Hypothesis	Alternative Hypothesis	Critical Region
$H_0 : \theta = \theta_0$	$H_+ : \theta > \theta_0$	$S_n \geq C_+$
$H_0^* : \theta \leq \theta_0$	$H_+ : \theta > \theta_0$	$S_n \geq C_+$
$H_0 : \theta = \theta_0$	$H_- : \theta < \theta_0$	$S_n \leq C_-$
$H_0^{**} : \theta \geq \theta_0$	$H_- : \theta < \theta_0$	$S_n \leq C_-$

We also need to consider the more general problem of testing

$$H_0^{**} : \theta \geq \theta_0 \tag{1.17}$$

against the alternative H_- of (1.16). An analysis similar to the above gives the test. This test is to

$$reject\ the\ null\ hypothesis\ if\ S_n \leq C_-, \tag{1.18}$$

where C_- is the largest integer such that

$$P(S_n \leq C_- \mid \theta_0) = P(Bin(n, \theta_0) \leq C_-) \leq \alpha. \tag{1.19}$$

A summary of the one-sided tests appears in Table 1.1.

A computer program for obtaining the critical values, C_+ and C_-, is given in Appendix B1. However, we can approximate the distribution of

$$Z(\theta) = \frac{S_n - n\theta}{\sqrt{n\theta(1 - \theta)}} \tag{1.20}$$

by the standard normal distribution, when $min\{n\theta, n(1 - \theta)\} \geq 5$. Using this result, we obtain approximations to the critical values C_+ and C_-. Starting from equation (1.13), and using the continuity correction, we have the condition

$$P(S_n \geq C_+ - 0.5 \mid \theta_0) \leq \alpha.$$

In turn, this condition is the same as

$$P\left(\frac{S_n - n\theta_0}{\sqrt{n\theta_0(1 - \theta_0)}} \geq C_+^* \right) \leq \alpha,$$

where

$$C_+^* = \frac{C_+ - 0.5 - n\theta_0}{\sqrt{n\theta_0(1 - \theta_0)}}.$$

The condition on the probability can be restated as

$$P(Z(\theta_0) \leq C_+^*) \geq 1 - \alpha.$$

Under H_0, a normal approximation can be used for the distribution of the statistic $Z(\theta_0)$. So we can satisfy the above condition by choosing C_+^* as

$z_{(1-\alpha)}$, the $100(1-\alpha)$ percentile of the standard normal distribution. Thus an approximation to C_+ is

$$C_+ \approx n\theta_0 + 0.5 + z_{(1-\alpha)}\sqrt{n\theta_0(1-\theta_0)}.$$

Since we want an integer value for C_+, this approximation is restated as

$$C_+ \approx \lfloor n\theta_0 + 0.5 + z_{(1-\alpha)}\sqrt{n\theta_0(1-\theta_0)}\rfloor + 1, \tag{1.21}$$

where $\lfloor x \rfloor$ denotes the integral part of x.

A similar analysis gives

$$C_- \approx \lfloor n\theta_0 - 0.5 + z_\alpha\sqrt{n\theta_0(1-\theta_0)}\,\rfloor. \tag{1.22}$$

Equations (1.21) and (1.22) give very good approximations, whenever $min\{n\theta_0, n(1-\theta_0\} \geq 5$. For example, when $n = 20$, $\theta = 0.25$, and $\alpha = 0.05$, the exact values are obtained using the computer program given in Appendix B1. These are $C_+ = 9$, and $C_- = 1$. From equations (1.21) and (1.22) the approximations are $C_+ \approx 9$ and $C_- \approx 1$. In this case the approximations are the same as the exact values.

1.2.3 P-values for one-sided tests

Instead of calculating the critical values and performing the test, one can compute the P-value (of the data), which is a measure of the strength of evidence against the null hypothesis, and compare it with the chosen α value. Let s be the observed value of the statistic S_n. The P-value for the test (1.12) is

$$P_+ = P(Bin(n, \theta_0) \geq s), \tag{1.23}$$

and when $min\{n\theta_0, n(1 - \theta_0\} \geq 5$, an approximation is

$$P_+ \approx \Phi[(n\theta_0 - s + 0.5)/\sqrt{n(\theta_0(1-\theta_0))}]. \tag{1.24}$$

The P-value for the test (1.18) is

$$P_- = P(Bin(n, \theta_0) \leq s), \tag{1.25}$$

and when $min\{n\theta_0, n(1 - \theta_0\} \geq 5$, an approximation is

$$P_- \approx \Phi[(s + 0.5 - n\theta_0)/\sqrt{n(\theta_0(1-\theta_0))}]. \tag{1.26}$$

In (1.24) and (1.26), $\Phi(.)$ is the cdf of the standard normal distribution. It is customary to give the P-value while reporting the results, and the computer programs usually report the P-values for tests.

We can also use the P-value for performing a test of hypothesis, as mentioned earlier. This method can be stated as follows:

Reject the null hypothesis if P-value $\leq \alpha$. $\qquad$ (1.27)

Example 1.1. The first example discussed at the beginning of Section 1.2 can be formulated as a problem of testing the null hypothesis

$$H_0^* : \theta \leq 0.2 \text{ against the alternative } H_+ : \theta > 0.2.$$

Let us take $\alpha = 0.05$. From the computer program we get $C_+ = 6$. Thus under the test (1.12), the researcher should reject H_0^* in favor of H_+ and develop the drug further when $S_{14} \geq 6$.

Suppose as indicated before that $S_{14} = 4$. The exact P-value from (1.23) is

$$P_+ = P(Bin(14, 0.2) \geq 4) = 0.3017,$$

which is obtained from the corresponding SAS function. Since $P_+ > \alpha = 0.05$, we do not reject $H_0^* : \theta \leq 0.2$.

Now we will study the power function of the test (1.12). It will be used for designing a study, which is the subject of Subsection 1.2.5.

1.2.4 Power function of one-sided tests

The power function of the test (1.12) is

$$\begin{aligned}
\pi_+(\theta) &= P(\text{rejecting } H_0 \mid \theta) \\
&= P(S_n \geq C_+ \mid \theta) \\
&= 1 - P(S_n \leq C_+ - 1 \mid \theta) \\
&= 1 - F(C_+ - 1; n, \theta) \\
&= 1 - I(1 - \theta; n - C_+ + 1, C_+) \\
&= I(\theta; C_+, n - C_+ + 1)
\end{aligned}$$

The last equality follows from the previous one by changing the variable of integration (see Appendix A1). From the integral representation, it is easy to see that this power function is an increasing function of θ. An approximation to the power function is useful for determining the size of an experiment. Using the normal approximation to the binomial distribution, an approximation to the power function is derived. Since the power function of test (1.12) is

$$\pi_+(\theta) = 1 - P(S_n \leq C_+ - 1 \mid \theta),$$

the normal approximation for the distribution of S_n, with continuity correction, gives the approximation

$$\pi_+(\theta) \approx 1 - \Phi[(C_+ - 0.5 - n\theta)/\sqrt{n\theta(1 - \theta)}].$$

Using the symmetry property of the normal cdf, we can simplify the right-hand-side expression and then we have

$$\pi_+(\theta) \approx \Phi[(n\theta + 0.5 - C_+)/\sqrt{n\theta(1 - \theta)}]. \tag{1.28}$$

Similarly, the power function of the test (1.18) can be seen to be

$$\pi_-(\theta) = I(1 - \theta; n - C_-, C_- + 1), \tag{1.29}$$

and it can be approximated as

$$\pi_-(\theta) \approx \Phi[(C_- + 0.5 - n\theta)/\sqrt{n\theta(1-\theta)}]. \tag{1.30}$$

1.2.5 Sample size

We want to set up a study with n trials to test $H_0 : \theta = \theta_0$ versus $H_+ : \theta > \theta_0$, at a significance level α. Consequently, the problem is to decide upon the sample size n so that the test, based on our study, has adequate power for all $\theta \geq \theta_1(> \theta_0)$. We want the power of the test (1.12) to be at least $1 - \beta$ for all $\theta \geq \theta_1$. In view of the monotone property of the power function, this requirement on the power is satisfied by requiring the power at θ_1 to be at least $1 - \beta$. Here, for convenience, we denote the critical value by c. Using the power function expression, the requirements are

$$\pi_+(\theta_0) \leq \alpha \,;\, \pi_+(\theta_1) \geq 1 - \beta.$$

These requirements are the same as

$$I(\theta_0; c, n - c + 1) \leq \alpha \,;\, I(\theta_1; c, n - c + 1) \geq 1 - \beta. \tag{1.31}$$

In the power function expression we had the constant C_+ and this is replaced by c for convenience. Thus we need to choose n and c so as to satisfy the inequalities (1.31).

An iterative technique is needed to find the required n and c. To start the iteration one can use approximations for n and c. Now we obtain a set of useful approximations.

Using the normal approximation with the continuity correction for the power function and changing the inequalities to equalities in (1.31), two equations are obtained. These are

$$\Phi\left[\frac{n\theta_0 + 0.5 - c}{\sqrt{n\theta_0(1-\theta_0)}}\right] = \alpha;\, \Phi\left[\frac{n\theta_1 + 0.5 - c}{\sqrt{n\theta_1(1-\theta_1)}}\right] = 1 - \beta.$$

These equations are the same as

$$\frac{n\theta_0 + 0.5 - c}{\sqrt{n\theta_0(1-\theta_0)}} = z_\alpha;\, \frac{n\theta_1 + 0.5 - c}{\sqrt{n\theta_1(1-\theta_1)}} = z_{1-\beta}.$$

Solving these equations, approximations for the required sample size and the critical value are obtained. The solution is (n^*, c^*), where

$$n^* = \frac{\left[z_\alpha\sqrt{\theta_0(1-\theta_0)} - z_{1-\beta}\sqrt{\theta_1(1-\theta_1)}\right]^2}{(\theta_1 - \theta_0)^2}$$

$$= \frac{\left[z_\alpha\sqrt{\theta_0(1-\theta_0)} + z_{(\beta)}\sqrt{\theta_1(1-\theta_1)}\right]^2}{(\theta_1 - \theta_0)^2}.$$

We can also rewrite the formula as

$$n^* = \frac{\left[z_{1-\alpha}\sqrt{\theta_0(1-\theta_0)} + z_{1-\beta}\sqrt{\theta_1(1-\theta_1)} \right]^2}{(\theta_1 - \theta_0)^2} \equiv \frac{A}{(\theta_1 - \theta_0)^2}, \qquad (1.32)$$

and

$$c^* = n^*\theta_0 + 0.5 - z_\alpha\sqrt{n^*\theta_0(1-\theta_0)}. \qquad (1.33)$$

Thus an integer approximation to the sample size is

$$n_u \approx \lfloor n^* \rfloor + 1. \qquad (1.34)$$

An integer approximation to c is

$$c_u \approx \lfloor c^* \rfloor + 1. \qquad (1.35)$$

A better approximation can be obtained using the results of Levin and Chen (1999) and these modified values will be given now. The n^* is modified as

$$n_L = \frac{n^*}{4}[1 + \sqrt{1 + 2(\theta_1 - \theta_0)/A}]^2,$$

where A is defined in (1.32) and the c^* is modified as

$$c_L = n_L\theta_0 + 0.5 - z_\alpha\sqrt{n_L\theta_0(1-\theta_0)}.$$

Using these values the modified integer approximations are

$$n_{LC} \approx \lfloor n_L \rfloor + 1 \qquad (1.36)$$

and

$$c_{LC} \approx \lfloor c_L \rfloor + 1. \qquad (1.37)$$

A computer program for doing these calculations is given in Appendix B1.
This sample size problem is the same as the problem of designing a *Phase II clinical trial* as discussed by Thall and Simon (1995). They give a table of n and c values that are solutions to (1.31). In connection with the Phase II trials, Thall and Simon indicate that reasonable values for the difference $(\theta_1 - \theta_0)$ are from 0.15 to 0.20. Now we illustrate the calculation of approximations (1.34) and (1.35) with an example.

Example 1.2. While testing $H_0^* : \theta \le 0.2$ against $H_+ : \theta > 0.2$ with $\alpha = 0.05$, the experimenter wants to have a power of 0.80 for the test, when $\theta = 0.35$. Then from (1.32), we have

$$n^* = [(1.645\sqrt{(0.2)(0.8)} + 0.84\sqrt{(0.35)(0.65)}]^2/(0.35 - 0.2)^2] = 49.81.$$

Using this value in (1.34), we get

$$n_u \approx \lfloor 49.81 \rfloor + 1 = 50.$$

Using the n^* value in (1.33) we get the c^* value and using this c^* in equation (1.35) we get $c_u \approx 16$. Thus the rejection region of an approximate 0.05-level test is $S_{50} \geq 15$. The modified approximations will turn out to be $n_{LC} = 57$, and $c_{LC} = 17$. These are taken from the output of a computer program given in Appendix B1. With the modified solution, the error probabilities are much closer to the specifications, compared to the unmodified solution.

Another approximation to the sample size can be obtained by using the arcsine transform of the statistic $\sqrt{(S_n/n)}$. The relevant details are given in Desu and Raghavarao (1990). This derivation is assigned as Problem 2. Using this transformation, Natrella (1963) prepared a table of the n-values.

1.2.6 Testing a two-sided hypothesis about θ

Suppose we want to test the simple null hypothesis (1.10) that $\theta = \theta_0$ against the two-sided composite alternative hypothesis

$$H_A : \theta \neq \theta_0. \tag{1.38}$$

The usual test is to

$$reject\ H_0\ if\ S_n \leq c_1 \quad or \quad S_n \geq c_2, \tag{1.39}$$

where c_1 is the largest integer and c_2 is the smallest integer such that

$$P(S_n \leq c_1 \mid H_0) \leq (\alpha/2); \ P(S_n \geq c_2 \mid H_0) \leq (\alpha/2). \tag{1.40}$$

This test can be derived from the *union-intersection principle*. The details of this derivation appear in Appendix A1. Using the normal approximation for the binomial distribution we can obtain approximations for the required c_1 and c_2 values. In particular, these approximations are

$$c_1 \approx \lfloor n\theta_0 - 0.5 + z_{(\alpha/2)} \sqrt{n\theta_0(1 - \theta_0)} \rfloor,$$

and

$$c_2 \approx \lfloor n\theta_0 + 0.5 + z_{(1-\alpha/2)} \sqrt{n\theta_0(1 - \theta_0)} \rfloor + 1. \tag{1.41}$$

An important special case is the one for which $\theta_0 = 0.5$, and it will be discussed in Subsection 1.3.2.

Remark 1.1. In this case, the P-value is usually computed as $2\,min(P_+, P_-)$, where P_+ and P_- are given by (1.23) and (1.25).

1.2.7 Confidence intervals for θ

In some applications the researcher may be interested in a confidence interval for θ. We need to find two functions $\theta_L(S_n)$ and $\theta_U(S_n)$ of the random variable S_n such that

$$P(\theta_L(S_n) < \theta < \theta_U(S_n)) \geq 1 - \alpha. \tag{1.42}$$

Then the interval $(\theta_L(S_n), \theta_U(S_n))$ is a confidence interval for θ with confidence coefficient $(1-\alpha)$. These limits are usually derived from the acceptance region of the two-sided test (1.39). As a prelude to this derivation, we consider the problem of finding one-sided confidence limits or confidence bounds. From these bounds we can get a confidence interval.

An *upper confidence bound* $\theta_{UB}(S_n)$ is a function of S_n such that

$$P(\theta < \theta_{UB}(S_n)) \geq 1 - \alpha, \tag{1.43}$$

and a *lower confidence bound* $\theta_{LB}(S_n)$ is a function of S_n such that

$$P(\theta_{LB}(S_n) < \theta) \geq 1 - \alpha. \tag{1.44}$$

From these bounds we get the one-sided confidence intervals $(\theta_{LB}(S_n), 1)$ and $(0, \theta_{UB}(S_n))$. In the case of Phase II trials one wants to ensure that the response rate is not too low. A lower bound for the response rate will enable a researcher to decide to proceed or not with the development of a new drug. An upper bound for the proportion of nonconforming units will enable an engineer to accept or reject manufactured items supplied by a vendor.

Upper confidence bound

To derive an upper confidence bound consider the lower tail α-level test (1.18). Let s be the observed value of S_n. Under this test we reject the null $H_0 : \theta = \theta_0$, if $s \leq c$, where $P(S_n \leq c|\theta_0) \leq \alpha$. In other words, we reject H_0 if

$$F(s; n, \theta_0) \leq \alpha,$$

where F is the cdf of S_n. Since the cdf is a decreasing function of θ_0, we can find θ_{UB} such that

$$F(s; n, \theta_{UB}) = \alpha.$$

Note that θ_{UB} is a function of s. We also have

$$F(s; n, \theta_0) \leq \alpha, \quad for \ \theta_0 \geq \theta_{UB},$$

and

$$F(s; n, \theta_0) > \alpha, \quad for \ \theta_0 < \theta_{UB}.$$

Thus we do not reject H_0 for $\theta_0 < \theta_{UB}(s)$, where s is the observed value of S_n. As the probability of not rejecting is at least $1 - \alpha$, we have

$$P(\theta < \theta_{UB}(S_n)|\theta) \geq 1 - \alpha.$$

Thus $\theta_{UB}(S_n)$ is a $(1-\alpha)$ upper confidence bound for θ. Using the incomplete beta function representation for the cdf of S_n, the bound θ_{UB} can be seen as the solution of the equation

$$I(\theta_{UB}; S_n + 1, n - S_n) = 1 - \alpha.$$

This leads to the formula

$$\theta_{UB}(S_n) = BINV(1 - \alpha; S_n + 1, n - S_n) \qquad (1.45)$$

for $S_n < n$, where $BINV(p; a, b)$ is the $100p$ percentile of the $Beta(a, b)$ distribution. The beta percentiles are standard SAS functions. When $S_n = n$, the upper bound is taken as one.

Lower confidence bound

A similar analysis will give an expression for the *lower confidence bound* as

$$\theta_{LB}(S_n) = BINV(\alpha; S_n, n - S_n + 1) \qquad (1.46)$$

for $S_n > 0$ and for $S_n = 0$ the lower bound is taken as zero.

For example, suppose an inspector examined a sample of 100 items and found that 3 of them are defective. To decide whether or not to accept the lot, an upper bound is calculated. For these data the 95% upper bound for θ is

$$\theta_{UB}(3) = BINV(0.95; 4, 97),$$

which turns out to be 0.0757, that is, 7.57%. So, in the worst case scenario, the percentage of nonconforming units could be as high as 7.57%. If this percentage is larger than the acceptable percentage, the lot would be rejected.

For example, suppose S_{14} is 4. Then the lower bound is

$$\theta_{LB}(4) = BINV(0.05; 4, 11) = 0.104.$$

The researcher will proceed further only if there is evidence that θ is at least 0.2. Because this lower bound is less than 0.2, further development of the drug will not be pursued.

Exact confidence limits

Let $\theta_L(S_n)$ be the lower $(1 - \alpha/2)$ confidence bound and $\theta_U(S_n)$ be the upper $(1 - \alpha/2)$ confidence bound. Then we have

$$P[\theta_L(S_n) < \theta < \theta_U(S_n)] = 1 - P(\theta \leq \theta_L(S_n)) - P(\theta_U(S_n) \geq \theta).$$

However,

$$P(\theta_L(S_n) < \theta) \geq (1 - \alpha/2) \Rightarrow -P(\theta \leq \theta_L(S_n)) \geq -(\alpha/2)$$

and

$$P(\theta < \theta_U(S_n)) \geq (1 - \alpha/2) \Rightarrow -P(\theta_U(S_n) \geq \theta) \geq -(\alpha/2).$$

Hence

$$P[\theta_L(S_n) < \theta < \theta_U(S_n)] \geq 1 - (\alpha/2) - (\alpha/2) = 1 - \alpha.$$

In other words $(\theta_L(S_n), \theta_U(S_n))$ is a $(1 - \alpha)$ confidence interval for θ. Thus, for $0 < S_n < n$ these confidence limits are given by

$$\theta_L(S_n) = BINV(\alpha/2; S_n, n - S_n + 1);$$

and

$$\theta_U(S_n) = BINV(1 - \alpha/2; S_n + 1, n - S_n). \tag{1.47}$$

If $S_n = 0$, the lower limit is taken as zero and if $S_n = n$, the upper limit is taken as 1. Several tabulations of these limits have been made. One reference is the set of tables edited by Lentner (1982).

Example 1.3. Suppose we observed 3 successes in 20 trials. We want to find a 95% confidence interval for the parameter θ, the success probability. The point estimate $\hat{\theta} = (3/20) = 0.15$. Using the formula (1.47) and a SAS program (see Appendix B1), we get the confidence limits $\theta_L = 0.0321$ and $\theta_U = 0.3789$. In other words, a 95% confidence interval for the parameter θ is $(0.0321, 0.3789)$.

Confidence limits using the asymptotic distribution

For large or moderate n, the confidence limits are usually derived using the normal approximation to the distribution of $\hat{\theta}$. In elementary textbooks the interval

$$(\hat{\theta} - z_{(1-\alpha/2)} \cdot v_1, \ \hat{\theta} + z_{(1-\alpha/2)} \cdot v_1) \tag{1.48}$$

is suggested as a confidence interval where $\hat{\theta}$ is given by (1.2) and $v_1^2 = [\hat{\theta}(1 - \hat{\theta})]/n$, a biased estimator of $var(\hat{\theta})$.

Samuels and Lu (1992) give a set of guidelines for deciding the situations when this interval provides a good answer.

Ghosh (1979) has investigated and recommended a method that is as simple as the above method for constructing a confidence interval and as good as the exact method. We give the result here and the method of derivation is relegated to Problem 1. This confidence interval for θ is

$$((\hat{\theta} + C - z_{(1-\alpha/2)} \cdot v_*)/(1 + 2C), (\hat{\theta} + C + z_{(1-\alpha/2)} \cdot v_*)/(1 + 2C)), \tag{1.49}$$

where

$$C = (z_{(1-\alpha/2)})^2/2n; \ v_*^2 = [\hat{\theta}(1 - \hat{\theta}) + (C/2)]/n = v_1^2 + (C/2n).$$

Recent studies of Agresti and Coull (1998) and Newcombe (1998) reinforced the recommendation of the interval (1.49). Also see Agresti and Caffo (2000) for further discussion on the confidence interval of θ. A computer program for calculating the interval (1.49) is given in Appendix B1.

For example, for $n = 20$ and $s = 3$ the 95% confidence interval (1.49) is $(0.0523, 0.3604)$.

Using a confidence interval for testing

A confidence interval can be used to test the simple null hypothesis (1.10) that $\theta = \theta_0$ against the two-sided alternative (1.38). The corresponding test is to

$$\text{reject } H_0 \text{ if } \theta_0 \text{ is not in the interval.} \qquad (1.50)$$

1.3 Complete data on continuous responses

In some studies the response variable can be viewed as a continuous random variable. In reliability studies and in clinical trials the response variable is *time to an event*. It may be the time to first breakdown of a machine or time to death of a patient with terminal cancer. In these studies we want to estimate some characteristics of the distribution of the variable of interest. Suppose we are interested in studying the properties of lifetime distributions.

Dunsmore (1974) obtained data on time to first breakdown for 20 machines. This set of 20 machines is viewed as a random sample. The time to first breakdown is the response variable. The data obtained here are observations on i.i.d. random variables $X_1, X_2, \ldots, X_n$, where the common probability distribution is defined by some probability density function $f(x)$. We have very limited knowledge about the density function. The objective is to estimate some characteristics of the (population) distribution of the time to first breakdown.

We assume that the probability distribution has a unique median. We want to estimate this median, which is usually used as a measure of location. In some cases we may want to test a hypothesis about the median. For example, a social scientist may be interested in testing that the median annual family income in a county is \$25,000 based on a random sample of family annual income data. This testing problem is also of interest in the evaluation of a cancer treatment, where the efficacy of a treatment is characterized by the median survival time. In clinical studies, observations on some subjects frequently are not complete, since different subjects enter the study at different times and for some subjects the event did not occur before the end of the study. These incomplete observations are called *right-censored observations*. In this section we discuss the results for complete data situations. Some generalizations for the censored data cases are discussed in Section 1.4.

1.3.1 Point estimation of the median

We have observations on $(X_1, X_2, \ldots, X_n)$, a random sample from the distribution defined by the pdf $f(.)$. The cdf of the population distribution is $F(.)$. Using these data we want to estimate the median, ξ, and test a hypothesis about the median.

The intuitive choice for the point estimator is the sample median. To give an expression for the sample median we need the order statistics of the sample. These are the sample values arranged in increasing order of magnitude.

We denote these order statistics by $X_{(1)}, X_{(2)}, \ldots, X_{(n)}$, where $X_{(1)} < X_{(2)} < \ldots < X_{(n)}$. A point estimate of ξ is $\hat{\xi}$, the sample median, and is defined as

$$\hat{\xi} = \begin{cases} X_{(k+1)}, & \text{if } n = 2k + 1, \\ (X_{(k)} + X_{(k+1)})/2, & \text{if } n = 2k. \end{cases}$$

The sampling distribution of this statistic depends on the population distribution in a complicated way. However, some properties of this estimator can be obtained by making certain assumptions about the population density function. Desu and Rodine (1969) showed that for symmetric densities, the sample median is an unbiased estimator of the population median, which is equal to the population mean. The interested reader is referred to their paper for the proofs and other details.

Sometimes one order statistic $X_{(s)}$ is used as an estimator of the median, where s is the integer $\lfloor (n/2) \rfloor + 1$. Further discussion along these lines appears in Subsection 1.3.6.

We first discuss the testing problem and then proceed to the problem of finding a confidence interval for the median ξ. This discussion can be carried out without any restrictions on the population distribution.

1.3.2 Sign test for testing a simple null hypothesis about the median

Let ξ be the population median. Consider the case of testing the null hypothesis

$$H_0 : \xi = \xi_0 \tag{1.51}$$

against the one-sided alternative

$$H_{A1} : \xi > \xi_0. \tag{1.52}$$

From the definition of the population median it is clear that $P(X_i > \xi) = 1/2$ or $P(X_i - \xi > 0) = 1/2$. Let $\theta = P(X_i - \xi_0 > 0)$. It follows that $\theta = 1/2$ or $> 1/2$ depending on whether (1.51) or (1.52) is true. Thus this hypothesis testing problem can be translated into a testing problem in relation to a binary data set. This translation will be explained now. We transform the data by defining

$$Z_i = \begin{cases} 1, & \text{if } X_i - \xi_0 > 0, \\ 0, & \text{otherwise.} \end{cases}$$

Denoting the $P(Z_i = 1)$ by θ, and using Zs, the statistical problem can be restated as that of testing the null hypothesis

$$H_0 : \theta = 1/2, \; \text{against the alternative } H_+ : \theta > 1/2. \tag{1.53}$$

From the discussion in Subsection 1.2.2, it is obvious that we can use the test defined by the critical region (1.12). Here the test statistic S_n is equal to $\sum_i Z_i$. In other words, S_n stands for the number of X values that are greater than ξ_0. The critical region of the test is

$$S_n \geq C_+,$$

where the constant C_+ is the smallest integer such that

$$P(Bin(n, 1/2) \geq C_+) \leq \alpha. \tag{1.54}$$

For $n \geq 10$, we can approximate C_+ of (1.54) as

$$C_+ \approx \lfloor (n/2) + 0.5 + z_{1-\alpha}\sqrt{(n/4)} \rfloor + 1. \tag{1.55}$$

Now let us consider the problem of testing the null hypothesis (1.51) against the other one-sided alternative,

$$H_{A2} : \xi < \xi_0. \tag{1.56}$$

This problem is equivalent to testing the null hypothesis of (1.53) against the other one-sided alternative,

$$H_- : \theta < 1/2. \tag{1.57}$$

Clearly this testing problem can be handled by the test defined by the critical region (1.18).

Suppose the alternative hypothesis is a two-sided one, namely,

$$H_A : \xi \neq \xi_0. \tag{1.58}$$

This testing problem is equivalent to testing

$$H_0 : \theta = 1/2 \text{ against the alternative } H_A : \theta \neq 1/2. \tag{1.59}$$

The relevant test for this two-sided alternatives case is to

$$\text{reject } H_0 \text{ if } S_n \leq C, \quad \text{or} \quad S_n \geq n - C, \tag{1.60}$$

because the null distribution of S_n is symmetrical under H_0. Here C is the largest integer such that

$$P(Bin(n, 1/2) \leq C) \leq (\alpha/2). \tag{1.61}$$

Using the table from MacKinnon (1964), we can obtain this C value. For $n \geq 10$, using the normal distribution approximation to the distribution of S_n, C can be approximated as

$$C \approx \lfloor (n/2) - 0.5 + z_{(\alpha/2)}\sqrt{(n/4)} \rfloor. \tag{1.62}$$

Example 1.4. For $n = 10$ and $\alpha = 0.05$, from (1.62) we have

$$C \approx \lfloor 5 - 0.5 - 1.96\sqrt{(2.5)} \rfloor = 1.$$

Using the computer program given in Appendix B1, we find that $C = 1$. Here the normal approximation and the exact value for C coincide.

These procedures can be adopted easily to test hypotheses about other percentiles. Since the Z is 1 or 0 depending on whether the difference $(X - \xi)$ is positive or negative, these tests are sometimes referred to as *sign tests*.

A distribution-free confidence interval for the median ξ

This interval can be obtained from the acceptance region of the two-sided test (1.60). Assume that for the given sample size n and the confidence coefficient $(1 - \alpha)$, the constant $C(> 0)$ satisfying (1.61) exists. Let $d = C + 1$. Then the acceptance region of the two-sided test (1.60) can be seen to be

$$\{\xi_0 : d \leq \Sigma_i Z_i \leq n - d\}. \tag{1.63}$$

This means that the number of X-values greater than ξ_0 is at least d and not more than $n - d$. Thus a $100\,(1 - \alpha)\%$ confidence interval for ξ is $[X_{(d)}, X_{(n-d+1)})$, where $X_{(i)}$'s are the order statistics of the sample. Van der Parren (1970) published a table of d-values, which can be used for constructing the confidence intervals. This table also gives the exact coverage probability, which is not available in the table of MacKinnon (1964).

Remark 1.2. In this discussion it is implicitly assumed that there are no ties. When there are tied values in the sample, a modification of this procedure is needed. This modification is given in Subsection 1.3.6.

Example 1.5. Dunsmore (1974) observed 20 machines and obtained data on times (in hours) to first breakdown. We consider only 10 observations. These are

$$18, 23, 29, 409, 24, 74, 13, 62, 46, \text{ and } 4.$$

The order statistics, $X_{(i)}$, of the sample can be seen to be

$$4, 13, 18, 23, 24, 29, 46, 62, 74, \text{ and } 409.$$

The sample median is $[X_{(5)} + X_{(6)}]/2 = 26.5$, which is a point estimate of ξ, the population median. In addition, we want a 95% confidence interval for the median ξ. Here $n = 10$, and $\alpha = 0.05$. In Example 1.4, we found that $C = 1$ and hence $d = 2$. Thus the confidence interval is $[X_{(2)}, X_{(9)})$. Hence a 95% confidence interval for the median ξ is $[13, 74)$.

1.3.3 Estimation of the cdf

Sample distribution function plays an important role in the analysis of continuous response data. It can be used to obtain estimates of certain probabilities of interest and from it we can also obtain a confidence band for the population distribution function.

Suppose that our sample is $(X_1, X_2, \ldots, X_n)$. The sample distribution function (or empirical distribution function) denoted by $F_n(x)$ is defined as

$$F_n(x) = \{number\ of\ X\ values\ which\ are\ \leq x\}/n$$

$$= \sum_{h=1}^{n} u(X_h, x)/n, \qquad (1.64)$$

where $u(a, b) = 1$, if $a \leq b$ and $= 0$, otherwise. It may be noted that this function depends on the sample values; however, our notation does not indicate this fact.

In general, if X is our response variable, the probability $P(X \leq x) = F(x)$ is estimated by $F_n(x)$, for each real x. In other words, for each real x,

$$\hat{F}(x) = F_n(x). \qquad (1.65)$$

Some properties of this estimator are noted for future use. For a fixed x, the statistic $nF_n(x)$ follows a *binomial distribution* with parameters n and $F(x)$. So it follows that

$$E(F_n(x)) = E(nF_n(x))/n = F(x), \quad var(F_n(x)) = F(x)(1 - F(x))/n.$$

By identifying $F(x)$ as θ, $nF_n(x)$ as S_n, and $F_n(x)$ as $\hat{\theta}$ in relation to the binary data setting of Section 1.2, we can find an exact or asymptotic confidence interval for $F(x)$. Let $v^2(x)$ be the unbiased estimator of the $var(F_n(x))$, so that

$$v^2(x) = F_n(x)(1 - F_n(x))/(n - 1). \qquad (1.66)$$

For large n, the distribution of $F_n(x)$ can be approximated by a normal distribution and using this approximate distribution, it can be seen that

$$(\,F_n(x) - z_{(1-\alpha/2)} \cdot v(x), F_n(x) + z_{(1-\alpha/2)} \cdot v(x)\,) \qquad (1.67)$$

is a confidence interval for $F(x)$ and the associated confidence coefficient is approximately equal to $(1 - \alpha)$.

Example 1.5 (cont'd.). From the Dunsmore data of Example 1.5, suppose we want to estimate the probability that the time to first breakdown is not greater than 46 hours. This probability is

$$P(X \leq 46) = F(46).$$

So it can be estimated by $F_n(46) = (7/10) = 0.7$. Now let us compute a confidence interval for $F(46)$. We first compute

$$v^2(46) = (7/10)(3/10)/9 = 0.0233,$$

and then

$$z_{0.975} \cdot v(46) = 1.96(.1527) = 0.2994.$$

Using the formula (1.67), we get a confidence interval for $F(46)$ as $(0.7 - 0.2994, 0.7 + 0.2994)$. In other words an approximate 95% confidence interval for $F(46)$ is $(0.4006, 0.9994)$.

1.3.4 Estimation of survival function

In reliability or survival studies, the researcher is interested in estimating the probability of surviving beyond x. This probability is

$$S(x) = P(X > x) = 1 - F(x),\qquad(1.68)$$

and this function is called the *survival function*. A natural estimator of $S(x)$ is

$$\hat{S}(x) = 1 - \hat{F}(x) = 1 - F_n(x) \equiv S_n(x),\qquad(1.69)$$

It is easy to verify that $S_n(x)$ in the above equation is the proportion of x values that are greater than x. This function is called the *sample survival function*.

Let us examine this estimator of $S(x)$ in more detail so that we can generalize this result for censored data. Let $Y_1 < Y_2 \ldots < Y_r$ be the distinct ordered values of the random sample of size n and let d_i be the number of times Y_i occurs in the sample. Recursively define $n_1 = n$ and $n_i = n_{i-1} - d_{i-1}$, for $i = 2, 3, \ldots r$. Note that n_i are the number of observations $\geq Y_i$. From (1.69), we have

$$\hat{S}(x) = \begin{cases} 1, & \text{for } x < Y_1, \\ 1 - \dfrac{\sum_{i=1}^{j} d_i}{n}, & \text{for } Y_j \leq x < Y_{j+1}, j = 1, 2, \ldots, r - 1, \\ 0, & \text{for } x \geq Y_r. \end{cases}\qquad(1.70)$$

Noting that

$$1 - \frac{d_1 + d_2}{n} = \left(1 - \frac{d_1}{n_1}\right)\left(1 - \frac{d_2}{n_2}\right),$$

we get

$$1 - \frac{\sum_{i=1}^{j} d_i}{n} = \prod_{i=1}^{j}\left(1 - \frac{d_i}{n_i}\right).$$

Thus the expression (1.70) can be rewritten as

$$\hat{S}(x) = \begin{cases} 1, & \text{for } x < Y_1, \\ \Pi_{\{i:Y_i \leq x\}}\left(1 - \frac{d_i}{n_i}\right), & \text{for } x \geq Y_1. \end{cases}\qquad(1.71)$$

This equation means that the estimated survival probability is the product of the probabilities of surviving in the Y-intervals preceding x.

It is easy to see that for fixed x,

$$E(\hat{S}(x)) = 1 - E(F_n(x)) = 1 - F(x) = S(x),$$

and

$$var(\hat{S}(x)) = var(F_n(x)) = F(x)[1 - F(x)]/n = S(x)(1 - S(x))/n. \quad (1.72)$$

For large n, a confidence interval for $S(x)$ with confidence coefficient $1 - \alpha$ is

$$\left(S_n(x) - z_{(1-\alpha/2)} \cdot v(x), S_n(x) + z_{(1-\alpha/2)} \cdot v(x)\right), \quad (1.73)$$

where $v^2(x)$ is given by (1.67).

Example 1.6. For the Dunsmore data of Example 1.5, suppose we want to estimate the probability that the first breakdown occurs after 46 hours. This probability is $S(46) = 1 - F(46)$. Thus $S_n(46) = 1 - F_n(46) = 0.3$. It is easy to see that an approximate 95% confidence interval for $S(46)$ is $(0.3 - 0.2994, 0.3 + 0.2994)$. In other words, the required confidence interval is $(0.0006, 0.5994)$.

Remark 1.3. Since $F(x)$ and $S(x)$ are probabilities we can use the exact methods of Subsection 1.2.7 for constructing the confidence intervals. Here we only give the large sample methods, since these generalize to the case of censored data.

1.3.5 Point estimation of population percentiles

For each positive fraction p, ξ_p is called the population $100p$ percentile if

$$P(X \le \xi_p) = p, \quad i.e. \ F(\xi_p) = p. \quad (1.74)$$

This percentile can also be defined as

$$S(\xi_p) = 1 - p. \quad (1.75)$$

It is easy to see that the population median is $\xi_{0.5}$. The above implicit definition can be reworded as

$$\xi_p = F^{-1}(p) = S^{-1}(1 - p).$$

The inverse function of F is called the population *quantile function* and is denoted by $Q(.)$. In other words, for $0 < p < 1$,

$$Q(p) \equiv F^{-1}(p) = \xi_p.$$

In connection with the estimation of population percentiles, the inverse of the sample distribution function is useful. This function is denoted by $Q_n(p)$ and is called the *sample quantile function*. For each positive fraction p, it is defined as

$$Q_n(p) \equiv F_n^{-1}(p) = \inf\{x : F_n(x) \ge p\}, \quad (1.76)$$

where $F_n(.)$ is the sample distribution function. This definition of the inverse function is needed since F_n is a step function. This simply means that $Q_n(p)$ is the smallest x-value such that $F_n(x)$ is not less than p for the first time. Let $j = \lfloor np \rfloor$. If there are *no ties* in the sample, it is easy to see that for $0 < p < 1$,

$$Q_n(p) = \begin{cases} X_{(j)}, & \text{if } np = j; \\ X_{(j+1)}, & \text{if } np > j, \end{cases}$$

where $X_{(j)}$ is the jth order statistic. This definition results in one order statistic and it is generally used in asymptotic discussions. This $Q_n(.)$ function is used for estimating the percentiles. A point estimate of the $100p$ percentile ξ_p is

$$\hat{\xi}_p = Q_n(p).$$

This estimate can also be expressed in terms of the sample survival function, $S_n(.)$. It is easy to see that

$$\begin{aligned} \hat{\xi}_p &= Q_n(p) \\ &= \inf\{x : F_n(x) \geq p\} \\ &= \inf\{x : 1 - S_n(x) \geq p\}. \end{aligned}$$

Finally, we have

$$\hat{\xi}_p = \inf\{x : S_n(x) \leq (1 - p)\}. \tag{1.77}$$

Remark 1.4. The last expression can easily be applied to cases where the data contain some right-censored observations.

In our breakdown time example, discussed in Subsection 1.3.3, the estimate of the first quartile $\xi_{0.25}$, is $X_{(3)} = 18$ hours and the estimate of the median $\xi_{0.50}$ is $X_{(5)} = 24$ hours.

1.3.6 Confidence intervals for percentiles

Suppose we want a $100(1 - \alpha)\%$ confidence interval for the $100p$ percentile ξ_p. Let us consider the order statistics $X_{(1)} < X_{(2)} < \cdots < X_{(n)}$ of the random sample. (We are assuming that there are no ties.) These order statistics partition the real line into $(n + 1)$ intervals. We first compute the probability that ξ_p belongs to the half open interval $[X_{(i)}, X_{(i+1)})$. We have

$$\begin{aligned} P(X_{(i)} \leq \xi_p < X_{(i+1)}) &= P(\text{exactly } i \text{ values are } \leq \xi_p) \\ &= \binom{n}{i} [F(\xi_p)]^i [1 - F(\xi_p)]^{n-i} \\ &= \binom{n}{i} p^i (1 - p)^{n-i} \end{aligned}$$

Now considering the union of such succesive intervals, we get the interval $[X_{(i)}, X_{(j)})$ and

$$P(X_{(i)} < \xi_p < X_{(j)}) = P(X_{(i)} \le \xi_p < X_{(j)})$$

$$= \sum_{l=i}^{j-1} \binom{n}{l} p^l (1-p)^{n-l} \equiv C(i,j). \qquad (1.78)$$

Thus the interval $(X_{(i)}, X_{(j)})$ is a confidence interval for ξ_p, with confidence coefficient $C(i,j)$. Generally, the confidence coefficient is chosen in advance. Thus we need to choose integers i and j such that $C(i,j)$ is at least $(1-\alpha)$, the chosen confidence coefficient. In other words, we choose integers i and j so as to satisfy the condition

$$P(i \le Bin(n,p) \le j-1) \ge 1 - \alpha. \qquad (1.79)$$

Now the interval $(X_{(i)}, X_{(j)})$ will be a $100(1-\alpha)$ confidence interval for ξ_p. It is obvious that more than one pair of integers (i,j) will satisfy the condition (1.79). For some additional remarks about the choice of i and j see Appendix A1.

One choice of i and j (as given in Appendix A1) is that

$$P(Bin(n,p) < i) \le (\alpha/2), \quad P(Bin(n,p) > j-1) \le (\alpha/2).$$

A computer program has been developed for this purpose and is given in Appendix B1.

A method for determining a lower confidence bound is also given in Appendix A1.

Example 1.7. Suppose we want a 95% confidence interval for the first (lower) quartile, $\xi_{0.25}$. With $n = 10$, from the output of the computer program, we have $i = clower + 1 = 1$ and $j - 1 = cupper - 1 = 5$. Hence $j = 6$ and a 95% confidence interval is $(X_{(1)}, X_{(6)})$. For the data of Example 1.5, this interval is (4, 29).

Remark 1.5. In this discussion we assumed that there are no tied observations. If there are tied observations, we proceed as follows. For the integers i and j determined to satisfy (1.79), find the quantiles $\xi_p^L = Q_n(i/n)$ and $\xi_p^U = Q_n(j/n)$, where $Q_n(.)$ is the sample quantile function. The resulting confidence interval is (ξ_p^L, ξ_p^U). Further details are available in Hutson (1999).

1.3.7 Kolmogorov's goodness-of-fit test

Sometimes we want to test a simple null hypothesis about the population distribution function. In other words, the null hypothesis is

$$H_0 : F(x) = F_0(x), \qquad (1.80)$$

where F_0 is a completely specified cdf and the two-sided alternative is

$$H_A : F(x) \neq F_0(x), for\ some\ x. \tag{1.81}$$

The test proposed by Kolmogorov tells us to evaluate the closeness of the sample distribution function $F_n(x)$ to the hypothesized cdf $F_0(x)$. The suggested closeness measure is

$$D_n = sup_x[|\ F_n(x) - F_0(x)\ |]. \tag{1.82}$$

This is the test statistic and an α-level test

$$rejects\ H_0\ if\ D_n \geq C_{1-\alpha}. \tag{1.83}$$

It should be noted that the null distribution of the test statistic D_n does not depend on $F_0(x)$. So this test is a distribution-free test. Birnbaum (1952) tabulated the distribution of D_n and gave a table of the critical values for $\alpha = 0.05$ and 0.01. An extensive table of percentage points is contained in Miller (1956).

To implement the test, a convenient formula for computing the test statistic is needed. This expression for the statistic will enable us to infer that the null distribution of the statistic is independent of the distribution F_0. For simplicity, let us assume that there are no ties. We observe that the order statistics $(X_{(1)} < \cdots < X_{(n)})$ partition the real line into $(n + 1)$ intervals and the sample distribution, F_n, is constant in each of these intervals. These $(n + 1)$ intervals, which constitute a partition of the real line, are $I_0 = (-\infty, X_{(1)}), I_j = [X_{(j)}, X_{(j+1)})$, for $j = 1, \ldots, n - 1$, and $I_n = [X_{(n)}, \infty)$. First we note that

$$D_n = max_j\{sup_{x \in I_j}|F_n(x) - F_0(x)|\}.$$

Next we calculate each of the supremums. It is easy to see that

$$sup_{x \in I_0}|F_n(x) - F_0(x)| = sup|0 - F_0(x)| = F_0(X_{(1)}),$$

and

$$sup_{x \in I_n}|F_n(x) - F_0(x)| = sup|1 - F_0(x)| = 1 - F_0(X_{(n)}).$$

For $j = 1, \ldots, n - 1$, we have

$$sup_{x \in I_j}|F_n(x) - F_0(x)| = max\{(j/n) - F_0(X_{(j)}), F_0(X_{(j+1)}) - (j/n)\}.$$

Using the supremums in the $(n + 1)$ intervals we have

$$D_n = max_{0 \leq j \leq n}[max\{(j/n) - F_0(X_{(j)}), F_0(X_{(j+1)}) - (j/n)\}]. \tag{1.84}$$

We note that $F_0(X_{(0)}) = 0$ and $F_0(X_{(n+1)}) = 1$. Under the null hypothesis the joint distribution of $(F_0(X_{(1)}), \ldots, F_0(X_{(n)}))$ is the same as the joint distribution of the order statistics of a sample of size n from the uniform distribution on the interval $(0, 1)$. Thus the statistic D_n does not depend on F_0, which implies that D_n is a distribution-free statistic. The above formula (1.84) for the statistic is equivalent to

$$D_n = max\{D_n^+, D_n^-\}, \tag{1.85}$$

where

$$D_n^+ = max_{0 \le j \le n}[(j/n) - F_0(X_{(j)})], \tag{1.86}$$

and

$$D_n^- = max_{0 \le j \le n}[F_0(X_{(j+1)}) - (j/n)]. \tag{1.87}$$

The expressions for the statistics D_n^+ and D_n^- can be simplified as follows:

$$D_n^+ = max\left[0, max_{1 \le j \le n}\left\{\frac{j}{n} - F_0(X_{(j)})\right\}\right],$$

and

$$D_n^- = max\left[max_{1 \le j \le n}\left\{F_0(X_{(j)}) - \frac{(j-1)}{n}\right\}, 0\right]$$

$$= max\left[max_{1 \le j \le n}\left\{F_0(X_{(j)}) - \frac{j}{n}\right\} + (1/n), 0\right].$$

It should be noted that these expressions are valid only for data sets with no ties.

Tests for one-sided alternatives

Even though these cases are of secondary importance, the test statistic D_n turned out to be a function of the two statistics D_n^+ and D_n^-. These two statistics, D_n^+ and D_n^- are useful for testing one-sided alternatives. For the alternative

$$H_+ : F(x) > F_0(x), \quad \text{for some } x, \tag{1.88}$$

the critical region is $D_n^+ \ge C_{1-\alpha}^+$ and for the alternative

$$H_- : F(x) < F_0(x), \quad \text{for some } x, \tag{1.89}$$

the critical region is $D_n^- \ge C_{1-\alpha}^-$.

Null distributions of D_n^+ and D_n^-

Birnbaum and Tingey (1951) derived the null distribution of D_n^+, and they showed that

$$P(D_n^+ \ge c|H_0) = c\sum_{j=0}^{J}\binom{n}{j}(1-c-(j/n))^{n-j}(c+(j/n))^{j-1} \equiv \pi(c), \tag{1.90}$$

where $J = \lfloor n(1-c) \rfloor$. So $C_{1-\alpha}^+$ is the solution of the equation

$$\pi(C_{1-\alpha}^+) = \alpha.$$

The null distributions of D_n^+ and D_n^- are the same. So we have

$$C_{1-\alpha}^+ = C_{1-\alpha}^-.$$

Thus the exact P-values for the one-sided tests are

$$P\text{-value}(D_n^+) = \pi(D_n^+(obs)) \quad and \quad P\text{-value}(D_n^-) = \pi(D_n^-(obs)),$$

where $D_n^+(obs)$ and $D_n^-(obs)$ are the observed values of the test statistics.
It can be shown that

$$\lim_{n\to\infty} P(\sqrt{n} \cdot D_n^+ \geq c) = \lim_{n\to\infty} \pi\left(\frac{c}{\sqrt{n}}\right) = exp(-2c^2).$$

Thus the P-values for one-sided tests can be approximated as

$$P\text{-value}(D_n^+) \approx exp(-2n(D_n^+(obs))^2), \tag{1.91}$$

where $D_n^+(obs)$ is the observed value of the test statistic. This limiting distribution can also be used to approximate the critical values. This approximation, suggested by Birnbaum (1952), is

$$C_{1-\alpha}^+ \approx \sqrt{\frac{-ln(\alpha)/2}{n}}. \tag{1.92}$$

An improved version, suggested by Stephens (1974), is

$$C_{1-\alpha}^+ \approx \frac{\sqrt{\frac{-ln(\alpha)}{2}}}{\sqrt{n} + 0.12 + \frac{0.11}{\sqrt{n}}}. \tag{1.93}$$

Null distribution of D_n

Kolmogorov (1933) derived a system of recursive relations, which enable us to tabulate the null distribution. Massey (1950) gave a simple method of calculating the $P(D_n < k/n|H_0)$ for finite n. Birnbaum (1952) tabulated the distribution using the results of Kolmogorov. It is known that

$$lim_{n\to\infty} P(\sqrt{n}D_n < k|H_0) = 1 - 2\sum_{j=1}^{\infty}(-1)^{j-1} exp(-2j^2k^2).$$

An approximation to the P-value is taken as

$$P\text{-value}((D_n) \approx 2exp(-2n(D_n(obs))^2), \tag{1.94}$$

where $D_n(obs)$ is the observed value of the test statistic.
Starting from equation (1.85), it can be shown (see Appendix A1) that the following relation between the critical values holds:

$$C_{1-\alpha} \approx C_{1-\alpha/2}^+ \approx \frac{\sqrt{\frac{-ln(\alpha/2)}{2}}}{\sqrt{n} + 0.12 + \frac{0.11}{\sqrt{n}}}.$$

Thus the tables of $C_{1-\alpha}^+$ or the approximations (1.92) or (1.93) can be used to implement the test defined by the critical region (1.83).

Table 1.2 *Quantities needed for computing Kolmogorov statistic*

j	$X_{(j)}$	j/n	$F_0(X_{(j)})$	$(j/n) - F_0(X_{(j)})$	$F_0(X_{(j)}) - (j/n)$
1	4	0.05	0.0555	−0.0055	0.0055
2	10	0.10	0.1331	−0.0331	0.0331
3	13	0.15	0.1695	−0.0195	0.0195
4	14	0.20	0.1813	0.0187	−0.0187
5	18	0.25	0.2267	0.0233	−0.0233
6	19	0.30	0.2377	0.0623	−0.0623
7	20	0.35	0.2485	0.1015	−0.1015
8	23	0.40	0.2800	0.1200	−0.1200
9	24	0.45	0.2903	0.1597	−0.1597
10	29	0.50	0.3392	0.1608	−0.1608
11	46	0.55	0.4817	0.0683	−0.0683
12	47	0.60	0.4890	0.1110	−0.1110
13	57	0.65	0.5570	0.0930	−0.0930
14	62	0.70	0.5876	0.1124	−0.1124
15	74	0.75	0.6526	0.0974	−0.0974
16	119	0.80	0.8173	−0.0173	0.0173
17	188	0.85	0.9318	−0.0818	0.0818
18	208	0.90	0.9488	−0.0488	0.0488
19	209	0.95	0.9495	0.0005	−0.0005
20	409	1.00	0.9971	0.0029	−0.0029

Example 1.7. Dunsmore (1974) gave a data set on times (hours) to first break-down of 20 machines. Two values have been modified. The modified observations are 18, 23, 29, 409, 24, 74, 13, 62, 46, 4, 57, 19, 47, 14, 20, 208, 119, 209, 10, and 188.

It is suspected that the time to first breakdown follows an exponential distribution. So we set up the null hypothesis as $H_0 : F_0(x) = 1 - e^{-x/70}$, for $x > 0$, and we want to test this null hypothesis. The quantities $F_0(X_{(j)})$ and (j/n) will be calculated and the necessary differences will be calculated. These quantities are given in Table 1.2.

From Table 1.2, we have $D_n^+ = max(max(d_j^+), 0) = 0.1608$ and $D_n^- = max(max(d_j^-) + (1/20), 0) = 0.0818 + 0.05 = 0.1318$. Hence the test statistic $D_n = max(D_n^+, D_n^-) = 0.1608$. From (1.93), $C_{.975}^+ = 0.294$, for $n = 20$, so that this value is equal to $C_{.95}$. Because the test statistic is less than the critical value we do not reject the null hypothesis.

Alternative distribution-free goodness-of-fit procedures have been proposed by Riedwyl (1967). Test statistics of two of the proposed procedures have the same asymptotic distribution as the Kolmogorov statistic. The reader is referred to Riedwyl's paper for further details. Subsection 1.3.9 gives another simple goodness-of-fit procedure.

Frequently, the null hypothesis is not simple. To deal with those cases, modifications of the above test have been proposed. Since these modified tests are *not* distribution-free procedures we will not discuss them. Some references in this connection are Lilliefors (1967, 1969) and Srinivasan (1970).

1.3.8 Confidence band for the population distribution function

We will now discuss how one can construct a confidence band for the population cdf using the acceptance region of the Kolmogorov test. This method uses the usual connection between a test of hypothesis and a confidence region. The acceptance region is $D_n < C_{1-\alpha}$. Thus the test will not reject any F_0 as long as $sup_x |F_n(x) - F_0(x)| < C_{1-\alpha}$. In other words, a $100(1-\alpha)\%$ confidence region for the population cdf, $F(x)$, is

$$\{F(x) : sup_x \mid F_n(x) - F(x) \mid < C_{1-\alpha}\}.$$

This statement is the same as

$$\{\mid F_n(x) - F(x) \mid < C_{1-\alpha}, \forall x\}.$$

This can be restated as

$$F_n(x) - C_{1-\alpha} < F(x) < F_n(x) + C_{1-\alpha}, \forall x.$$

Using the fact that $0 \le F(x) \le 1$, the above confidence region can be seen to be

$$max(0, F_n(x) - C_{1-\alpha}) < F(x) < min(1, F_n(x) + C_{1-\alpha}), \forall x. \qquad (1.95)$$

In other words, the lower and upper bounds for $F(x)$ are

$$F_L(x) = max(0, F_n(x) - C_{1-\alpha}); F_U(x) = min(1, F_n(x) + C_{1-\alpha}).$$

These bounds are step functions since F_n is a step function. It is easy to see that, for $j = 1, 2, \ldots, n$,

$$F_L = max(0, F_n(X_{(j)}) - C_{1-\alpha}), \quad for \ X_{(j)} \le x < X_{(j+1)},$$

and

$$F_U = min(1, F_n(X_{(j)}) + C_{1-\alpha}), \quad for \ X_{(j)} \le x < X_{(j+1)}.$$

Also $F_L = 0$ for $x < X_{(1)}$ and $F_U = 1$ for $x > X_{(n)}$. These expressions are useful for computational purposes. Now we illustrate the details of this procedure using the data on failure times. The calculations are restricted to the positive real line, since the failure time is a positive-valued random variable.

Example 1.7 (cont'd.). Let us compute the confidence band (1.95) for $F(x)$, with 95% as the confidence coefficient. We use the data of Example 1.6. Here $n = 20$, $\alpha = 0.05$, and the critical value $C_{0.95} = 0.294$. The bounds for various intervals are given in Table 1.3. The lower and upper bounds are denoted by F_L and F_U, respectively.

Table 1.3 *Confidence bounds*

Interval of x	F_L	F_U
$[0, 4)$	0	0.294
$[4, 10)$	0	0.344
$[10, 13)$	0	0.394
$[13, 14)$	0	0.444
$[14, 18)$	0	0.494
$[18, 19)$	0	0.544
$[19, 20)$	0.006	0.594
$[20, 23)$	0.056	0.644
$[23, 24)$	0.106	0.694
$[24, 29)$	0.156	0.744
$[29, 46)$	0.206	0.794
$[46, 47)$	0.256	0.844
$[47, 57)$	0.306	0.894
$[57, 62)$	0.356	0.944
$[62, 74)$	0.406	0.994
$[74, 119)$	0.456	1.000
$[119, 188)$	0.506	1.000
$[188, 208)$	0.556	1.000
$[208, 209)$	0.606	1.000
$[209, 409)$	0.656	1.000
$[409, \infty)$	0.706	1.000

A computer program for obtaining the quantities used in Table 1.2 and Table 1.3 is given in Appendix B1.

1.3.9 A plotting procedure

The main idea behind the test discussed in Subsection 1.3.7 is the comparison of the sample distribution function with the hypothesized distribution function. By comparing the two curves one has to decide whether or not they are close to each other. This task of comparison is much easier if we transform the data using the probability integral transformation. We know that $F(X)$ follows the uniform distribution on $(0,1)$. Let $U_i = F_0(X_i)$, for $i = 1, \ldots, n$. The transformed data, $(U_1, \ldots, U_n)$, is a random sample from the uniform distribution under the null hypothesis. The cdf of this uniform distribution is the $45°$ line through the origin. Now we want to compare the sample distribution function with the hypothesized population distribution.

A graphical method can be used to make this comparison. This graphical method consists of comparing the P-P plot with the $45°$ line through the origin. Let $(U_{(1)}, \ldots, U_{(n)})$ be the order statistics of the sample. The P-P plot is the plot of $F_0(X_i)$ against $F_n(X_i)$, which is the same as the plot of $U_{(i)}$ against (i/n). This P-P plot for the data of Example 1.6 is given in Figure 1.1.

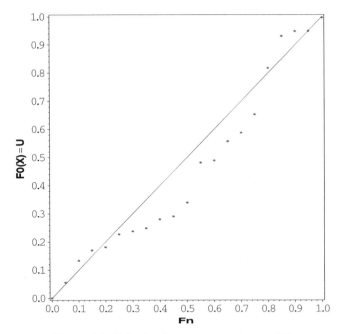

Figure 1.1 *P-P plot for testing goodness of fit.*

It is not always easy to assess the closeness of the plot to the $45°$ line through the origin visually. An objective way to make this assessment is to compute a measure of closeness and assess the significance of the measure. It has been suggested that we consider the vertical distances of the plotted points from the objective line and construct a summary statistic. These distances are the differences

$$d_j^- = F_0(X_{(j)}) - (j/n) = U_{(j)} - (j/n).$$

Two possible summary measures are

$$D^* = max(d_j^-) \quad and \quad W^* = \sum_i d_i^-.$$

Now testing for significance of D^* is equivalent to performing Kolmogorov's test for the alternatives (1.89).

Since $W^* = \sum_i U_{(i)} - (n+1)/2$ and as the distribution of $\sum_i U_{(i)}$ is the same as the distribution of $W = \sum_i U_i$, the test based on W^* is equivalent to the test based on W, proposed by Moses (1964).

Since $W/n = (1/n)\sum U_i$ is the average of n independent uniform random variables its distribution can be approximated by a normal distribution with mean $(1/2)$ and variance $(1/12n)$. This approximation is adequate if $n \geq 5$. So approximate α-level critical regions can be constructed using the

Table 1.4 *Goodness-of-fit tests based on W*

Alternative	Critical Region
$H_{A1} : F > F_0$	$Z_W < z_\alpha$
$H_{A2} : F < F_0$	$Z_W > z_{1-\alpha}$
$H_{A3} : H_{A1}$ or H_{A2}	$|Z_W| > z_{(1-\alpha/2)}$

statistic

$$Z_W = \frac{(1/n) \sum_i F_0(X_i) - (1/2)}{\sqrt{\frac{1}{12n}}}.$$

The tests of the null hypothesis $H_0 : F = F_0$ are given in Table 1.4.

It is suggested that the visual inspection of the plot be supplemented by the Kolmogorov test and the W test. It should be pointed out that the tests based on W are relatively easy to implement and a computer program for implementing these tests is given in Appendix B1.

For the data of Example 1.7, the statistic $W = 9.5743$ with a P-value 0.7416. So we do not reject the null hypothesis.

For an interesting review of graphical methods in nonparametric statistics see Fisher (1983).

1.4 Procedures for censored data

In life-testing experiments, sometimes the experiment is stopped before all items have failed. Two types of stopping rules have been used: (1) stopping after fixed time τ and (2) stopping after r items have failed. In both these cases the observations on the lifetimes of the items that have not failed are incomplete. In clinical trials, which are undertaken to study effects of some treatments, we also come across incomplete observations on some individuals. Typically, in clinical trials different patients enter the study at different times and they are given some treatment. A continuous response variable is used to evaluate the treatments. A commonly used response variable is the time to some event, such as recurrence of cancer or death of the patient, etc. This time to an event variable is called the *survival time*. Thus we want to observe the survival times of a group of patients. As the study will be terminated after a fixed duration, for some patients the event has not happened, so the survival times for these patients are incomplete or censored. For these patients we note for how long they were under observation. This type of censoring is called *random censoring*. The following discussion deals only with random censoring.

Suppose that n individuals are participating in a clinical trial and their survival times are $X_1, X_2, \ldots, X_n$. These X variables are considered as i.i.d. random variables with some common continuous cdf $F(.)$. The associated survival function is $S(.) = 1 - F(.)$. To accommodate the possibility of censoring, it is assumed that a censoring variable C_i is associated with the ith individual. The censoring variable C_i represents the period for which the ith individual is under observation. In other words, this individual will not be under observation after time C_i; if the event has not happened by then, the observation is C_i and it is considered a censored observation. Thus instead of observing X_i we observe the pairs (T_i, δ_i), where

$$T_i = min(X_i, C_i), \qquad (1.96)$$

and

$$\delta_i = \begin{cases} 1, & \text{if } X_i \leq C_i \\ 0, & \text{if } X_i > C_i. \end{cases} \qquad (1.97)$$

Observations on T_i for which $\delta_i = 1$ are called *uncensored times* or *failure times*, and observations on T_i for which $\delta_i = 0$ are called *censored times*. A typical one-sample data set is $(t_i, \delta_i)(i = 1, \ldots, n)$. Sometimes for reporting convenience t_i is used for uncensored observation and t_i+ or t_i^* is used for censored observation. For example, the following data on remission times (in weeks) of leukemia patients treated with the drug 6-MP have been reported by Gehan (1965):

$$6+, 6, 6, 6, 7, 9+, 10+, 10, 11+, 13, 16, 17+, 19+, 20+, 22, 23, 25+,$$
$$32+, 32+, 34+, 35 + .$$

Here, for example, the observation 25+ means that the event of interest has not happened during the observational period of 25 weeks.

1.4.1 Kaplan-Meier estimate of the survival function

To derive estimates of the survival function, certain assumptions about the censoring variables are needed. The censoring variables are assumed to be i.i.d. continuous random variables with some common cdf G(.). These C-variables are assumed to be independent of the survival time variables. It is also assumed that the data on C-variables do not provide any information on the survival function $S(.)$. With these assumptions, using a general definition of maximum likelihood estimator, Kaplan and Meier (1958) obtained an estimator of the survival function $S(t) = P(X > t)$, which we discuss now.

Our data are $(t_i, \delta_i)(i = 1, 2, \ldots, n)$. We order the t's. The resulting ordered values are $t'_{(1)} \leq t'_{(2)} \leq \cdots \leq t'_{(n)}$. Let $\delta_{(i)}$ be the δ associated with $t'_{(i)}$. From the $t'_{(i)}$, we obtain

$$t_{(0)} < t_{(1)} < t_{(2)} < \cdots < t_{(k)},$$

the distinct ordered uncensored observations. Here, $t_{(0)} = 0$.

Table 1.5 *Computation of Kaplan-Meier estimate*

j	1	2	3	4	5	6	7
$t_{(j)}$	6	7	10	13	16	22	23
d_j	3	1	1	1	1	1	1
n_j	21	17	15	12	11	7	6

Let d_j be the number of uncensored observations that are equal to $t_{(j)}$, and n_j be the number of observations (both censored and uncensored) that are $\geq t_{(j)}$, for $j = 0, 1, 2, \ldots, k$. Now, generalizing the discussion leading to Equation (1.71), the *Kaplan-Meier estimator*, $\hat{S}(t)$, is

$$\hat{S}(t) = \Pi_{\{j:t_{(j)} \leq t\}} \{1 - (d_j/n_j)\}, \quad for\ t \leq t_{(k)}. \tag{1.98}$$

This estimator is undefined for values of $t > t_{(k)}$, the largest uncensored observation. This estimator is also known as the *product limit estimator*. Equation (1.98) leads to the recurrence relation

$$\hat{S}(t_{(j)}) = \hat{S}(t_{(j-1)})[1 - (d_j/n_j)]. \tag{1.99}$$

Calculation of this estimator will be illustrated with the data on remission times given earlier.

Example 1.7. In the data of remission times reported in Gehan (1965) there are no observations that are equal to zero. The $t_{(j)}, d_j$, and n_j values, which are needed to compute the estimator (1.98), are given in Table 1.5.

The estimator is constant between two successive uncensored values. Thus it is sufficient to compute the distinct values of the estimator. These are the values $\hat{S}(t_{(j)})$, which are obtained using the recurrence relation (1.99).

Now

$$\hat{S}(0) = 1;$$

$$\hat{S}(6) = \hat{S}(0)[1 - (3/21)] = 1 \times (18/21) = 0.8571;$$

$$\hat{S}(7) = \hat{S}(6)[1 - (1/17)] = \hat{S}(6) \times (16/17) = 0.8067;$$

$$\hat{S}(10) = \hat{S}(7)[1 - (1/15)] = \hat{S}(7) \times (14/15) = 0.7529.$$

Following the same steps, we get

$$\hat{S}(13) = 0.6902, \qquad \hat{S}(16) = 0.6275,$$

and

$$\hat{S}(22) = 0.5378, \qquad \hat{S}(23) = 0.4482.$$

A plot of the Kaplan-Meier estimate is given in Figure 1.2.

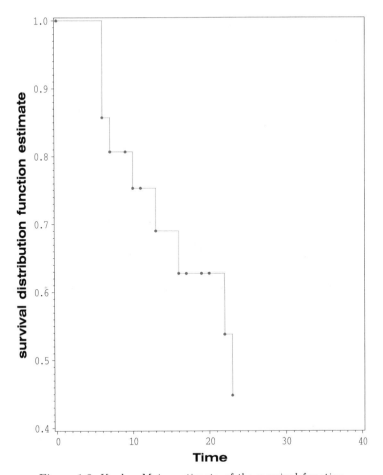

Figure 1.2 *Kaplan-Meier estimate of the survival function.*

In order to get a confidence interval for the survival probability $S(t)$, we can use the asymptotic normality of the estimator. The asymptotic mean of $\hat{S}(t)$ is $S(t)$ and the asymptotic variance is denoted by $AV(\hat{S}(t))$. An estimator of the asymptotic variance is

$$\hat{A}V(\hat{S}(t)) = [\hat{S}(t)]^2 \sum_{j=0}^{i} \frac{d_j}{n_j(n_j - d_j)} \quad \text{for } t_{(i)} \leq t < t_{(i+1)}. \tag{1.100}$$

This formula is known as *Greenwood's formula*. Now, for a given t, a confidence interval for $S(t)$ is

$$(\hat{S}(t) - z_{(1-\alpha/2)}\sqrt{\hat{A}V(\hat{S}(t))}, \qquad \hat{S}(t) + z_{(1-\alpha/2)}\sqrt{\hat{A}V(\hat{S}(t))}), \tag{1.101}$$

with an approximate confidence coefficient $(1 - \alpha)$.

Some mathematical details about the derivation of the asymptotic variance of $\hat{S}(t)$ appear in Appendix A1, and a computer program for obtaining Kaplan-Meier estimates and the confidence interval (1.101) is given in Appendix B1.

1.4.2 Estimation of the quartiles

We will discuss the estimation of the median in detail and these results can be easily extended to the case of other quartiles. As in the case of complete data, we can obtain a point estimator of the median $\xi_{0.5}$. Since $\xi_{0.5} = F^{-1}(0.5)$, a natural estimator is

$$\hat{\xi}_{0.5} = \hat{F}^{-1}(0.5) = \inf\{t : \hat{F}(t) \geq 0.5\} = \inf\{t : \hat{S}(t) \leq 0.5\}. \qquad (1.102)$$

This means that $\hat{\xi}_{0.5}$ is the smallest uncensored t for which $\hat{S}(t)$ is or falls below 0.5. So for the leukemia data considered in Example 1.7, an estimate of the median is 23 weeks.

Now we consider the problem of constructing a confidence interval for the median. As usual we start with the problem of testing the null hypothesis

$$H_0 : \xi_{0.5} = t_0 \ against \ H_A : \xi_{0.5} \neq t_0. \qquad (1.103)$$

Noting that

$$\frac{\{\hat{S}(t) - S(t)\}^2}{\hat{A}V(\hat{S}(t))} \approx \chi_1^2, \qquad (1.104)$$

an approximate α-level test of $H_0 : \xi_{0.5} = t_0$ is to

$$reject \ H_0 \ if \ (\hat{S}(t_0) - 0.5)^2 \geq \chi_1^2(1 - \alpha)\hat{A}V(\hat{S}(t_0)). \qquad (1.105)$$

where $\chi_1^2(1-\alpha)$ is the $100(1-\alpha)$ percentile of the chi-square distribution with 1 degree of freedom. Thus the null hypothesis $H_0 : \xi_{0.5} = t_0$ is not rejected at an approximate level α whenever

$$(\hat{S}(t_0) - 0.5)^2 \leq \chi_1^2(1 - \alpha)\hat{A}V(\hat{S}(t_0)). \qquad (1.106)$$

The set of values of t_0 satisfying (1.106) constitutes a confidence region for the median with approximate confidence coefficient $(1 - \alpha)$. This set reduces to an interval, which is called the *test based interval*. This method is from Brookmeyer and Crowley (1982) and they call the test (1.105) a *sign test for censored data*.

Several authors have considered this problem of constructing a confidence interval for the median. Slud et al. (1984) compared the performance of six proposals, including the above interval. They recommended the use of the *simple reflected interval*. This interval is

$$\{t : [\hat{S}(t) - 0.5]^2 \leq \chi_1^2(1 - \alpha)\hat{A}V(\hat{S}(\hat{\xi}_{0.5})]\}, \qquad (1.107)$$

and it is easy to compute. It is recommended for use with censored data.

In some situations the upper limit is not defined and in those cases it is possible to obtain a lower bound. It is obtained as the smallest t for which

$$\hat{S}(t) > 0.5 + z_{(1-\alpha)}\sqrt{\hat{AV}(\hat{S}(\hat{\xi}_{.05}))}. \tag{1.108}$$

This entire discussion can be carried out in relation to any other quartile. In addition to the median, the upper quartile (75th percentile) and the lower quartile (25th percentile) are usually of interest. Some computer programs give (point) estimates and confidence intervals of all the three quartiles. The SAS package uses the method proposed by Brookmeyer and Crowley (1982).

In relation to the leukemia data, we have already noted that a point estimate of the median is 23.0. Since there are no uncensored values above 23, we cannot find an upper confidence limit for the median. Here

$$\hat{AV}(\hat{S}(\hat{\xi}_{0.5})) = \hat{AV}(\hat{S}(23.0)) = 0.1346,$$

and a lower 95% confidence bound is the smallest t at which

$$\hat{S}(t) > 0.5 + z_{0.95}(0.1346) = 0.5 + (1.645)(0.1346) = 0.7214.$$

From the survival function calculations, it can be seen that such a t-value is 13. In other words, a 95% lower confidence bound for the median is 13 weeks.

The computer program mentioned earlier for calculating the estimate of the survival function also gives estimates of the three quartiles and their confidence intervals. It can be found in Appendix B1.

1.5 Appendix A1: Mathematical supplement

A1.1 Binomial cdf expressed as a beta integral

Here we prove the result (1.5). For fixed positive integer n and integer x such that $0 \le x \le n$, the cdf, $F(x; n, \theta)$, of the binomial distribution is a function of θ only and so let us denote it by $G(\theta)$. Since

$$G(\theta) = \sum_{j=0}^{x} \binom{n}{j} \theta^j (1-\theta)^{n-j},$$

the derivative of G with respect to θ is

$$G'(\theta) = -n(1-\theta)^{n-1} + \sum_{j=1}^{x} \binom{n}{j} [j\theta^{j-1}(1-\theta)^{n-j}$$
$$- (n-j)\theta^j (1-\theta)^{n-j-1}].$$

The sum on the right side reduces to a single expression. Denoting this derivative by $g(\theta)$, we have

$$g(\theta) = -(n-x)\binom{n}{x}\theta^x (1-\theta)^{n-x-1}.$$

Now using the fundamental theorem of calculus, we have

$$\int_0^\theta g(u)\,du = G(\theta) - G(0) = G(\theta) - 1.$$

Hence

$$G(\theta) = 1 - \int_0^\theta (n-x)\binom{n}{x}u^x(1-u)^{n-x-1}du$$

$$= (n-x)\binom{n}{x}\int_\theta^1 u^x(1-u)^{n-x-1}du$$

$$= (n-x)\binom{n}{x}\int_0^{1-\theta} v^{n-x-1}(1-v)^x dv$$

$$= I(1-\theta; n-x, x+1),$$

where I is the incomplete beta function.

In some problems we need $P[Bin(n,\theta) \geq c]$. This probability can be expressed in terms of an incomplete beta function, using the result we just proved. In other words,

$$P[Bin(n,\theta) \geq c] = 1 - F(c-1; n, \theta)$$

$$= 1 - I(1-\theta; n-c+1, c)$$

$$= \int_{1-\theta}^1 (n-c+1)\binom{n}{c-1}v^{n-c}(1-v)^{c-1}dv$$

$$= \int_0^\theta (n-c+1)\binom{n}{c-1}w^{c-1}(1-w)^{n-c}dw$$

$$= I(\theta; c, n-c+1).$$

A1.2 Union-intersection principle

In some problems, the null hypothesis H_0 can be expressed as an intersection of several subhypotheses and the alternative hypothesis can be expressed as a union of several subhypotheses. For example, in Section 1.1 our null hypothesis is $H_0 : \theta = \theta_0$. This can be viewed as an intersection, namely,

$$H_0 = [H_{01} : \theta \leq \theta_0] \cap [H_{02} : \theta \geq \theta_0].$$

The alternative hypothesis can be seen to be a union, namely,

$$H_A = [H_{A1} : \theta > \theta_0] \cup [H_{A2} : \theta < \theta_0].$$

We know how to test H_{01} against H_{A1} and also to test H_{02} against H_{A2}. If we conduct these two tests, we will reject H_0 only when either H_{01} or H_{02} is rejected. So a critical region for the original testing problem is the union of the two critical regions for the tests of H_{01} against H_{A1} and H_{02} against H_{A2}.

In this binomial problem, the critical region is

$$S_n \leq c_1 \quad or \quad S_n \geq c_2.$$

Now we will state the details of the general method of test construction using the union and intersection principle. We want to test the null hypothesis H_0 against H_A. It is possible to express these hypotheses as

$$H_0 = \bigcap_{i=1}^{k} H_{0i}, \quad and \quad H_A = \bigcup_{i=1}^{k} H_{Ai},$$

where H_{Ai} is an appropriate alternative to H_{0i}. A test of H_{0i} against H_{Ai} is defined by the critical region C_i. Then a test of H_0 against H_A is defined by the critical region

$$C = \bigcup_{i=1}^{k} C_i.$$

A1.3 Distribution of the rth order statistic

Let $X_{(r)}$ be the rth order statistic of the random sample $(X_1, X_2, \ldots, X_n)$ from the cdf $F(.)$. The cdf of $X_{(r)}$ is

$$
\begin{aligned}
G_r(y) &= Pr(X_{(r)} \leq y) \\
&= Pr(\text{at least } r \text{ X-values are } \leq y) \\
&= \sum_{i=r}^{n} \binom{n}{i} [F(y)]^i [1 - F(y)]^{n-i} \\
&= P[Bin(n, F(y)) \geq r] \\
&= I(F(y); r, n - r + 1).
\end{aligned}
$$

The last equality follows from the relation between the binomial cdf and the incomplete beta function. Let $f(.)$ be the population pdf, which is

$$f(x) = \frac{d}{dx} F(x).$$

Then the pdf of $X_{(r)}$ is

$$g_r(y) = \frac{d}{dy} G_r(y).$$

Differentiating each term in the sum and simplifying, we get

$$g_r(y) = n \binom{n-1}{r-1} [F(y)]^{r-1} [1 - F(y)]^{n-r} f(y).$$

In the context of computing confidence intervals for percentiles the following result is useful. This result makes use of the cdf we just derived. We want to find the following probability:

$$P[X_{(r)} \leq y < X_{(r+1)}] = P[X_{(r)} \leq y] - P[X_{(r+1)} \leq y]$$

$$= \sum_{i=r}^{n} \binom{n}{i} [F(y)]^i [1 - F(y)]^{n-i}$$

$$- \sum_{i=r+1}^{n} \binom{n}{i} [F(y)]^i [1 - F(y)]^{n-i}$$

$$= \binom{n}{r} [F(y)]^r [1 - F(y)]^{n-r}.$$

Thus for integers r and s such that $1 \leq r < s \leq n$, we have

$$P[X_{(r)} \leq y < X_{(s)}] = \sum_{i=r}^{s-1} \binom{n}{i} [F(y)]^i [1 - F(y)]^{n-i}.$$

Distribution of $U_{(k)}$

Let $(U_1, \ldots, U_n)$ be a random sample from the uniform distribution on $(0,1)$ and $U_{(k)}$ be the kth order statistic of this sample. The pdf of this order statistic is

$$g_k(u) = n \binom{n-1}{k-1} u^{k-1} [1 - u]^{n-k}.$$

This is the pdf of a beta distribution. In other words, $U_{(k)}$ follows $Beta(k, n - k + 1)$ distribution.

Remark. The distribution of $F(X_{(k)})$ is the same as the distribution of $U_{(k)}$.

A1.4 Confidence intervals for percentiles

Let $X_{(1)}, X_{(2)}, \ldots, X_{(n)}$, be the order statistics associated with the random sample $X_1, X_2, \ldots, X_n$ from a continuous distribution with the $100p$ percentile ξ_p. It has been indicated that $[X_{(i)}, X_{(j)})$ is a distribution-free confidence interval for the percentile ξ_p, with the confidence coefficient

$$C(i, j) = \sum_{l=i}^{j-1} \left\{ \binom{n}{l} p^l (1 - p)^{n-l} \right\}.$$

Usually we want a confidence interval with a specified confidence coefficient $(1 - \alpha)$. Thus, for given n and p, we need to find an integer pair (i, j) such that the resulting confidence coefficient $C(i, j)$ is at least $(1 - \alpha)$. In general, more than one pair (i, j) will satisfy this requirement. So we may have to impose another condition to come up with one pair to use. It is reasonable to minimize $j - i$. Table 12.3 of Owen (1962) gives the (A, B) pairs chosen in this manner for the 25th percentile ($\xi_{0.25}$) and the 75th percentile ($\xi_{0.75}$).

Wilks (1959) gave large sample approximations to the required values of i and j, so that $j - i$ is as small as possible. These values are the *nearest integers* to the two values of r, which are the roots of the equation

$$(r - np)^2 = z^2_{(1-\alpha/2)} \cdot np(1 - p).$$

Clearly these roots r_1 and r_2 are

$$r_1 = np - z_{(1-\alpha/2)} \cdot \sqrt{np(1 - p)},$$

and

$$r_2 = np + z_{(1-\alpha/2)} \cdot \sqrt{np(1 - p)}.$$

Alternatively, we can choose i and j so as to satisfy the following inequalities:

$$P(Bin(n, p) < i) \leq (\alpha/2); P(Bin(n, p) > j - 1) \leq (\alpha/2).$$

It can be seen that if $i(\alpha/2; n, p)$ is the solution for the first constraint, then the j solution of the second constraint is given by

$$j(\alpha/2; n, p) = n + 1 - i(\alpha/2; n, 1 - p).$$

The proof of this result is Problem 7.

A computer program for obtaining these indices, i and j, is given in Appendix B1.

Determination of a lower bound for ξ_p

Suppose that we want to find a lower confidence bound for ξ_p, the population $100p$ percentile. An order statistic $X_{(i)}$ will serve as a lower bound, if i is chosen so as to satisfy the condition

$$P[X_{(i)} \leq \xi_p] = P[Bin(n, p) \geq i] = 1 - \alpha.$$

In other words, i is chosen so as to satisfy the condition

$$P[Bin(n, p) < i] = \alpha.$$

Thus $(i - 1)$ is a 100α percentile of $Bin(n, p)$ distribution.

A1.5 Delta method

Here we obtain approximations to the mean and variance of a function of a statistic. Let X be a statistic with mean μ and variance σ^2. At times we want to find approximations to the mean and variance of a function of X, say $T = h(X)$. By considering a Taylor expansion of $h(X)$ around μ the required approximations are derived. By assuming that the function $h(x)$ is twice differentiable around μ and the derivatives are not zero at μ, the variable $h(X)$ can be approximated as

$$T = h(X) \approx h(\mu) + h'(\mu)(X - \mu).$$

This means we can use the right-side variable to obtain approximations to the mean and variance of $T = h(X)$. Now taking expectations of both sides we get

$$E(h(X)) \approx h(\mu),$$

and then

$$var(h(X)) \approx \{h'(\mu)\}^2 \sigma^2.$$

This method of getting approximations to the mean and the variance is called the *delta method*. We now give three applications.

1. Applying this method to the statistic $T = 2Arcsin(\sqrt{\hat{\theta}})$, we have

$$E(T) \approx 2Arcsin(\sqrt{\theta}),$$

and

$$var(T) \approx (1/n).$$

2. Another function of interest is $T = ln(X)$. From the general result, we have

$$E(T) \approx ln(\mu) \quad and \quad var(T) \approx \frac{1}{\mu^2}\sigma^2.$$

3. Now we apply this method to $ln\hat{S}(t)$ to gain an insight into the derivation of *Greenwood's formula* (1.100) for the variance of $\hat{S}(t)$. From the formula (1.98), taking logarithms of both sides, we get

$$ln(\hat{S}(t)) = \sum_{\{j:t_{(j)}\leq t\}} ln\{1 - (d_j/n_j)\}.$$

Thus

$$var[ln(\hat{S}(t))] \approx \sum_j var[ln\{1 - (d_j/n_j)\}],$$

where the covariance terms are omitted. Now

$$var[ln(\hat{S}(t)] \approx \frac{1}{S^2(t)} AV(\hat{S}(t)),$$

and

$$var[ln\{1 - (d_j/n_j)\}] \approx \frac{1}{(1 - (d_j/n_j))^2} \frac{(d_j/n_j)(1 - (d_j/n_j))}{n_j}$$

$$= \frac{d_j}{n_j(n_j - d_j)}.$$

Using these approximate variances in the expression for $var[ln(\hat{S}(t))]$, we get

$$\frac{1}{S^2(t)} AV(\hat{S}(t)) \approx \sum \frac{d_j}{n_j(n_j - d_j)},$$

which is the same as

$$AV(\hat{S}(t)) \approx S^2(t) \sum \frac{d_j}{n_j(n_j - d_j)}.$$

So an estimate of the asymptotic variance is

$$\hat{AV}(\hat{S}(t)) = (\hat{S}(t))^2 \sum \frac{d_j}{n_j(n_j - d_j)},$$

which is the formula (1.100).

A1.6 Relation between percentiles of Kolmogorov tests

Note that

$$\alpha = P_0(D_n \geq C_{(1-\alpha)}) = P_0(D_n^+ \geq C_{(1-\alpha)}, \ or \ D_n^- \geq C_{(1-\alpha)}).$$

However, the probability on the right side has an upper bound. In other words,

$$P_0(D_n \geq C_{(1-\alpha)}) \leq P_0(D_n^+ \geq C_{(1-\alpha)}) + P_0(D_n^- \geq C_{(1-\alpha)}).$$

Since the null distribution of D_n^+ is equal to that of D_n^-, the upper bound is $2P_0(D_n^+ \geq C_{(1-\alpha)})$. Finally, we have

$$\alpha = P_0(D_n \geq C_{(1-\alpha)}) \leq 2P_0(D_n^+ \geq C_{(1-\alpha)}).$$

Thus using the definitions of the quantiles we have

$$C_{(1-\alpha)} \approx C_{(1-\alpha/2)}^+.$$

1.6 Appendix B1: Computer programs

B1.1 $(1 - \alpha)$ quantile and α quantile of Bin(n, θ) distribution

Here we calculate the critical value for the test of $H_0 : \theta \leq \theta_0$ vs. $H_+ : \theta > \theta_0$, where θ is the probability of a binomial distribution with n as the other parameter. The test statistic is S. The test rejects H_0, *if $S \geq C_+$*. We calculate the critical value C_+, which is the $(1 - \alpha)$ quantile. Because of the discrete nature of the distribution, the actual size is usually different from α. The output gives the attained size, which is less than or equal to α. A normal approximation to C_+ is also calculated. This approximation uses the continuity correction. In the program for convenience we use p0 for θ_0. In the output we use theta0. We also calculate C_-, the α quantile and the normal approximation to C_-. One can choose the desired value. The output results are given at the end of Subsection 1.2.2.

```
data one;
n=20;*sample size*;
alpha=0.05;
p0=0.25;
mu=n*p0;
mv=sqrt(mu*(1-p0));
czp=mu+mv*probit(alpha)-0.5;
cz1=int(czp);
czu=mu+mv*probit(1-alpha)+0.5;
cz2=int(czu)+1;
ut=1-probbnml(p0,n,cz2-1);
lt=probbnml(p0,n,cz1);
theta0=p0;
proc print;
title1 '(1-alpha) quantile and alpha quantile';
title2 'specifications, expected(S) and std(S)';
var n theta0  alpha mu mv;
run;
```

In the following we calculate C_+^* and C_+.

```
data two;
set one;
p00=1-p0;
c=0;
cum=probbnml(p00,n,c);
do while (cum le alpha);
c+1;
cum=probbnml(p00,n,c);
put c cum;
end;
if c-1 le 0 then c2=0;
else c2=c-1;
cum2=probbnml(p00,n,c2);
cplus=n-c2;
asize=cum2;
proc print;
title 'approximate & exact critical values and tail area';
var  cz2 cplus asize;
run;
```

Here we calculate the exact value of C_-.

```
data three;
set one;
c=0;
cum=probbnml(p0,n,c);
```

```
do while (cum le alpha);
c+1;
cum=probbnml(p0,n,c);
end;
if c-1 le 0 then c2=0;
else c2=c-1;
cum2=probbnml(p0,n,c2);
cminus=c2;
asize=cum2;
proc print;
title 'approximate & exact critical value and tail area';
var cz1 cminus asize;
run;
```

Output

<pre>
        (1-alpha) quantile and alpha quantile
        specifications, expected(S) and std(S)

   n   theta0   alpha   mu   mv

   20   0.25   0.05   5   1.93649

        approximate & exact critical values
                  and tail area

           cz2   cplus   asize

            9   9   0.040925

        approximate & exact critical value
                  and tail area

           cz1   cminus   asize

            1   1   0.024313
</pre>

B1.2 Sample size calculation

This is a program for computing sample size for a one-sample binomial testing problem. The null hypothesis is $\theta = \theta_0$ and the alternative is $\theta > \theta_0$. The requirements are that $P(type\,I\,error) \leq \alpha$ and the power at $\theta = \theta_1$ should be $1 - \beta$. To get the critical region of the test, first we obtain the solutions (1.32) and (1.33), and then obtain the modified version (1.36), suggested by Levin and Chen (1999). We use the specifications of Example 1.1.

```
data one;
p0=0.20;
p1=0.35;
alpha=0.05;
power=0.80;
beta=1-power;
z11=probit(alpha);
z22=probit(power);
sdn=sqrt(p0*(1-p0));
sda=sqrt(p1*(1-p1));
del=(p0-p1)**2;
nnum=(z11*sdn-z22*sda)**2;
nnor=nnum/del;
cnor=nnor*p0+0.5-z11*sqrt(nnor*p0*(1-p0));
c=int(cnor)+1;
n=int(nnor)+1;
m=c-1;
error2=probbnml(p1,n,m);
error1=1-probbnml(p0,n,m);
power1=1-error2;
proc print;
title 'specifications';
var p0 p1 alpha power ;
proc print;
title 'Sample size using normal approximation
                          (1.34) and (1.35)';
var n c;
proc print;
title 'Resultant error probabilities and power';
var error1 error2 power1;
data two;
set one;
a=nnum;
b=1+2*((p1-p0)/a);
b1=sqrt(b);
ncf=((1+b1)/2)**2;
ncc=nnor*ncf;
nlc=int(ncc)+1;
cnc=ncc*p0+0.5-z11*sqrt(ncc*p0*(1-p0));
clc=int(cnc)+1;
mc=clc-1;
error2c=probbnml(p1,nlc,mc);
error1c=1-probbnml(p0,nlc,mc);
powerc=1-error2c;
proc print;
```

```
title 'Levin and Chen modified sample size';
var nlc clc;
proc print;
title 'Attained errors and power with new sample size';
var error1c error2c powerc;
run;
```

Output
___

```
                      specifications
               p0      p1      alpha      power

              0.2     0.35     0.05       0.8

          Sample size using approximation (1.34)
                      and (1.35)
                    Obs     n     c

                     1     50     16
          Resultant error probabilities and power
                error1        error2       power1

              0.030803      0.28010      0.71990



          Levin and Chen modified sample size

                       nlc     clc

                       57      17
          Attained errors and power with new
                      sample size
                error1c       error2c      powerc

              0.050466      0.16930      0.83070
```

B1.3 Confidence limits for θ using beta percentiles

This is a program for computing confidence limits for the success probability using the data from a binomial experiment with n trials. Here, n is the sample size, s is the number of successes, opr is the observed proportion of successes, and cc is the confidence coefficient.

```
data one;
n=20;*sample size*;
s=3;*number of successes*;
```

```
opr=s/n;*observed prop. of successes*;
op=100*opr;*observed percentage of successes*;
cc=0.95;*confidence coefficient*;
alpha=1 - cc;
op=100*opr;
alpha1=(1-cc)/2;
if s=0 then f1=0;
else f1=betainv(alpha1,s,n-s+1);
thetal=f1;
if s=n then f2=1;
else f2=betainv(1-alpha1,s+1,n-s);
thetau=f2;
proc print;
title 'Data and Specifications';
var cc alpha alpha1 n s  opr;
proc print ;
title 'Confidence limits using Beta Percentiles';
var thetal thetau;
run;
```

Output
<hr>

Data and Specifications

cc	alpha	alpha1	n	s	opr
0.95	0.05	0.025	20	3	0.15

Confidence limits using Beta Percentiles

Obs	thetal	thetau
1	0.032071	0.37893

B1.4 Large sample confidence limits for θ (Ghosh's method)

This is a program for computing a confidence interval for the success probability using the data from a binomial experiment with n trials. Here, n is sample size, s is number of successes, opr is the observed proportion of successes, and cc is the confidence coefficient (enter as a fraction). The confidence limits (1.49) are computed. This method is the same as Wilson's method.

```
data one;
n=20;*sample size*;
s=3;*number of successes*;
```

```
opr=s/n;*observed prop. of successes*;
thetahat=opr;
cc=0.95;*confidence coefficient*;
alpha=1 - cc;
op=100*opr;*observed percentage of successes*;
z=probit(1-alpha/2);
cg=z*z/(2*n);
vg=(opr*(1-opr)+(cg/2))/n;
sg=sqrt(vg);
dg=z*sg;
If opr=0 then lbg=0 ;
else lbg=(opr+cg-dg)/(1+2*cg);
if opr=1 then ubg=1;
else ubg=(opr+cg+dg)/(1+2*cg);
proc print;
title1 'Confidence limits for Theta';
title2 'Data and Specifications';
var cc n s  thetahat;
proc print;
title 'Confidence limits (1.49) using large sample theory';
var n s thetahat cc lbg ubg;
run;
```

Output

Confidence limits for Theta

Data and Specifications

cc	n	s	thetahat
0.95	20	3	0.15

Confidence limits (1.49) using large sample theory

n	s	thetahat	cc	lbg	ubg
20	3	0.15	0.95	0.052369	0.36042

B1.5 Critical values for a two-sided test for the median

Here we calculate the critical values for the two-sided test of $H_0 : \theta = 1/2$ *vs.* $H_1 : \theta \neq 1/2$, where θ is the probability of the binomial distribution with n as the other parameter. The tails are size $\leq \alpha/2$. It is sufficient to get the lower critical value C. The upper critical value is $n - C$. Because of the discrete nature of the distribution, the actual size is usually different from the nominal value α. The output gives the attained level. A normal approximation is also calculated. This approximation uses the continuity correction. This program is especially designed for the sign test and calculation of confidence limits for the median. The null value $1/2$ is denoted by p0.

```
data one;
n=10;*sample size*;
alpha=0.05;
p0=0.50;
c=0;
mu=n*p0;
mv=sqrt(mu*(1-p0));
czp=mu+mv*probit(1-alpha/2)+0.5;
cz2=int(czp)+1;
czm=mu+mv*probit(alpha/2)-0.5;
cz1=int(czm);
az=probbnml(p0,n,cz1)+1-probbnml(p0,n,cz2-1);
cum=probbnml(p0,n,c);
do while (cum le alpha/2);
c+1;
cum=probbnml(p0,n,c);
end;
if c-1 le 0 then c2=0;
else c2=c-1;
cum2=probbnml(p0,n,c2);
clower=c2;
cupper=n-c2;
atsize=cum2+1-probbnml(p0,n,cupper-1);
proc print;
title ' Specifications , expected(S) and std(S)';
var n p0    alpha mu mv;
proc print;
title 'approximate critical values & exact size with czs';
var cz1 cz2 az;
proc print;
title 'exact  critical value and attained size';
var   alpha    clower cupper atsize;
run;
```

Output

<pre>
          Specifications , expected(S) and std(S)

          n   p0   alpha   mu   mv

          10   0.5   0.05   5   1.58114

     approximate critical values & exact size with czs
                  cz1   cz2   az
                   1   9   0.021484

      exact critical value and attained size
          alpha   clower   cupper   atsize
          0.05   1   9   0.021484
</pre>

B1.6 Critical values for a two-sided test for a quantile

Here we calculate the critical values for the two-sided test of $H_0 : \theta = \theta_0$ vs. $H_1 : \theta \neq \theta_0$, where $\theta_0 \neq 1/2$. We want to construct a confidence interval for the pth quantile ξ_p. The critical values for the two-sided test will give us the indices of the order statistics, which are the confidence limits. The test statistic S is the number of values less than or equal to ξ_0, the hypothesized value of ξ_p. We consider the test with equal tails, so that each tail size is less than or equal to $\alpha/2$. Because of the discrete nature of the distribution, the actual size is usually different from the nominal value α. The output gives the attained level. A normal approximation is also calculated. This approximation uses the continuity correction. First the lower critical value $C_-(\theta_0)$ is obtained. Using the same steps we get the critical value $C_-(1 - \theta_0)$. The upper critical value is $C_+(\theta_0) = n - C_-(1 - \theta_0)$. From these critical values indices of order statistics for getting a confidence interval are obtained. These are i and $j(= i + k)$.

```
data one;
  n=10;*sample size*;
  alpha=0.05;
  p=0.25;
  q=1-p;
  c=0;
  mu=n*p;
```

```
mv=sqrt(mu*q);
czp=mu+mv*probit(1-alpha/2)+0.5;
cz2=int(czp)+1;
czm=mu+mv*probit(alpha/2)-0.5;
cz1=int(czm);
az=probbnml(p,n,cz1)+1-probbnml(p,n,cz2-1);
cum=probbnml(p,n,c);
do while (cum le alpha/2);
c+1;
cum=probbnml(p,n,c);
end;
if c-1 le 0 then c1=0;
else c1=c-1;
cum1=probbnml(p,n,c1);
ltail=cum1;
clower=c1;
c=0;
cum=probbnml(q,n,c);
do while (cum le alpha/2);
c+1;
cum=probbnml(q,n,c);
end;
if c-1 le 0 then c3=0;
else c3=c-1;
c2=n-c3;
cupper=c2;
cum2=1-probbnml(p,n,cupper-1);
utail=cum2;
atsize=cum1+cum2;
i=clower+1;
j=cupper;
cc=100(1-alpha).
proc print;
title ' specifications, expected(S) and std(S)';
var n p    alpha mu mv;
proc print;
title 'approximate critical values & exact size with czs';
var cz1 cz2 az;
proc print;
title 'exact  critical values and attained size';
var  alpha   clower cupper ltail utail atsize;
proc print;
title 'Indices of order statistics for getting a CI ';
var p cc i j;
run;
```

Output

---

specifications, expected(S) and std(S)

n	p	alpha	mu	mv
10	0.25	0.05	2.5	1.36931

approximate critical values & exact size with czs

cz1	cz2	az
0	6	0.076041

exact critical value and attained size

alpha	clower	cupper	ltail	utail	atsize
0.05	0	6	0.056314	0.019728	0.076041

Indices of order statistics for getting a CI

p	cc	i	j
0.25	95	1	6

B1.7 K-S goodness of fit

We calculate the K-S goodness-of-fit statistic D. We also obtain the confidence bounds FL and FU. The constant c3 is the approximate critical value. We also calculate the Moses statistic and approximate P-value.

```
data one;
n=20;
alpha=0.05;
c1=sqrt(0.5*log(2/alpha));
c2=sqrt(n)+0.12+(0.11/sqrt(n));
c3=c1/c2;
proc print data=one;
var n alpha c3;
```

```
data two;
input id x @@;
lines;
1 18 2 23 3 29 4 409 5 24 6 74 7 13 8 62
9 46 10 4 11 57 12 19
13 47 14 14 15 20 16 208 17 119 18 209 19 10
20 188
;
proc rank data=two  out=three;
var x;
ranks rx;
run;
proc  sort data=three out=four;
by rx;
run;
data five;
set four;
n=20;
c=0.294;
fn=rx/n;
F00=1 - exp(-x/70);
F0=round(F00,.0001);
d1=(fn - F0);
d2=-d1;
FL=max(0, fn - c);
FU=min(1, fn+c);
u=F0;
proc print data=five;
title 'Table1.2. Quantities needed for Kolmogorov test';
var x fn f0 d1 d2;
run;
proc iml;
use five;
read  all variables{d1} into M1;
read  all variables{d2} into M2;
read all variables{F0} into M3;
n=nrow(M1);
D11=max(M1);
D21=max(M2);
D1=max(D11,0);
D2=max(D21+(1/n),0);
D=max(D1,D2);
Dplus=D1;
Dminus=D2;
W1=sum(M3);
```

```
W=round(W1,.0001);
num=(W/n) - 0.5;
den=sqrt(1/(12*n));
z=num/den;
p11=probnorm(-abs(z));
p1=round(p11,.0001);
p2=2*p1;
print 'Various statistics' Dplus Dminus D;
print 'Moses statistic' W ;
print '2sided Pvalue for W' p2;
exit ;
run;
proc print data=five;
title 'Table1.3. Confidence Band';
var x FL FU;
proc univariate data=five edf;
beta(a=1 b=1) ;
run;
```

Output

n	alpha	c3
20	0.05	0.29417

Table1.2. Quantities needed for Kolmogorov test

x	fn	F0	d1	d2
4	0.05	0.0555	-0.0055	0.0055
10	0.10	0.1331	-0.0331	0.0331
13	0.15	0.1695	-0.0195	0.0195
14	0.20	0.1813	0.0187	-0.0187
18	0.25	0.2267	0.0233	-0.0233
19	0.30	0.2377	0.0623	-0.0623
20	0.35	0.2485	0.1015	-0.1015
23	0.40	0.2800	0.1200	-0.1200
24	0.45	0.2903	0.1597	-0.1597
29	0.50	0.3392	0.1608	-0.1608
46	0.55	0.4817	0.0683	-0.0683
47	0.60	0.4890	0.1110	-0.1110
57	0.65	0.5570	0.0930	-0.0930
62	0.70	0.5876	0.1124	-0.1124
74	0.75	0.6526	0.0974	-0.0974

119	0.80	0.8173	-0.0173	0.0173
188	0.85	0.9318	-0.0818	0.0818
208	0.90	0.9488	-0.0488	0.0488
209	0.95	0.9495	0.0005	-0.0005
409	1.00	0.9971	0.0029	-0.0029

	DPLUS	DMINUS	D
Various statistics	0.1608	0.1318	0.1608

	W
Moses statistic	9.5742

	P2
Pvalue for W	0.7416

Table1.3. Confidence Band

x	FL	FU
4	0.000	0.344
10	0.000	0.394
13	0.000	0.444
14	0.000	0.494
18	0.000	0.544
19	0.006	0.594
20	0.056	0.644
23	0.106	0.694
24	0.156	0.744
29	0.206	0.794
46	0.256	0.844
47	0.306	0.894
57	0.356	0.944
62	0.406	0.994
74	0.456	1.000
119	0.506	1.000
188	0.556	1.000
208	0.606	1.000
209	0.656	1.000
409	0.706	1.000

B1.8 Kaplan-Meier estimation

This program computes the Kaplan-Meier estimate using a sample with right-censored observations. We use the data considered in Example 1.7. When all observations are complete the estimate is the sample survival function. Each observation is a pair (T, Delta). The OUTSURV option gives confidence limits for $S(t)$. The PLOT option gives the plot of the survival function.

```
data luke;
input T Delta @@;
lines;
6 0 6 1 6 1 6 1 7 1 9 0 10 0 10 1 11 0 13 1 16 1 17 0
19 0 20 0 22 1 23 1
25 0 32 0 32 0 34 0 35 0
;
proc print ;
title2 'Data';
proc lifetest data=luke outsurv=a;
time T*Delta(0);
proc print data=a;
title 'Confidence limits for S(t)';
run;
```

Output

	Data	
Obs	T	Delta
1	6	0
2	6	1
3	6	1
4	6	1
5	7	1
6	9	0
7	10	0
8	10	1
9	11	0
10	13	1
11	16	1
12	17	0
13	19	0
14	20	0
15	22	1
16	23	1
17	25	0
18	32	0
19	32	0
20	34	0
21	35	0

The LIFETEST Procedure
Product-Limit Survival Estimates

T	Survival	Failure	Survival Standard Error	Number Failed	Number Left
0.00	1.0000	0	0	0	21
6.00	.	.	.	1	20
6.00	.	.	.	2	19
6.00	0.8571	0.1429	0.0764	3	18
6.00*	.	.	.	3	17
7.00	0.8067	0.1933	0.0869	4	16
9.00*	.	.	.	4	15
10.00	0.7529	0.2471	0.0963	5	14
10.0	.	.	.	5	13
11.00*	.	.	.	5	12
13.00	0.6902	0.3098	0.1068	6	11
16.00	0.6275	0.3725	0.1141	7	10
17.00*	.	.	.	7	9
19.00*	.	.	.	7	8
20.00*	.	.	.	7	7
22.00	0.5378	0.4622	0.1282	8	6
23.00	0.4482	0.5518	0.1346	9	5
25.00*	.	.	.	9	4
32.00*	.	.	.	9	3
32.00*	.	.	.	9	2
34.00*	.	.	.	9	1
35.00	.	.	.	9	0

NOTE: The marked survival times are censored observations.

Summary Statistics for Time Variable T

Quartile Estimates

Percent	Point Estimate	95% Confidence Interval [Lower	Upper)
75	.	23.0000	.
50	23.0000	13.0000	.
25	13.0000	6.0000	23.0000

Mean	Standard Error
17.9092	1.6474

NOTE: The mean survival time and its standard error were underestimated because the largest observation was censored and the estimation was restricted to the largest event time.

Summary of the Number of Censored and
Uncensored Values

Total	Failed	Censored	Percent Censored
21	9	12	57.14

In the following, CENSOR is 1 for censored observation and 0 for uncensored observation.

Confidence limits for S(t)

T	CENSOR	SURVIVAL	SDF_LCL	SDF_UCL
0	0	1.00000	1.00000	1.00000
6	0	0.85714	0.70748	1.00000
6	1	0.85714	.	.
7	0	0.80672	0.63633	0.97711
9	1	0.80672	.	.
10	0	0.75294	0.56410	0.94178
10	1	0.75294	.	.
11	1	0.75294	.	.
13	0	0.69020	0.48084	0.89955
16	0	0.62745	0.40391	0.85099
17	1	0.62745	.	.
19	1	0.62745	.	.
20	1	0.62745	.	.
22	0	0.53782	0.28648	0.78915
23	0	0.44818	0.18439	0.71197
25	1	.	.	.
32	1	.	.	.
32	1	.	.	.
34	1	.	.	.
35	1	.	.	.

1.7 Problems

1. *Large sample theory confidence interval for θ, the probability of success*: In obtaining an approximation for the critical value C_- of the test (1.18), we used the approximate normal distribution of the statistic S_n. This statistic is approximately normally distributed with mean $n\theta$ and variance $n\theta(1-\theta)$.

Using this approximate distribution we can find an interval for θ by forming the quadratic equation

$$n(\hat{\theta} - \theta)^2 = \theta(1 - \theta)z_{(1-\alpha/2)}^2$$

and solving it for θ. Let θ_1 and θ_2 be the solutions, where $\theta_1 < \theta_2$. Then the interval (θ_1, θ_2) is a confidence interval with an approximate confidence coefficient $(1 - \alpha)$. Determine this interval (θ_1, θ_2). (Ghosh (1979) investigated the properties of this interval for various values of n and found that the coverage probability is close to the nominal coverage probability $1 - \alpha$. Chen (1990) also recommended this interval over the other approximate interval (1.23). Newcombe (1998) and Agresti and Coull (1998) also recommended this interval.)

2. In some discussions, the statistic $T = 2Arcsin(\sqrt{\hat{\theta}})$ is considered as a normally distributed random variable with mean $2Arcsin(\sqrt{\theta})$ and variance $1/n$. Using this approximate normal distribution of the statistic T and the formulation considered in Subsection 1.2.5, show that an approximation to the sample size (a result analogous to (1.32)) is

$$n = (z_\alpha + z_\beta)^2/d^2,$$

where $d = 2Arcsin(\sqrt{\theta_1}) - 2Arcsin(\sqrt{\theta_0})$.

3. Suppose that in a binomial experiment with $n = 10$, you observe $S_n = 6$. Compute the confidence limits given by (1.49) by choosing $(1 - \alpha) = 0.90$.

4. Suppose we denote the lower confidence limit θ_L of (1.22) by $\theta_L(1-\alpha, n, S_n)$. Show that the upper confidence limit $\theta_U(1 - \alpha, n, S_n)$ is given by

$$\theta_U(1 - \alpha, n, S_n) = 1 - \theta_L(1 - \alpha, n, n - S_n).$$

5. *A confidence interval in the context of a process control:* Usually the quality of a manufacturing process is conveyed by the proportion of units that do not conform to specifications. Ideally, we want this proportion to be zero. However, we want to ensure that this proportion is as small as possible. To estimate this proportion, θ, a sample of n items is examined. The number of nonconforming items, S, is ascertained. From the value of S we can estimate θ. This data gathering method is modeled as a binomial experiment and the statistical inferences about θ can be handled by the methods for binary data. In some applications, as indicated by Hahn and Meeker (1991), confidence intervals for certain probabilities are of interest. One such probability is π_{LE}; this is the probability that the number of nonconforming units in a future sample of size m is less than or equal to J, for given values m and J. A confidence interval for this probability can be obtained from a set of confidence

limits for θ. To see this connection, we need to observe that

$$\pi_{LE} = P[Bin(m, \theta) \le J] = I(1 - \theta; m - J, J + 1),$$

which is a decreasing function of θ. Using this observation, show that a $100(1 - \alpha)\%$ confidence interval for π_{LE} is

$$[I(1 - \theta_U; m - J, J + 1), \ I(1 - \theta_L; m - J, J + 1)],$$

where $[\theta_L, \theta_U]$ is a $100(1 - \alpha)\%$ confidence interval for θ. (This discussion has been adopted from Hahn and Meeker (1991).)

6. Using the pdf of the rth order statistic derived in Appendix A1, show that

$$E[F(X_{(r)})] = \frac{r}{(n + 1)}.$$

Using this result and the delta method, show that

$$E(X_{(r)}) \approx F^{-1}\left(\frac{r}{n + 1}\right) \approx F^{-1}(p),$$

where $p = r/n$. (Thus $X_{(r)}$ can be viewed as an approximately unbiased estimator of the quantile $\xi_p = F^{-1}(p)$.)

7. Using the second method of choosing the integers i and j in the discussion on the construction of confidence intervals for ξ_p, if $i(\alpha/2; n, p)$ is the solution for the first constraint, then show that the solution of the second constraint is given by

$$j(\alpha/2; n, p) = n + 1 - i(\alpha/2; n, 1 - p).$$

8. Schafer and Angus (1979) gave the following data on the operation life (in hours) of 20 bearings: 6278, 3113, 9350, 5236, 11584, 12628, 7725, 8604, 14266, 6215, 3212, 9003, 3523, 12888, 9460, 13431, 17809, 2812, 11825, 2398.
Find a point estimate for the quantile $\xi_{0.10}$. Find also a 95% lower confidence bound for $\xi_{0.10}$. (Hint: Refer to Appendix A1.4.)

9. In a random sample of 50 families in a community, 30 families were observed to have an income of more than \$50,000 per annum. Does this information constitute statistical evidence at 0.05 level that more than 50% of the families in that community have an income of more than \$50,000? Based on these data, find a 95% confidence interval for the probability that a family has an income of more than \$50,000.

10. For the data given in Example 1.7, test the null hypothesis that the median $\xi = 30$ against a two-sided alternative hypothesis $\xi \ne 30$. Use $\alpha = 0.05$. Also, find a 95% confidence interval for ξ. Does 60 lie in this confidence interval?

11. The following data are a sample of test scores: 18, 27, 49, 43, 46, 34, 30, 37, 48, 31. Test the hypothesis that the scores have a normal distribution with mean 38 and variance 49.

12. The following data from Pike (1966) are the times (days) from insult with a carcinogen to mortality from vaginal cancer in rats: 143, 164, 188, 190, 192, 206, 209, 213, 216, 220, 227, 230, 234, 246, 265, 304, 216+, 244+.
(a) Obtain the Kaplan-Meier estimate of the survival probability $S(250)$. Using this estimate obtain a 95% confidence interval for $S(250)$.
(b) Obtain a point estimate of the median survival time and construct a 95% confidence interval for the median survival time.

1.8 References

Agresti, A. and Caffo, B. (2000). Simple and effective confidence intervals for proportions and differences of proportions result from adding two successes and two failures. *Amer. Statist.*, **54**, 280–288.

Agresti, A. and Coull, B.A. (1998). Approximate is better than exact for interval estimation of binomial proportion. *Amer. Statist.*, **52**, 119–126.

Birnbaum, Z.W. (1952). Numerical tabulation of the distribution of Kolmogorov's statistic for finite sample size. *J. Amer. Statist. Assoc.*, **47**, 425–441.

Bradley, J.V. (1969). A survey of sign tests based on the binomial distribution. *J. Qual. Tech.*, **1**, 89–101.

Birnbaum, Z.W. and Tingey, F.H. (1951). One-sided confidence contours for the probability distribution functions. *Ann. Math. Statist.*, **22**, 592–596.

Brookmeyer, R. and Crowley, J. (1982). A confidence interval for the median survival time. *Biometrics*, **38**, 29–41.

Casella, G. and Berger, R.L. (1990). *Statistical Inference*, Duxbury Press, Wadsworth Publishing Company, Belmont, California.

Chen, H. (1990). The accuracy of approximate intervals for a binomial parameter. *J. Amer. Statist. Assoc.*, **85**, 514–518.

Desu, M.M. and Raghavarao, D. (1990). *Sample Size Methodology*, Academic Press, Boston.

Desu, M.M. and Rodine, R.H. (1969). Estimation of population median. *Skan. Aktuar.*, **52**, 67–70.

Dunsmore, I.R. (1974). The Bayesian predictive distribution in life testing models. *Technometrics*, **16**, 455–460.

Fisher, N.I. (1983). Graphical methods in nonparametric statistics: a review and annotated bibliography. *Inter. Statist. Rev.*, **51**, 25–58.

Gehan, E.A. (1965). A generalized Wilcoxon test for comparing arbitrarily simply-censored samples. *Biometrika*, **52**, 203–223.

Ghosh, B.K. (1979). A comparison of some approximate confidence intervals for the binomial parameter. *J. Amer. Statist. Assoc.*, **74**, 894–900.

Hahn, G.J. and Meeker, W.Q. (1991). *Statistical Intervals: A Guide for Practitioners*, John Wiley & Sons, New York.

Hutson, A.D. (1999). Calculating nonparametric confidence intervals for quantiles using fractional order statistics. *J. Appl. Statist.*, **26**, 343–353.

Kaplan, E.L. and Meier, P. (1958). Nonparametric estimation from incomplete observations. *J. Amer. Statist. Assoc.*, **53**, 457–481.

Kolmogorov, A.N. (1933). Sulla determinazione empirica di una legge di distribuzione. *Giornale dell' Istituto Italianio degli Attuari*, **4**, 83–91.

Lentner, C. (Ed.) (1982). *Geigy Scientific Tables (Vol. 2)*, Ciba-Geigy Corp., West Caldwell, New Jersey.

Levin, B. and Chen, X. (1999). Is the one-half continuity correction used once or twice to derive a well known approximate sample size formula to compare two independent binomial distributions? *Amer. Statist.*, **53**, 62–66.

Lilliefors, H.W. (1967). On the Kolmogorov-Smirnov test for normality with mean and variance unknown. *J. Amer. Statist. Assoc.*, **62**, 399–402.

Lilliefors, H.W. (1969). On the Kolmogorov-Smirnov test for the exponential distribution with mean unknown. *J. Amer. Statist. Assoc.*, **64**, 387–389.

MacKinnon, W.J. (1964). Table for both the sign test and distribution-free confidence intervals of the median for sample sizes to 1000. *J. Amer. Statist. Assoc.*, **59**, 935–956.

Massey, F.J. (1950). A note on the estimation of a distribution function by confidence limits. *Ann. Math. Statist.*, **21**, 116–119.

Miller, L.H. (1956). Table of percentage points of Kolmogorov statistics. *J. Amer. Statist. Assoc.*, **51**, 111–121.

Moses, L.E. (1964). One sample limits of some two-sample rank tests. *J. Amer. Statist. Assoc.*, **59**, 645-651.

Natrella, M.G. (1963). *Experimental Statistics*, National Bureau of Standards Handbook 91.

Newcombe, R.G. (1998). Two-sided confidence intervals for the single proportion: comparison of seven methods. *Statist. Med.*, **17**, 857–872.

Owen, D.B. (1962). *Handbook of Statistical Tables*, Addison-Wesley, Reading, Massachusetts.

Pike, M.C. (1966). A method of analysis of certain class of experimental carcinogenesis. *Biometrics*, **22**, 142–161.

Riedwyl, H. (1967). Goodness of fit. *J. Amer. Statist. Assoc.*, **62**, 390–398.

Samuels, M.L. and Lu, T. (1992). Sample size requirements for the back-of-the-envelope binomial confidence interval. *Amer. Statist.*, **46**, 228–231.

Schafer, R.E. and Angus, J.E. (1979). Estimation of Weibull quantiles with minimum error in the distribution function. *Technometrics*, **21**, 367–370.

Slud, E.V., Byar, D.P., and Green, S.B. (1984). A comparison of reflected versus test-based confidence intervals for the median survival time, based on censored data. *Biometrics*, **40**, 587–600.

Srinivasan, R. (1970). An approach to testing the goodness of fit of incompletely specified distributions. *Biometrika*, **57**, 605–611.

Stephens, M.A. (1974). EDF statistics for goodness of fit and some comparisons. *J. Amer. Statist. Assoc.*, **69**, 730–737.

Thall, P.F. and Simon, R.M. (1995). Recent developments in the design of phase II clinical trials. *Recent Advances in Clinical Trial Design and Analysis*, Edited by Peter F. Thall, Kluwer Academic Publishers, Boston, pp. 49–71.

Van der Parren, J.L. (1970). Tables for distribution-free confidence limits for the median. *Biometrika*, **57**, 613–617.

Wilks, S.S. (1959). Nonparametric statistical inference. *Probability and Statistics: The Harald Cramér Volume*, Edited by Ulf Grenander, John Wiley & Sons, New York, pp. 331–354.

CHAPTER 2

Procedures for two independent samples

2.1 Introduction

Often researchers are interested in comparing two processes, or two fertilizers, or two detergents, or two therapies, etc. Data are collected on a suitable response variable under two settings and are used to compare the characteristics. In some cases data are gathered by conducting an experiment using two treatments on homogeneous units or subjects. In these experiments treatments are randomly assigned to the units and data are collected on some response variable. This type of experiment is called a *simple comparative study* or a *parallel study*.

The two samples are denoted by $(X_1, X_2, \ldots, X_m)$ and $(Y_1, Y_2, \ldots, Y_n)$. The X's constitute a random sample on the random variable X, with cdf F(.), and the Y's make up a random sample on the random variable Y, with cdf G(.). We consider two situations: (1) X and Y are discrete variables and (2) X and Y are continuous variables. The testing problem is usually formulated as testing the null hypothesis $H_0 : F(z) = G(z)$, for all real z. This hypothesis is called the *homogeneity hypothesis*. We will also consider the clinical equivalence of two therapies with discrete and continuous response variables. As in Chapter 1, the results for discrete responses are given first, followed by the results for the continuous responses, with complete and censored data. Derivation of optimal linear rank statistics and ARE of tests are also discussed in this chapter.

2.2 Two-sample problem with binary responses

Here the response variables X and Y can assume only two values, which are designated as 1 and 0. The probability models are

$$P(X = 1) = \theta_1, \quad P(X = 0) = 1 - \theta_1;$$
$$P(Y = 1) = \theta_2, \quad P(Y = 0) = 1 - \theta_2, \tag{2.1}$$

where $0 < \theta_1 < 1$ and $0 < \theta_2 < 1$. As the response variable can take only two values, the data are usually summarized by the number of 1's. In other words, the data can be displayed as a 2×2 table, such as Table 2.1.

Let $N = m + n$, $A = \sum X_i$, and $B = \sum Y_j$. We are interested in testing whether or not $\theta_1 = \theta_2$. One measure that indicates the difference between

Table 2.1 *Results of a parallel study*

Value	X-sample	Y-sample	Total
1	A	B	$A + B$
0	$m - A$	$n - B$	$N - (A + B)$
Total	m	n	N

θ's is the *risk difference*, defined as

$$\Delta = \theta_1 - \theta_2. \tag{2.2}$$

Sometimes the comparison problem uses a different measure, the *risk ratio*, which is

$$\psi = (\theta_1/\theta_2). \tag{2.3}$$

In the following subsections we consider not only the testing of hypothesis problems but also the confidence interval estimation problems.

2.2.1 *Testing the homogeneity hypothesis*

We formulate the homogeneity hypothesis in terms of the parameter Δ. Thus the null hypothesis $H_0 : \theta_1 = \theta_2$ can be restated as

$$H_0 : \Delta = 0. \tag{2.4}$$

Clearly one would consider the maximum likelihood estimator of Δ for constructing a test of H_0. This estimation is done assuming that $0 < A < m$ and $0 < B < n$ and the estimator is

$$\hat{\Delta} = \hat{\theta}_1 - \hat{\theta}_2 = (A/m) - (B/n) \equiv \frac{N}{mn}(A - m(T/N)), \tag{2.5}$$

where $T = A + B$. It is easy to verify that

$$E(\hat{\Delta}|H_0) = 0; \quad var(\hat{\Delta}|H_0) = \Theta(1 - \Theta)(N/mn). \tag{2.6}$$

where Θ is the common value of θ_1 and θ_2, under H_0. A natural estimator of Θ is

$$\hat{\Theta} = (A + B)/(m + n) = (T/N). \tag{2.7}$$

Using this estimator, the null variance of $\hat{\Delta}$ is estimated by

$$v^2 = \hat{\Theta}(1 - \hat{\Theta})(N/mn) = (T/N)(1 - (T/N))(N/mn), \tag{2.8}$$

and we can also write this estimator as

$$v^2 = (A + B)[N - (A + B)]/(Nmn). \tag{2.9}$$

Now a test statistic for testing the null hypothesis (2.4) is

$$Z = \hat{\Delta}/v. \tag{2.10}$$

This statistic is asymptotically a $N(0,1)$ variable and this result is used to construct critical regions of approximate size α. This method of test construction has been proposed by Wald (1943).

One-sided alternatives

When the alternative is

$$H_+ : \Delta > 0 (i.e., \theta_1 > \theta_2),\tag{2.11}$$

the proposed test is to

$$reject\, H_0\; if\; Z > z_{1-\alpha}.\tag{2.12}$$

The associated P-value is $P(N(0,1) \geq z(obs))$, where $z(obs)$ is the observed value of the test statistic (2.10).

This formulation has been used by Rodary et al. (1989) in the context of an efficacy trial. Now for the other one-sided alternative

$$H_- : \Delta < 0 (i.e., \theta_1 < \theta_2),\tag{2.13}$$

the proposed test is to

$$reject\, H_0\; if\; Z < z_\alpha.\tag{2.14}$$

The associated P-value is $P(N(0,1) \leq z(obs))$, where $z(obs)$ is the observed value of the test statistic (2.10).

Two-sided alternatives

For the two-sided alternatives

$$H_A : \Delta \neq 0 (i.e., \theta_1 \neq \theta_2),\tag{2.15}$$

the test is to

$$reject\, H_0\; if\; |Z| > z_{(1-\alpha/2)}.\tag{2.16}$$

The associated P-value is $2P(N(0,1) > |z(obs)|)$.

Sometimes the critical region of test (2.16) is stated in terms of Z^2 and it is

$$Z^2 > \chi_1^2(1-\alpha),\tag{2.17}$$

where $\chi_1^2(1-\alpha)$ is the $100(1-\alpha)\%$ percentile of the chi-square distribution with 1 degree of freedom, and

$$Z^2 = \frac{N(nA - mB)^2}{mn(A+B)(N-A-B)} = \frac{[A - m(T/N)]^2}{[\frac{mn}{N}(T/N)(1-(T/N))]}.\tag{2.18}$$

In some applications the problem is formulated in terms of ψ, the risk ratio. In terms of the risk ratio the null hypothesis under test is

$$H_0 : \psi = 1.$$

The relevant test for the two-sided alternatives under the risk ratio formulation is given in Subsection 2.2.5.

Example 2.1. In a study to evaluate two treatments for AIDS patients, one group of 225 patients, treated with treatment 1, showed a response rate of 69%. In the other group of the same number of patients, treated with treatment 2, the response rate was 52%. Is there enough evidence to conclude that treatment 1 has a higher response rate compared to treatment 2?

This problem is formulated as a testing problem and the one-sided alternative H_+ of (2.11) is appropriate. So we calculate

$$\hat{\Delta} = 0.69 - 0.52 = 0.17.$$

Further,

$$\hat{\Theta} = (0.69 + 0.52)/2 = 0.605,$$

and

$$v^2 = (0.605)(0.395)(2/225) = 0.002124.$$

Hence $Z = 0.17/0.0461 = 3.6876$. For $\alpha = 0.05$, $z_{1-\alpha} = 1.645$. From the test (2.12), we reject the null hypothesis. We conclude that treatment 1 has a higher response rate than treatment 2.

2.2.2 Fisher's exact test

The earlier discussion is based on the asymptotic distributional results. When the sample sizes are not large, a different procedure is used. To understand the motivation behind this procedure, we examine the joint distribution of the variables A and B, which are independent binomial random variables. Their joint distribution is defined by the probability function

$$P(A = a, B = b | \theta_1, \theta_2) = \binom{m}{a} \theta_1^a (1 - \theta_1)^{m-a} \binom{n}{b} \theta_2^b (1 - \theta_2)^{n-b}$$

for integers a and b, where $0 \le a \le m$ and $0 \le b \le n$. Under the null hypothesis $H_0 : \theta_1 = \theta_2 = \Theta$, this probability function simplifies to

$$P(A = a, B = b | \Theta) = \binom{m}{a} \binom{n}{b} \Theta^t (1 - \Theta)^{N-t}$$

$$= \binom{N}{t} \Theta^t (1 - \Theta)^{N-t} \frac{\binom{m}{a}\binom{n}{b}}{\binom{N}{t}}$$

$$= P(T = t | H_0) \frac{\binom{m}{a}\binom{n}{b}}{\binom{N}{t}},$$

where $T = A + B$, $t = a + b$, and $N = m + n$. Since the first part on the right side depends on the unknown parameter Θ, we use the second part for constructing tests. This second part can be seen as the probability function of the conditional distribution of A, given $T = t$. It is easy to see that

$$f(a|t, H_0) = P(A = a|T = t, H_0) = \frac{\binom{m}{a}\binom{n}{b}}{\binom{N}{t}},$$

and we can rewrite this function as

$$f(a|t, H_0) = \frac{\binom{t}{a}\binom{N-t}{m-a}}{\binom{N}{m}} \tag{2.19}$$

for integers a such that $max(t - n, 0) \le a \le min(m, t)$. The critical regions are constructed in relation to this conditional distribution, which is a hypergeometric distribution.

The critical region of the proposed test for testing the null hypothesis (2.4) against the alternative hypothesis H_+ of (2.11) is

$$A \ge a_1, \tag{2.20}$$

where a_1 is the smallest integer such that

$$P(A \ge a_1|t, H_0) \le \alpha. \tag{2.21}$$

The critical region of the proposed test for the alternative H_- of (2.13) is

$$A \le a_2, \tag{2.22}$$

where a_2 is the largest integer such that

$$P(A \le a_2|t, H_0) \le \alpha. \tag{2.23}$$

Finally, the critical region of the proposed test for the two-sided alternative H_A of (2.15) is

$$A \le a_3 \quad or \quad A \ge a_4, \tag{2.24}$$

where a_3 is the largest integer such that

$$P(A \le a_3|t, H_0) \le (\alpha/2),$$

and a_4 is the smallest integer such that

$$P(A \ge a_4|t, H_0) \le (\alpha/2). \tag{2.25}$$

Various tables of these critical values are available. One such table by Finney et al. (1963) is included in *Biometrika Tables for Statisticians*, Volume 1. *Geigy Scientific Tables* (1982) also contains a table of these critical values.

Remark 2.1. Some further details about the conditional distribution appear in Appendix A.

Here it is easier to compute the P-values rather than the critical values. Often the computer programs give the P-values. The appropriate P-values can be expressed in terms of the cumulative probabilities of a hypergeometric distribution. Let a be the observed value of A. The P-values for the one-sided alternatives (2.11) and (2.13) are, respectively,

$$P_+ = P(A \geq a|t, H_0); \quad P_- = P(A \leq a|t, H_0), \tag{2.26}$$

For the two-sided alternatives (2.15), some packages calculate the P-value as the sum of the probabilities of all configurations, which are more extreme than the observed configuration. In other words

$$P\text{-value} = \sum_{i \in S} P(A = i|t, H_0), \tag{2.27}$$

where $S = \{j|P(j|t) \leq P(a|t)\}$.

Example 2.2. Two samples of manufactured items have been examined for imperfections. Each item has been classified as conforming (0) or nonconforming (1). The samples come from different lots produced at different plants. The results are given in Table 2.2.

Can we conclude that plant 1 is producing more nonconforming items compared to plant 2?

Here the one-sided alternative H_+ of (2.11) is appropriate. The relevant P-value is

$$P\text{-value} = P(A \geq 7|T = 12, H_0) = 0.355.$$

This P-value is obtained from a SAS program where this is referred to as the *right tail* (see Appendix B2). Since this P-value is greater than 0.05 (the chosen α-value), the data are not significant. In other words, there is support for the null hypothesis that the product quality is the same in both plants.

Table 2.2 *Results of sample inspection*

	Plant		
Value	1	2	Total
1	7	5	12
0	8	10	18
Total	15	15	30

2.2.3 Establishing clinical equivalence

In some cases instead of testing the homogeneity hypothesis, we want to test the clinical (or therapeutic) equivalence hypothesis. This objective is relevant when one of the treatments is the standard one and the other is a new treatment. The new treatment is considered as *equivalent* to the standard treatment if it is only negligibly inferior. This problem of establishing therapeutic equivalence is formulated as a test of hypothesis problem where the null hypothesis corresponds to nonequivalence. The θ's defined earlier are called the *effect rates*. The standard treatment is identified as treatment 1 and the new treatment is identified as treatment 2. Usually the hypotheses are stated in terms of the risk difference $\Delta = \theta_1 - \theta_2$ or the risk ratio $\psi = \theta_1/\theta_2$. We will indicate the results for the risk difference formulation. In some cases larger effect rates are desirable. It is suspected that the effect rate of the new treatment is less than the effect rate of the standard treatment. However, the experiment is undertaken to establish the noninferiority nature of the new treatment. The new treatment is considered noninferior to the standard treatment if $\Delta = \theta_1 - \theta_2$ is less than Δ_0, where Δ_0 is a specified positive number.

This problem is formulated (see Dunnett and Gent (1977) and Rodary et al. (1989)) as testing

$$H_0 : \Delta = \Delta_0, \quad against \quad H_A : \Delta < \Delta_0, \qquad (2.28)$$

where Δ_0 is a specified positive constant. In this discussion we assume that $0 < A < m$ and $0 < b < n$, as in Subsection 2.2.1. The proposed test is based on the asymptotic normal distribution of $\hat{\Delta}$, where

$$\hat{\Delta} = (A/m) - (B/n) \equiv \hat{\theta}_1 - \hat{\theta}_2.$$

Recall that the variance of $\hat{\Delta}$ is

$$var(\hat{\Delta}) = var(\hat{\theta}_1) + var(\hat{\theta}_2)$$
$$= [\theta_1(1 - \theta_1)/m] + [\theta_2(1 - \theta_2)/n].$$

This expression can be rewritten as

$$var(\hat{\Delta}) = [\theta_1(1 - \theta_1)/m] + [(\theta_1 - \Delta)(1 - \theta_1 + \Delta)/n] \equiv V(\theta_1, \Delta). \qquad (2.29)$$

The parameter θ_1 is estimated using the maximum likelihood method under the restriction $\Delta = \Delta_0$. To get this estimator one needs to solve a third-degree equation. Farrington and Manning (1990) gave the maximum likelihood equation and its solution. In Appendix A2 the maximum likelihood equation is derived and the solution is given. A computer program for getting this estimator, $\hat{\theta}_{1D}$, is given in Appendix B2. This restricted maximum likelihood estimator is used in place of the parameter θ_1 and Δ is set to Δ_0 to get the estimator, v_D^2, of the variance $V(\theta_1, \Delta_0)$. In other words,

$$v_D^2 = V(\hat{\theta}_{1D}, \Delta_0).$$

Table 2.3 *Results of an equivalence study*

Treatment	Response		Total
	1	0	
Standard	120	80	200
New	57	43	100

The test statistic is

$$Z_D = (\hat{\Delta} - \Delta_0)/v_D,$$

and the test

declares equivalence if $Z_D < z_\alpha$.

Example 2.3. In a study, a new treatment is compared with a standard treatment and the data collected appear in Table 2.3.

We are interested in testing for the equivalence of the new treatment. Using $\Delta_0 = 0.15$, test the equivalency hypothesis.

Here $\hat{\theta}_1 = (120/200) = 0.6$ and $\hat{\theta}_2 = (57/100) = 0.57$. Hence

$$\hat{\Delta} = 0.60 - 0.57 = 0.03$$

From the output of a program, given in Appendix B1, the restricted ML estimate $\hat{\theta}_{1D} = 0.6377$ and the standard error $v_D = 0.0604$. Further, $Z_D = -1.9852$, with an approximate P-value 0.0236. Therefore we declare that the new treatment is equivalent to the standard treatment.

2.2.4 Confidence interval for the risk difference $\Delta = \theta_1 - \theta_2$

One of the commonly used methods is based on the asymptotic normal distribution of $\hat{\Delta}$. To use this method we need to estimate the variance of $\hat{\Delta}$. When $0 < A < m$ and $0 < B < n$, an unbiased estimator of this variance is

$$v^2 = [\hat{\theta}_1(1 - \hat{\theta}_1)/(m - 1)] + [\hat{\theta}_2(1 - \hat{\theta}_2)/(n - 1)].$$

Now using the result that the variable

$$\frac{(\hat{\Delta} - \Delta)}{v},$$

is asymptotically distributed as a $N(0, 1)$ variable, we can get a confidence interval for Δ. This interval is

$$(\hat{\Delta} - z_{(1-\alpha/2)}\, v, \hat{\Delta} + z_{(1-\alpha/2)}\, v). \tag{2.30}$$

In addition to this method several other methods were proposed. Newcombe (1998) evaluated 11 noniterative methods for constructing a confidence interval for the difference. A method that combines the intervals for the two θ's (see (1.48) of Chapter 1) has been recommended. This method will be described now.

Combining Wilson intervals for θ_1 and θ_2

Let (l_i, u_i) be the $100(1 - \alpha)$ percent confidence intervals for $\theta_i, (i = 1, 2)$ recommended by Wilson (1927) (see Newcombe (1998)). These are the intervals given by (1.48) of Chapter 1. Using these limits and the estimators $\hat{\theta}_i$, compute

$$\delta = \sqrt{(\hat{\theta}_1 - l_1)^2 + (u_2 - \hat{\theta}_2)^2},$$

$$\epsilon = \sqrt{(u_1 - \hat{\theta}_1)^2 + (\hat{\theta}_2 - l_2)^2}.$$

Now a $100(1 - \alpha)\%$ confidence interval for Δ is

$$(\hat{\Delta} - \delta, \hat{\Delta} + \epsilon). \tag{2.31}$$

This interval has good coverage probability and can be implemented very easily. In a later paper Newcombe (2001) reinforced the recommendation of this interval. A program for computing this interval is given in Appendix B2.

Thomas and Gart (1977) developed a table of confidence limits for the risk differences and the risk ratios. They use the non-null conditional distribution considered in the exact test. This distribution depends on the odds ratio, $\rho = [\theta_1/(1-\theta_1)]/[\theta_2/(1-\theta_2)]$. Using an iterative method they obtain limits for the odds ratio, ρ. From these limits they obtain limits for the risk difference Δ and for the risk ratio, ψ. Santner and Snell (1980) indicate some deficiencies of the results of Thomas and Gart. They also give an algorithm for constructing the intervals.

Remark 2.2. An upper bound for Δ is also useful in devising a test for a clinical equivalence problem. Further details about this method are available in Rodary et al. (1989).

2.2.5 Confidence interval for the risk ratio $\psi = (\theta_1/\theta_2)$

Katz et al. (1978) investigated three methods for obtaining this confidence interval and recommended Method C, which is relatively simple to use. This method is given now.

Here a confidence interval for $\phi = \log \psi$ is constructed and using this interval, an interval for ψ is obtained. In the ensuing discussion we assume that $0 < A < m$ and $0 < B < n$. A natural estimator of $\phi = \log \psi = \log(\theta_1/\theta_2)$ is

$$\hat{\phi} = log(\hat{\theta}_1/\hat{\theta}_2) = log(\hat{\theta}_1) - log(\hat{\theta}_2), \tag{2.32}$$

where $\hat{\theta}_1 = (A/m)$ and $\hat{\theta}_2 = (B/n)$. Using the Delta method (see Appendix A1.5) we see that the mean of this statistic is approximately equal to $\phi(= log\psi)$ and

$$var(\hat{\phi}) = var(log\hat{\theta}_1) + var(log(\hat{\theta}_2)) \approx \frac{(1 - \theta_1)}{m\theta_1} + \frac{(1 - \theta_2)}{n\theta_2} \equiv V^2. \tag{2.33}$$

An estimator of this approximate variance is

$$v^2 = \frac{1 - \hat{\theta}_1}{A} + \frac{1 - \hat{\theta}_2}{B} = \frac{1}{A} - \frac{1}{m} + \frac{1}{B} - \frac{1}{n}.$$

Now using the approximate normal distribution of $\hat{\phi}$, the end points of a confidence interval for ϕ are $\hat{\phi} \pm z_{(1-\alpha/2)}v$. These limits are converted into limits for ψ, using exponentiation. The resulting confidence limits for ψ are

$$exp(\hat{\phi} \pm z_{(1-\alpha/2)}v).$$

In other words, an approximate $100(1 - \alpha)\%$ confidence interval for the risk ratio, ψ, is

$$(\hat{\psi}.exp(-z_{(1-\alpha/2)}v), \hat{\psi}.exp(z_{(1-\alpha/2)}v)). \tag{2.34}$$

Remark 2.3. A slightly different method, which is related to a chi-square test, has been proposed by Koopman (1984). In Appendix A2, details of the chi-square test and the associated confidence interval are given. This method needs an iterative technique.

The variance estimate is zero when $A = m$ and $B = n$. Also, the variance estimate does not exist when $A = 0$ or $B = 0$. In these cases the confidence limits need modification, and the modifications are given in Table 2.4 (see Koopman (1984)).

Testing the homogeneity hypothesis

It may be noted that the confidence interval (2.34) can be used to test the homogeneity hypothesis

$$H_0 : \psi = 1, \quad against \quad H_A : \psi \neq 1. \tag{2.35}$$

The relevant test is to

reject $H_0 : \psi = 1$ if the interval (2.34) does not include 1. (2.36)

Table 2.4 *Methods for handling extreme values of A and B*

Value of A, B	Lower Bound	Upper Bound
$A = 0, B = 0$	0	∞
$A = 0, B \neq 0$	0	Substitute $A = 1/2$
$A \neq 0, B = 0$	Substitute $B = 1/2$	∞
$A = m, B = n$	Substitute $A = m - (1/2)$ $B = n - (1/2)$	Substitute $A = m - (1/2)$ $B = n - (1/2)$

Example 2.4. We use the data of Example 2.1 and obtain an interval estimate for the risk ratio. We have

$$\hat{\psi} = 0.69/0.52 = 1.3269 \quad and \quad v^2 = 0.0060996.$$

Further, $v = 0.0781$. We want a 95% confidence interval, so $z_{(1-\alpha/2)} = 1.96$. Also $exp(z_{(1-\alpha/2)}.v) = 1.1654$. Thus a confidence interval for the risk ratio is

$$[(1.3269/1.1654), (1.3269)(1.1654)] = (1.1386, 1.5464).$$

As this interval does not include 1, we reject the homogeneity hypothesis at an approximate significance level of 5.

2.2.6 Designing a parallel study

We want to design a study with an equal number of units in the two samples. The objective is to test the homogeneity hypothesis against the one-sided alternative $H_+ : \Delta > 0$. We want to control the power of an α-level test at the alternative $\Delta = \Delta_1 (> 0)$ to be at least $1 - \beta$. One of the earliest investigations was by Cochran and Cox (1957). They provide two tables of the required sample sizes. These correspond to one-sided alternatives and two-sided alternatives. These tables are based on the arcsine transform. The details of this method are given in Desu and Raghavarao (1990). It should be noted that the tables of Cochran and Cox have been revised by Gehan and Schneiderman (1973).

Gail and Gart (1973) determined the sample size in relation to Fisher's exact test. Haseman (1978) also discussed this problem in relation to the conditional test. Many other papers have discussed this problem. A survey of the various proposals was made by Sahai and Khurshid (1996). Before the availability of computer software, there was a need for a good approximation formula. Casagrande et al. (1978) came up with a good approximation that depends on the asymptotic normal distribution with the continuity correction. Recently Levin and Chen (1999) gave a careful explanation for the good accuracy of the approximation proposed by Casagrande et al. (1978). We now give this formula. Let θ_1 and θ_2 be values of the probabilities anticipated under the alternative and let $\bar{\theta} = (\theta_1 + \theta_2)/2$. Let

$$v_0 = \sqrt{2\bar{\theta}(1-\bar{\theta})}, \quad and \quad v_1 = \sqrt{\theta_1(1-\theta_1) + \theta_2(1-\theta_2)},$$

$$A = [z_{1-\alpha}.v_0 + z_{1-\beta}.v_1]^2. \tag{2.37}$$

The asymptotic normal method leads to the sample size

$$n' = A/[(\theta_1 - \theta_2)^2]. \tag{2.38}$$

Now the continuity-corrected sample size formula is

$$n_{cps} = (n'/4)\left[1 + \sqrt{(1 + (4/(n'.\Delta_1)))}\right]^2, \tag{2.39}$$

where $\Delta_1 = \theta_1 - \theta_2$, under the alternative. A computer program for calculating this sample size is given in Appendix B2.

Table 2.5 *A 2 × k contingency table of sample frequencies*

	Response Category						
Sample	1	2	$\cdots$	i	$\cdots$	k	Total
X	A_1	A_2	$\cdots$	A_i	$\cdots$	A_k	m
Y	B_1	B_2	$\cdots$	B_i	$\cdots$	B_k	n
Total	C_1	C_2	$\cdots$	C_i	$\cdots$	C_k	N

Remark 2.4. This sample size is very close to the size needed when we use Fisher's exact test. The determination of the sample size for the exact test is a relatively complicated problem.

2.3 Studies with categorical responses

Here we assume that the response variables X and Y are categorical variables with $k(> 2)$ categories. The probability distributions are defined as follows. For $i = 1, 2, \ldots, k$,

$$P(X\varepsilon \ category \ i) = \pi_{xi}, P(Y\varepsilon \ category \ i) = \pi_{yi}.$$

We want to test the null hypothesis

$$H_0 : \pi_{xi} = \pi_{yi}, \quad for \ i = 1, 2, \ldots, k. \tag{2.40}$$

against the two-sided alternatives

$$H_A : not \ H_0. \tag{2.41}$$

The sample data are given in Table 2.5. The test to be described is a version of Pearson's goodness-of-fit test. We estimate the probabilities using the combined sample and obtain the expected frequencies for each category. Then we compute a discrepancy measure and test it for significance. Under the null hypothesis $\pi_{xi} = \pi_{yi} = \pi_i$, say. The estimates of π_i are

$$\hat{\pi}_i = (A_i + B_i)/N = C_i/N.$$

The expected frequency for the ith category in the X-sample is $m.\hat{\pi}_i$, whereas the expected frequency in the Y-sample is $n.\hat{\pi}_i$. The test criterion is

$$T = \sum_{i=1}^{k}[(A_i - m\hat{\pi}_i)^2/(m\hat{\pi}_i) + (B_i - n\hat{\pi}_i)^2/(n\hat{\pi}_i)]. \tag{2.42}$$

Using the fact that $A_i - m\hat{\pi}_i = -(B_i - n\hat{\pi}_i)$, we can simplify the expression for T. This simplification leads to

$$T = \frac{1}{mn}\sum_{i=1}^{k}[(nA_i - mB_i)^2/(C_i)]. \tag{2.43}$$

The null distribution of this statistic can be approximated by a chi-square distribution with $(k-1)$ degrees of freedom. Thus an approximate α-level test is to

$$reject\ H_0\ if\ T > \chi^2_{(k-1)}(1-\alpha) \qquad (2.44)$$

and the associated P-value is $P(\chi^2_{(k-1)} > T(obs))$, where $T(obs)$ is the observed value of the test statistic T of (2.43).

Remark 2.5. Whenever any one of the expected frequencies is less than 5, it is suggested that adjacent categories be combined so that in the modified table expected frequencies are greater than 5. Then the test should be performed on the adjusted table.

Remark 2.6. An extension of Fisher's test is available for this problem. Statistical packages such as SAS and StatXact give the P-values. Some details about the computation of the exact P-values for Fisher's test and the test (2.44) are given in Appendix A2.

Remark 2.7. When we have only two categories, the test (2.44) is equivalent to the test (2.17).

Remark 2.8. If there is an ordering among the categories, we recommend the use of the Wilcoxon-Mann-Whitney procedure. This procedure will be discussed later.

Example 2.5. Among applicants to different colleges in a university, a sample of 100 male applicants and a sample of 75 female applicants showed their preferences as indicated in Table 2.6.

Can we conclude that there are no differences between the genders regarding the preferences for admission to various colleges?

The estimates of π's under the homogeneity hypothesis are

$$\hat{\pi}_1 = 35/175 = 0.2, \quad \hat{\pi}_2 = 0.486, \quad and \quad \hat{\pi}_3 = 0.314.$$

Using these estimates, the expected frequencies and contributions of various cells to T are calculated using a computer program and they are given in Table 2.7. The numbers in the cells are the expected frequencies and the

Table 2.6 *Preferences of applicants*

Gender	Arts (1)	Business (2)	Engineering (3)	Total
Males	20	45	35	100
Females	15	40	20	75
Total	35	85	55	175

Table 2.7 *Quantities needed for calculating T*

Gender	Response Category		
	1	2	3
Males	20(0)	48.57(0.2626)	31.43(0.4058)
Females	15(0)	36.43(0.3501)	23.57(0.5411)

quantities in parentheses are the contributions to the statistic T. Adding all the contributions we get $T = 1.56$. For $\alpha = 0.05$, the critical value is 5.991. Since the value of the statistic is less than the critical value, we do not reject the null hypothesis. In other words the data do not indicate any gender differences with regard to the preferences for admission to various colleges. The P-value in relation to the chi-square test is 0.458. The P-value in relation to Fisher's exact test is 0.440. These P-values are from the SAS program output (see Appendix B2).

2.4 Methods for continuous responses

Here we consider the two-sample problem in relation to continuous response variables. Let X and Y be two independent continuous random variables with cdf's $F(.)$ and $G(.)$, respectively. We have random samples on X and Y. The problem of interest is to test the null hypothesis

$$H_0 : F(x) = G(x), \forall x. \tag{2.45}$$

This problem is usually called the *two-sample problem*. Starting in the early 1940s several procedures have been proposed. In 1943, Mathisen proposed a test, which is currently called the *control median test*. In 1945, Wilcoxon proposed the rank sum procedure for samples of equal size. This procedure was extended by Mann and Whitney in 1947. They established a linear relationship between their statistic and the Wilcoxon statistic, thereby showed the equivalence of the two tests, so this test now is known as the *Wilcoxon-Mann-Whitney (WMW) test*. In 1948, Brown and Mood proposed a procedure that depends on the median of the combined sample. Mood (1950) described this method in detail. The median tests are simple to use and the control median test is related to precedence tests, which are popular in life test studies.

Before we go into the details of the various procedures, we will introduce a general class of alternatives to the null hypothesis H_0. It is instructive to start with a model that is used in the problem of comparing means of two normal distributions. The model assumes that X follows $N(\mu_1, \sigma^2)$ distribution, whereas Y follows $N(\mu_2, \sigma^2)$ distribution, so that the null hypothesis (2.45) is the same as $H_0 : \mu_1 = \mu_2$ or $\delta \equiv \mu_2 - \mu_1 = 0$. Clearly the alternatives of interest are $\delta > 0$, or $\delta < 0$, or $\delta \neq 0$. In this case the two cdf's are related to each other and the relationship is

$$G(x) = F(x - \Delta), \tag{2.46}$$

where $\Delta = \delta/\sigma$. We also say that F and G are related through a *shift model* and Δ is called the *shift parameter*.

Under this model there is an ordering between the two distribution functions. In particular,

$$F(x) \geq G(x), \forall x \quad when \quad \Delta > 0,$$

and

$$F(x) \leq G(x), \forall x \quad when \quad \Delta < 0.$$

These orderings do not depend on the normality assumption, but only on the shift model assumption. There are models other than (2.46) that lead to the orderings such as the above. Some discussion on the other models appears in Appendix A2. These ordered alternatives can be generalized as

$$F(x) \geq G(x), \quad and \quad F \not\equiv G, \tag{2.47}$$

and

$$F(x) \leq G(x), \quad and \quad F \not\equiv G. \tag{2.48}$$

In the first case (2.47), we say that *F is stochastically smaller than G or X is stochastically smaller than Y* and is denoted as $F\ st < G$ or $X\ st < Y$. Loosely speaking, in this case X-values tend to be smaller than Y-values. In the second case (2.48), we say that *F is stochastically larger than G or X is stochastically larger than Y* and this is denoted as $F\ st > G$ or $X\ st > Y$. Here X-values tend to be larger than Y-values. These are the one-sided alternatives of interest and they are denoted by H_{A1} and H_{A2}, respectively. The two-sided alternatives that we consider will be denoted by H_{A3}, which is equal to $H_{A1} \cup H_{A2}$.

We first discuss the median tests and then the discussion focuses on the Wilcoxon-Mann-Whitney test. In all three procedures we first consider the stochastic alternatives H_{A1}, H_{A2}, and H_{A3}. Later we specialize the discussion to the shift model case. In particular we look into the confidence interval estimation of the shift parameter, which can be interpreted as the difference between the medians as well as the difference between the means.

We then discuss the proportional hazard model analysis, which also induces a stochastic ordering between the distributions F and G. Finally, we consider Smirnov's test, which can be used for all alternatives.

2.4.1 Precedence tests — control median test (Mathisen's test)

Mathisen (1943) was concerned with devising a simple method for testing the hypothesis that the two samples are from the same population, based on medians and quantiles. One of the methods suggested depends on the statistic T_M, the number of Y-observations that are less than or equal to the median

of the X-sample. This method is called the *method of two intervals*. Gart
(1963) further investigated this test, calling it the *control median test*. Epstein
(1954) studied the statistic U, which denotes the number of values in the Y-
sample that exceed a given order statistic of the X-sample. A generalization of
Mathisen's test statistic T_M is T_a, which denotes the number of values in the
Y-sample that precede the ath order statistic of the X-sample. Nelson (1963)
proposed a test using this generalized statistic in the context of life testing
studies and it is called a *precedence test*. As all these tests are based on the
number of Y-sample values that precede or exceed a specified order statistic
of the X-sample, we consider a test in general terms and present Mathisen's
test as studied by Gart as a particular case.

Let $X_{(a)}$ be the ath order statistic of the X-sample, and

$$T_a = number\ of\ Y\text{-}values\ \leq X_{(a)}.$$

The distribution of this statistic under the null hypothesis (2.45) is independent of the common cdf and it is shown that (see Appendix A2), for
$j = 0, 1, \ldots, n$,

$$P(T_a = j|H_0) = \frac{m}{N} \frac{\binom{m-1}{a-1}\binom{n}{j}}{\binom{N-1}{a+j-1}}, \tag{2.49}$$

where $N = m+n$. To determine the critical regions for various alternatives, let
us evaluate $E(T_a)$. This calculation makes use of the fact that the conditional
distribution of T_a, given $X_{(a)}$, is the binomial distribution with parameters n
and $G(X_{(a)})$. Thus we have

$$E(T_a) = E(E(T_a|X_{(a)})) = E(nG(X_{(a)})). \tag{2.50}$$

Using the result $E(F(X_{(a)})) = a/(m+1)$, from (2.50) we get

$$E(T_a|H_0) = E(nF(X_{(a)})) = \frac{na}{m+1}. \tag{2.51}$$

We note that $E(T_a) >$ *or* $< E(T_a|H_0)$ depending on whether $Xst > Y$
or $Xst < Y$. Thus extreme values of the difference $T_a - E(T_a|H_0)$ indicate
deviations from the null hypothesis.

Before we express these tests as *precedence tests*, we want to point out some
similarities of these tests to the sign tests that we discussed in Chapter 1. The
sign tests of Chapter 1 use the number of observations below a hypothesized
quantile. Here the statistic $X_{(a)}$ is an estimator of the pth quantile, $\xi_p(X)$, of
the X-distribution, where $p = a/(m+1)$. The statistic T_a counts the number
of Y-values below an estimated quantile of the X-distribution, which is the
same as the Y-distribution under H_0. Thus the tests based on T_a are analogous
to the one-sample sign tests. Thus we call them *two-sample sign tests*. The
alternative hypotheses, critical regions and P-values for the two-sample sign
tests are shown in Table 2.8. Here the constants c_i's are chosen so that the tests
are of size α, $T_a(obs)$ is the observed value of T_a, and $S = \{k|P(T_a = k|H_0) \leq P(T_a = T_a(obs)|H_0)\}$.

Table 2.8 *Two-sample sign tests*

Alternative	Critical Region	P-Value
$H_{A1} : X st < Y$	$T_a \leq c_1$	$P_1 = P(T_a \leq T_a(obs)|H_0)$
$H_{A2} : X st > Y$	$T_a \geq c_2$	$P_2 = P(T_a \geq T_a(obs)|H_0)$
$H_{A3} : H_{A1}$ or H_{A2}	$T_a \leq c_3$ or $T_a \geq c_4$	$P_3 = \sum_{j \in S} P(T_a = j|H_0)$

Remark 2.9. In the case of the two-sided alternatives, we suggest choosing the tails to be of equal size.

Some properties of the null distribution of T_a

(1) The probability function satisfies the recurrence relation

$$P(T_a = j + 1|H_0) = P(T_a = j|H_0)\frac{(n-j)(a+j)}{(j+1)(N-a-j)},$$

which can be used to generate the null distribution.

(2) The variance (see Appendix A2) is

$$var(T_a|H_0) = \frac{na(m+1-a)(N+1)}{(m+1)^2(m+2)}.$$

(3) It can be shown that (see Appendix A2)

$$P(T_a \geq t|H_0) = \sum_{u=0}^{a-1}\left\{ \frac{\binom{m}{u}\binom{n}{a+t-u}}{\binom{m+n}{a+t}} \right\}.$$

The right-side sum is the cdf of a hypergeometric distribution and it is a SAS function.

(4) In large samples, the distribution of

$$Z = \frac{T_a - E(T_a|H_0)}{\sqrt{var(T_a|H_0)}},$$

can be approximated by the standard normal distribution, and this approximation can be used to find approximations to the critical values and the P-values.

Remark 2.10. In the case of two-sided alternatives, we suggested an equal tail test. Thus for large samples the P-value, P_3, can be approximated as

$$P_3 \approx P\big(\chi_1^2 > Z^2(obs)\big).$$

Precedence life tests

For applications to life testing experiments, we rephrase the two-sample sign tests as precedence tests. In certain applications like life testing, observations come in order. To take advantage of this feature, we look at the critical regions in a different way. Let us examine the test for the one-sided alternative

Table 2.9 *Precedence tests*

Alternative	Critical Region
$H_{A1} : Xst < Y$	$X_{(a)} < Y_{(c_1+1)}$
$H_{A2} : Xst > Y$	$X_{(a)} \geq Y_{(c_2)}$
$H_{A3} : H_{A1}\ or\ H_{A2}$	$X_{(a)} < Y_{(c_3+1)}\ or\ X_{(a)} \geq Y_{(c_4)}$

$H_{A1} : Xst < Y$. For this case the critical region is $T_a \leq c_1$. This region corresponds to the configuration that at most c_1 of Y-values are below $X_{(a)}$. In other words, the order statistic $Y_{(c_1+1)}$ is greater than $X_{(a)}$ in this region. Thus the reformulated test is to

reject H_0 in favor of H_{A1} if $X_{(a)} < Y_{(c_1+1)}$,

where c_1 is the largest integer such that

$$P(T_a \leq c_1|H_0) = P(X_{(a)} < Y_{(c_1+1)}|H_0) \leq \alpha.$$

In the context of a life testing experiment to implement this test, one need not wait until all items have failed. It is enough to wait until the $min(X_{(a)}, Y_{(c_1+1)})$ is observed, and then one can decide whether the test will accept or reject the null hypothesis. This early stopping of the life test will result in considerable saving of testing costs. For future reference, the critical regions in this format for various alternatives are given in Table 2.9.

The power of the precedence test for the alternatives H_{A1} has been studied by Eilbott and Nadler (1965), when the two distributions are exponentials. Later Shorack (1967) showed the validity of these results for a larger class of distributions of interest in life testing and related areas.

Exceedence statistic of Epstein

The statistic U_r^n of Epstein (1954) is related to T_r and the relationship is $U_r^n = n - T_r$. In this study, Epstein assumed $m = n$, and gave a tabulation of $P(U_r^n \leq u|H_0)$.

Exceedence statistic of Rosenbaum

The two-sample location statistic of Rosenbaum (1954) is the number of Y-values that exceed the largest X. In other words, this statistic is $n - T_m$. This has been suggested in the context of shift alternatives, with $\Delta > 0$. The two-sample scale statistic is $n - T_m + T_1$; that is, it is the number of Y-values outside the range of X-values. It is suggested as a test statistic for detecting scale differences.

Mathisen's test (control median test)

Mathisen's test is one of the two-sample sign tests discussed earlier. Because of its special importance, we describe the test procedure in detail. Later a confidence interval procedure based on this test will be presented.

Mathisen assumed that $m = 2k + 1$, and set $a = k + 1$; thus the statistic, T_M, of Mathisen is the same as T_{k+1}. Gart (1963) studied this test further and suggested the statistic T_s, where $s = \lfloor (m/2) \rfloor + 1$. He also indicated that the normal approximation to the null distribution of the test statistic T_s is quite accurate when $min(m, n) \geq 10$. We recall that the probability function of T_s is

$$P(T_s = j | H_0) = \frac{m}{N} \frac{\binom{m-1}{s-1}\binom{n}{j}}{\binom{N-1}{s+j-1}} \tag{2.52}$$

for $j = 0, 1, \ldots, n$. Further

$$E(T_s | H_0) = n \cdot \frac{s}{m+1} \approx \frac{n}{2},$$

and

$$var(T_s | H_0) = \frac{ns(m+1-s)(N+1)}{(m+1)^2(m+2)} \approx \frac{nN}{4m}.$$

Gart (1963) indicated that the null distribution of

$$Z = \frac{T_s - (n/2)}{\sqrt{\frac{nN}{4m}}}$$

can be approximated by the standard normal distribution, which is adequate whenever $min(m, n) \geq 10$.

Control median tests for shift model

One of the important models is the shift model. The following discussion deals with this case. We assume that the two distribution functions F and G are related to each other in the following manner:

$$G(x) = F(x - \Delta).$$

The quantity Δ is called the *shift parameter*. This relationship means that the distribution of $Y - \Delta$ is equal to the distribution of X. Let ξ_X and ξ_Y be the medians of the X-distribution and the Y-distribution, respectively. It has been noted (see Appendix A2) that

$$\Delta = \xi_Y - \xi_X,$$

and when the expectations exist,

$$\Delta = E(Y) - E(X).$$

Thus under the shift model the null hypothesis (2.45) can be restated as

$$H_0 : \Delta = 0 \quad or \quad \xi_Y = \xi_X \quad or \quad E(Y) = E(X).$$

Further, the alternatives restated in terms of Δ and the corresponding critical regions are presented in Table 2.10. These tests are based on T_s (where $s = \lfloor (m/2) \rfloor + 1$), and are stated as in Table 2.8. We also express the critical

Table 2.10 *Control median tests*

	Critical Regions	
Alternatives	Median Tests	Precedence Tests
$H_{A1} : \Delta > 0$	$T_s \leq c_1$	$X_{(s)} < Y_{(c_1+1)}$
$H_{A2} : \Delta < 0$	$T_s \geq c_2$	$X_{(s)} \geq Y_{(c_2)}$
$H_A : \Delta \neq 0$	$T_s \leq c_3$ or $T_s \geq c_4$	$X_{(s)} < Y_{(c_3+1)}$ or $X_{(s)} \geq Y_{(c_4)}$

regions as in Table 2.9. This representation of the critical regions will help us in the derivation of the confidence interval.

Remark 2.11. For two-sided alternatives, Gastwirth (1968) proposed a different version of the control median test, and called it the *first median test*.

Confidence interval for Δ from control median test
     We use the acceptance region of the test for the alternatives H_A, to obtain a two-sided confidence interval for Δ. This interval is

$$(Y_{(c_3)} - X_{(s)}, Y_{(c_4)} - X_{(s)}), \tag{2.53}$$

where c_3 and c_4 are chosen to satisfy the condition

$$P(X_{(s)} < Y_{(c_3+1)}|H_0) \leq \alpha/2, P(X_{(s)} \geq Y_{(c_4)}|H_0) \leq \alpha/2.$$

The integers c_3 and c_4 can be approximated as follows. Let

$$c^* = (n/2) + 0.5 - z_{(1-\alpha/2)}\{nN/4m\}^{1/2}.$$

Then

$$c_3 \approx \lfloor c^* \rfloor, \quad and \quad c_4 \approx \lfloor (n - c^*) \rfloor + 1. \tag{2.54}$$

     The detailed derivation of the confidence interval is given in Appendix A2.
     Albers and Löhnberg (1984) indicated the usefulness of a confidence interval for the difference $\xi_Y - \xi_X$ between the medians in relation to a biomedical problem. They gave a method for obtaining an interval *without the shift assumption*, which is useful in large samples. This will be discussed in a later section.

2.4.2 Combined sample percentile tests — Mood's median test

We consider a general class of test statistics, of which one member can be related to the Mood's median statistic. Let $Z_{(s)}$ be the sth order statistic of the combined sample. Let

$$A_s = number\ of\ X\text{-}values\ less\ than\ or\ equal\ to\ Z_{(s)}.$$

Table 2.11 *Combined sample percentile tests*

Alternative	Critical Region
$H_{A1} : X st < Y$	$A_s \geq a_1$
$H_{A2} : X st > Y$	$A_s \leq a_2$
$H_{A3} : H_{A1} \text{ or } H_{A2}$	$A_s \leq a_3 \text{ or } A_s \geq a_4$

It can be shown (see Appendix A2) that

$$P[A_s = k|H_0] = \frac{\binom{s}{k}\binom{N-s}{m-k}}{\binom{N}{m}},$$

for integers k between $max(s - n, 0)$ and $min(s, m)$. If B_s denotes the number of Y-values among the smallest s values of the combined sample, we have

$$A_s + B_s = s.$$

Intuitively, if $A_s > B_s$, we would infer that X-values are tending to be smaller than Y-values. In other words, $A_s \geq B_s - c$, is an appropriate critical region of a test for the one-sided alternative $H_{A1} : X st < Y$. This critical region can be restated as $A_s \geq a_1$. A similar argument gives an appropriate critical region for the alternative H_{A2}. Combining these two critical regions for the two one-sided alternatives we get the critical region for the alternative H_{A3}. Now we summarize the alternatives and the corresponding critical regions in Table 2.11.

In the case of two-sided alternatives, we suggest an equi-tail test. The constants a_i's are the appropriate percentiles of the null distribution of the statistic A_s.

Remark 2.12. Smirnov's test depends on the differences $d_s = F_m(Z_{(s)}) - G_n(Z_{(s)})$, and it is easy to see that $d_s = \frac{N}{mn} A_s - \frac{s}{n}$. This relationship is invoked to infer that the null distribution of Smirnov's statistic, to be discussed later in this chapter, does not depend on the population distributions.

The critical regions can be reworded as the critical regions of precedence-type tests. For this purpose let us consider the critical region of the test for the alternative H_{A1}. This set is $A_s \geq a_1$. This set corresponds to the event that the s smallest values include at least a_1, X-values and hence at most $s - a_1$, Y-values. In other words, in this set, we have

$$Z_{(s)} \geq X_{(a_1)} \quad and \quad Z_{(s)} < Y_{(s-a_1+1)},$$

where $Z_{(s)}$, is the sth order statistic of the combined sample. This set is the same as

$$X_{(a_1)} < Y_{(s-a_1+1)}.$$

Thus this set is the critical region for the alternatives H_{A1}. Arguing in a similar manner other critical regions can be restated. The critical regions of the precedence test versions are listed in Table 2.12.

Table 2.12 *Precedence test versions of tests of Table 2.11*

Alternative	Critical Region
$H_{A1} : Xst < Y$	$X_{(a_1)} < Y_{(s-a_1+1)}$
$H_{A2} : Xst > Y$	$X_{(a_2+1)} > Y_{(s-a_2)}$
$H_{A3} : H_{A1} \, or \, H_{A2}$	$X_{(a_3+1)} > Y_{(s-a_3)} \, or \, X_{(a_4)} < Y_{(s-a_4+1)}$

From these critical regions, confidence intervals for the shift parameter can be derived. We do this confidence interval estimation using Mood's median test.

Mood's median test

Mood's median test is a member of the class of tests considered thus far. Because of its importance we describe the details here. Later a confidence interval procedure for the shift parameter will be described. Let $\tilde{z}$ be the median of the combined sample and let $N = m + n$. This is the sth order statistic in the combined sample, where

$$s = \begin{cases} (N+1)/2, & \text{if } N \text{ is odd,} \\ N/2, & \text{if } N \text{ is even.} \end{cases}$$

Let V be the random variable that denotes the number of X-observations that *exceed* $\tilde{z}$. It is easy to see that $V = m - A_s$. Thus the null distribution of V is given by

$$P(V = v|H_0) = P(A_s = m - v|H_0) = \frac{\binom{s}{m-v}\binom{N-s}{v}}{\binom{N}{m}},$$

and rewriting the right-side expression, we have

$$P(V = v|H_0) = \frac{\binom{m}{v}\binom{n}{N-s-v}}{\binom{N}{N-s}}, \qquad (2.55)$$

for all integers v between $max(m - s, 0)$ and $min(m, N - s)$. Identifying this distribution as a hypergeometric distribution, we have

$$E(V|H_0) = (N - s)(m/N) \approx (m/2),$$

and

$$var(V|H_0) = \frac{mns(N - s)}{(N - 1)N^2} \approx \frac{mn}{4N}. \qquad (2.56)$$

Various alternatives and the corresponding critical regions are listed in Table 2.13, where the critical values are appropriate percentiles of the null distribution of V. We suggest a test with equal tails for the two-sided alternatives H_{A3}. A computer program for finding the critical values is given in Appendix B2. It is convenient to compute the P-values since the cdf of a hypergeometric distribution is a SAS function.

Table 2.13 *Mood's median tests*

Alternative	Critical Region
$H_{A1} : Xst < Y$	$V \leq v_1,$
$H_{A2} : Xst > Y$	$V \geq v_2,$
$H_{A3} : H_{A1}$ or H_{A2}	$V \leq v_3$ or $V \geq v_4,$

Gastwirth (1968) showed that the control median test reaches a decision before the Mood's median test, when one is testing H_0 against the one-sided alternative $H_{A1} : \Delta > 0$, under the shift model. In this testing situation when early stopping is important, one should use the control median procedure rather than Mood's median procedure.

Confidence interval for Δ from Mood's median test

Here, like in the previous section, we assume that

$$G(x) = F(x - \Delta),$$

and consider the problem of testing $H_0 : \Delta = 0$ against the two-sided alternatives $H_{A3} : \Delta \neq 0$. The critical region of the relevant test, based on the statistic V, is

$$V \leq v_3 \quad or \quad V \geq v_4, \qquad (2.57)$$

where v_3 is the largest integer such that

$$P(V \leq v_3 | H_0) \leq (\alpha/2),$$

and v_4 is the smallest integer such that

$$P(V \geq v_4 | H_0) \leq (\alpha/2).$$

Using the acceptance region, the complement of the critical region (2.57), we obtain a confidence interval for the shift parameter, Δ. This interval is

$$\left(Y_{(s-m+v_3+1)} - X_{(m-v_3)}, Y_{(s-m+v_4)} - X_{(m-v_4+1)}\right), \qquad (2.58)$$

where v_3 and v_4 are chosen as in the definition of the critical region (2.57). The derivation of this confidence interval appears in Appendix A2.

Large sample approximations to v_3 and v_4

In large samples we can approximate the distribution of

$$Z = [V - (m/2)]/(mn/4N)^{1/2}$$

by the standard normal distribution. Thus an approximate α-level test for H_{A3} is to

reject H_0 when $|Z| > z_{(1-\alpha/2)}.$

Comparing the critical region of this two-sided test with the critical region (2.57), we get the approximations. Let

$$v^* = (m/2) - 0.5 - z_{(1-\alpha/2)}(mn/4N)^{1/2}.$$

The approximations are

$$v_3 \approx \lfloor v^* \rfloor, \quad v_4 \approx \lfloor (m - v^*) \rfloor + 1. \tag{2.59}$$

These approximations are quite accurate as long as $min(m, n)$ is at least 12.

2.4.3 Wilcoxon-Mann-Whitney procedure

Here X stands for the response variable under *control treatment*, whereas Y stands for the response under *active treatment*. If the treatment has no effect, the variables X and Y are identically distributed, so that $F(x) = G(x)$ for every x. In some cases the treatment tends to increase the response level. This situation is described as $P(X > x) \le P(Y > x)$, for all x with strict inequality for some x. This condition means that Y *is stochastically larger than X*. When the treatment tends to decrease the response level, we say that Y *is stochastically smaller than X*. Thus we consider the problem of testing the homogeneity hypothesis against stochastically ordered alternatives.

A parameter that can be used to describe the stochastic ordering between the two random variables X and Y is

$$\theta = P(X < Y) = \int_{-\infty}^{\infty} F(y)g(y)dy, \tag{2.60}$$

where $g(.)$ is the pdf of Y. It is easy to see that

$$\theta = (1/2) \quad \text{if } F = G,$$
$$> (1/2) \quad \text{if } Gst > F (i.e. Y st > X),$$
$$< (1/2) \quad \text{if } Gst < F (i.e. Y st < X). \tag{2.61}$$

In view of the above observation, one is tempted to use an estimate of θ as a test criterion. This is essentially the idea behind the proposal of Mann and Whitney. Their test is defined in terms of the statistic U_{YX}, which denotes the number of times a y *precedes* an x. However, we consider a related statistic U_{XY}, which denotes the number of times an x *precedes* a y. To compute this number we need to pair each X with each Y and then count the number of pairs (X_i, Y_j) for which $X_i < Y_j$. In other words,

$$U_{XY} = \# \text{ of pairs } (X_i, Y_j) \text{ for which } X_i < Y_j.$$

We can express this number as a sum, which is

$$U_{XY} = \sum_{i=1}^{m}\sum_{j=1}^{n} \psi(X_i, Y_j), \tag{2.62}$$

Table 2.14 *Mann-Whitney tests*

Alternative	θ-Version	Critical Region
$H_{A1} : Xst < Y$	$\theta > (1/2)$	$U_{XY} \geq c_1$
$H_{A2} : Xst > Y$	$\theta < (1/2)$	$U_{XY} \leq c_2$
$H_{A3} : H_{A1}$ or H_{A2}	$\theta \neq (1/2)$	$U_{XY} \leq c_3$ or $U_{XY} \geq c_4$

where

$$\psi(a,b) = \begin{cases} 1, & \text{if } a < b, \\ 0, & \text{otherwise.} \end{cases}$$

When there are *no ties* in the combined sample, the number of pairs (X_i, Y_j) for which $X_i = Y_j$ is zero. Then the mn possible pairs (X_i, Y_j) fall into two categories: (1) those for which $X_i > Y_j$ and (2) those for which $X_i < Y_j$. As the statistic U_{YX} denotes the *number of pairs* (X_i, Y_j) for which $Y_j < X_i$, we have the relationship

$$U_{XY} + U_{YX} = mn.$$

In other words,

$$U_{YX} = mn - U_{XY}. \tag{2.63}$$

We now compute the expected value of U_{XY}. It is

$$E(U_{XY}) = \sum_{i=1}^{m} \sum_{j=1}^{n} E[\psi(X_i, Y_j)] = \sum_{i=1}^{m} \sum_{j=1}^{n} P(X_i < Y_j),$$
$$= mn\theta.$$

Hence $U_{XY}/(mn)$ is an unbiased estimator of θ, which is equal to $(1/2)$ under the null hypothesis. Thus extreme values of $U_{XY} - (mn/2)$ suggest the rejection of the null hypothesis

$$H_0 : F(x) = G(x), \text{ for every } x; \quad \text{or} \quad \theta = (1/2). \tag{2.64}$$

The various alternatives of interest and the critical regions of the tests are given in Table 2.14. The constants c_i's are chosen so that the tests are of size α. For the two-sided alternative the two tails are chosen to be of size $\alpha/2$.

Mann and Whitney (1947) gave some tabulations of the null distribution of U_{YX}. Several others have done some tabulations. Milton (1964) published an extended table of critical values for the statistic U_{YX}, which can be used to obtain the critical values needed for implementing the tests mentioned earlier. Mann and Whitney observed that the null distribution of U_{YX} differs very little from the normal distribution when $min(m, n)$ is at least 8. To compute this approximation we need an expression for the null variance of U_{XY}. It can be shown that

$$E(U_{XY}|H_0) = mn/2 \quad and \quad var(U_{XY}|H_0) = \frac{mn(m + n + 1)}{12}.$$

Table 2.15 *Wilcoxon rank sum tests*

Alternative	Critical Region	P-Value
$H_{A1} : X st < Y$	$W_Y \geq c_1$	$P_1 = P(W_Y \geq W_Y(obs)|H_0)$
$H_{A2} : X st > Y$	$W_Y \leq c_2$	$P_2 = P(W_Y \leq W_Y(obs)|H_0)$
$H_{A3} : H_{A1} \ or \ H_{A2}$	$W_Y \leq c_3 \ or \ W_Y \geq c_4$	$P_3 = 2min(P_1, P_2)$

The derivation of the variance is given in Appendix A2. In the next sub-section these critical regions will be defined in terms of the Wilcoxon rank sum statistic and are given in Table 2.15. This is accomplished by finding the relationship between the statistic U_{XY} and the Y-ranks sum statistic.

Wilcoxon rank sum procedure

For the testing problem considered earlier, Wilcoxon (1945) proposed a different test statistic, which depends on the ranks of the observations in the combined sample. In other words the two samples are combined and the observations are ranked, so that the smallest value will get rank 1 and the largest value will get rank $N(= m+n)$. We will describe the remaining details of this procedure by assuming that there are no tied values in the combined sample. Of course, the necessary modifications when there are ties will be mentioned later in this section.

Let $R_1 < R_2 < \cdots < R_n$ be the ordered ranks of the Y-observations in the combined sample. This means R_1 is the rank of the smallest Y-observation, R_2 is the rank of second smallest Y-observation, and so on. The Wilcoxon rank sum statistic is defined as

$$W_Y = \ sum \ of \ the \ Y\text{-}ranks = \sum_{i=1}^{n} R_i. \tag{2.65}$$

Here again large or small values of $W_Y - E(W_Y|H_0)$ will lead to the rejection of the null hypothesis (2.64). The reasoning behind this characterization of the critical regions follows from the relationship between the statistics W_Y and U_{XY}, which will be established now.

Since there are no ties in the combined sample, it is easy to see that $(R_1 - 1)$ X-observations are below the Y-observation that received the rank R_1. Likewise there are $(R_2 - 2)$ X-observations below the Y-observation that received the rank R_2 and so on. Since U_{XY} stands for the number of X-observations below each Y-observation, we have

$$U_{XY} = \sum_{i=1}^{n}(R_i - i) = \sum_{i=1}^{n} R_i - \frac{n(n+1)}{2}.$$

In other words,

$$U_{XY} = W_Y - [n(n+1)/2]. \tag{2.66}$$

Now it follows that

$$E(W_Y|H_0) = E(U_{XY}|H_0) + [n(n+1)/2] = n(N+1)/2.$$

In view of the linear relationship (2.66), the critical regions of the tests for stochastically ordered alternatives can be expressed in terms of the statistic W_Y.

Remark 2.13. A relation similar to (2.66) exists between U_{YX} and W_X, the sum of X-ranks. It is

$$U_{YX} = W_X - [m(m+1)/2].$$

In view of the relations among the four statistics U_{YX}, U_{XY}, W_X and W_Y, the critical regions can be expressed in terms of any one of these statistics.

The null distribution of W_Y can be developed from the distribution of the ranks, which is uniform over all possible values. Thus by enumeration we can develop this distribution and hence can determine the critical values. These calculations will be demonstrated for the case $m = 2$ and $n = 3$.

Example 2.5. The null distribution of W_Y for $m = 2, n = 3$: The values of W_Y for possible values for the ranks are given in the following table.

(R_1, R_2, R_3)	W_Y	(R_1, R_2, R_3)	W_Y
(1,2,3)	6	(1,4,5)	10
(1,2,4)	7	(2,3,4)	9
(1,2,5)	8	(2,3,5)	10
(1,3,4)	8	(2,4,5)	11
(1,3,5)	9	(3,4,5)	12

All these possible values for the rank vector are equally likely, so the probability for each value is $(1/10)$. This information will enable us to derive the distribution of W_Y, which is given in the following table.

w	6	7	8	9	10	11	12
$P(W_Y = w)$	0.1	0.1	0.2	0.2	0.2	0.1	0.1

From this table it is easy to calculate the P-values and also get the critical values to perform the tests. We observe that the distribution is symmetric about 9, which is the mean of the distribution.

In general, we have

$$E(W_Y|H_0) = n(N+1)/2, \quad var(W_Y|H_0) = nm(N+1)/12. \tag{2.67}$$

A detailed derivation of the variance appears in Appendix A2. Using these quantities we obtain the standardized random variable,

$$Z_W = (W_Y - E(W_Y|H_0))/\sqrt{var(W_Y|H_0)}, \tag{2.68}$$

whose distribution can be approximated by the standard normal distribution. This approximate normal distribution of Z_W is useful for approximating the critical values or for approximating the P-values.

For future reference, we express the critical regions, given in Table 2.14, in terms of W_Y in Table 2.15. In this table $W_Y(obs)$ stands for the observed value of W_Y. The critical values are chosen so that the tests are of size α. The null distribution of W_Y is symmetric about its mean (see Appendix A2). So, for any integer c, we have

$$P(W_Y \leq \mu - c|H_0) = P(W_Y \geq \mu + c|H_0),$$

where $\mu = E(W_Y|H_0) = n(N+1)/2$. In view of this fact, we use an equi-tailed test for the two-sided alternative case. In other words, the critical region is taken as

$$|W_Y - n(N + 1)/2| \geq c_5,$$

and thus the P-value is

$$P_3 = P(|W_Y - n(N + 1)/2| \geq |W_Y(obs) - n(N + 1)/2||H_0).$$

It can be shown that $P_3 = 2min(P_1, P_2)$.

Critical values for tests based on W_Y

To implement the test for the alternative H_{A2} we have to find the constant c, which is the largest integer such that

$$P(W_Y \leq c(m, n)|H_0) \leq \alpha.$$

Verdooren (1963) published tables of critical values for Wilcoxon's test statistic. The tables prepared by Wilcoxon, Katti and Wilcox (1973) also can be used to implement the tests defined by the statistic W_Y. The tables of Wilcoxon et al. enable us to choose a critical region of size close to the chosen α. From the tables prepared by Milton (1964), we also can obtain the required critical values. The computer packages usually give the P-values.

Large sample null distribution of W_Y

Mann and Whitney derived a recurrence relation that can be used to generate the null distribution of U_{XY} and observed that this null distribution differs very little from the standard normal distribution when $min(m, n) \geq 8$. Since

$$U_{XY} = W_Y - \frac{n(n + 1)}{2},$$

the null distribution of W_Y can also be approximated by a normal distribution, whenever $min(m, n)$ is at least 8. This result will enable us to get approximations for the critical values needed to implement tests based on U_{XY} or W_Y. In an example we examine the closeness of this approximation. A continuity correction usually improves the approximation. In other words, the cdf is approximated as

$$P(W_Y \le c(m, n)|H_0) \approx \Phi \left[\frac{c(m, n) + 0.5 - E(W_Y|H_0)}{\sqrt{var(W_Y|H_0)}} \right].$$

This leads to the approximation for the α quantile, namely,

$$c(\alpha; m, n) \approx E(W_Y|H_0) - 0.5 + z_\alpha \sqrt{var(W_Y|H_0)}.$$

Example 2.6. Calculation of approximate critical values: Let $m = n = 8$ and $\alpha = 0.05$. For this case,

$$z_{0.05} = -1.645$$

and

$$E(W_Y|H_0) = 8 \cdot (17)/2 = 68, \quad var(W_Y|H_0) = 90.67.$$

Hence

$$c(0.5; 8, 8) \approx \lfloor 68 - 0.5 + (-1.645)\sqrt{90.67} \rfloor = 51.$$

From the table of Wilcoxon et al. we have that

$$P(W_Y \le 52|H_0) = 0.0524 \quad and \quad P(W_Y \le 51|H_0) = 0.0415.$$

So the exact value of c is 51, which is the same as the approximation.

Example 2.7. Let the X-sample be $\{69, 61, 76\}$ and the Y-sample be $\{73, 85, 79, 72\}$.Thus $m = 3$, and $n = 4$. We want to test H_0 of (2.64) against the alternative H_{A1} at level $\alpha = 0.05$. The ordered sample values, their sample ID and the ranks are displayed below.

Values	61	69	72	73	76	79	85
ID	x	x	y	y	x	y	y
Ranks	1	2	3	4	5	6	7

Now the sum of Y-ranks is $W_Y = 20$. Using a computer program (see Appendix B2), the exact P-value is found to be 0.1143. As this quantity is greater than 0.05, we do not reject H_0.

Remark 2.14. In Example 2.7, it is easy to calculate the P-value by enumerating all possible values for the Y-ranks whose sum is greater than or equal to 20, the observed value. As mentioned in the worked-out example, each of these possible values is associated with probability $1/\binom{7}{4} = 1/35$. These cases are

$$7 + 6 + 5 + 4 = 22; \quad 7 + 6 + 5 + 2 = 20;$$
$$7 + 6 + 5 + 3 = 21; \quad 7 + 6 + 4 + 3 = 20.$$

Thus the required P-value is $(4/35) = 0.1143$, as given by the program output.

Extension of the WMW procedure, when there are ties

Even when the variable under observation is a continuous variable, due to the limitations of the measuring instruments, we do get tied values. An extreme case is the ordered categorical data setting. We now discuss the modified definitions of the statistics W_Y and U_{XY} and the corresponding testing procedures.

We first consider a method of assigning ranks, which will lead to a natural definition for the statistic U_{XY}. When there are ties, one method is to assign the average of the ranks that the tied group would have received if the values were distinct. We will explain this concept by considering an example. Suppose that our combined sample is 80, 85, 85, 85, 90, 90. The value 80 will receive rank 1. In the next group we have three values, each equal to 85. If these values were distinct they would have received the ranks 2, 3, and 4. The average of these three numbers is 3. Thus each of the three values of 85 will be assigned the average rank of 3. Using a similar argument, we will assign an average rank of 5.5 to the two values that are equal to 90.

Formulas for ranks

It is advantageous to have formulas for computing the average ranks. The values in the combined sample are denoted by $(Z_1, Z_2, \ldots, Z_N)$. To find the rank of Z_h, we need to compare this value with every other value in the group and find out the number of values less than Z_h and the number of values equal to Z_h. Then the rank of Z_h is

$$r(Z_h) = \#\ less\ than\ Z_h + \frac{1}{2}\big(\#\ equal\ to\ Z_h] + 1\big) \qquad (2.69)$$

We can express the quantity on the right in terms of the function

$$c(a, b) = \begin{cases} 0, & \text{if } a < b, \\ (1/2), & \text{if } a = b, \\ 1, & \text{if } a > b. \end{cases} \qquad (2.70)$$

Now the rank of the observation Z_h is seen to be

$$r(Z_h) = \frac{1}{2} + \sum_{k=1}^{N} c(Z_h, Z_k). \qquad (2.71)$$

This expression gives the usual ranks when there are no ties and the average ranks if there are some ties.

w-ranks

Noether (1967) suggested the use of quantities called w-ranks in place of the ranks. They are defined as

$$w(Z_h) = \#\ less\ than\ Z_h - \#\ greater\ than\ Z_h. \qquad (2.72)$$

These w-ranks can be expressed as sums of *sign* functions. In other words,

$$w(Z_h) = \sum_{k=1}^{N} sign(Z_h, Z_k), \qquad (2.73)$$

where the sign function is

$$sign(a, b) = \begin{cases} -1, & \text{if } a < b, \\ 0, & \text{if } a = b, \\ 1, & \text{if } a > b. \end{cases}$$

It is easy to see that $sign(Z_h, Z_k) = 2c(Z_h, Z_k) - 1$; hence we can relate the (usual) ranks with the w-ranks. In fact,

$$w(Z_h) = 2\left[r(Z_h) - \frac{(N+1)}{2}\right]. \qquad (2.74)$$

Thus the w-ranks are twice the deviations of the ranks from $(N+1)/2$, which is the average of all ranks.

Some properties of the w-ranks:
(1) The w-ranks are always integers.
(2) The w-ranks can be negative.
(3) The sum of the w-ranks is zero.
(4) Variance of the w-ranks $= (1/(N-1)) \sum_h w^2(Z_h) = 4\times$ variance of the ranks.

These w-ranks are generalized to w-scores, when the data contain some right-censored values. This generalization will be discussed in a later section.

Ranks as functions of the sample distribution function
Let $F_N(.)$ be the distribution function of the sample $(Z_1, \ldots, Z_N)$. Recall that

$$F_N(Z_h) \equiv F_N(Z_h + 0) = (\# \text{ of values less than or equal to } Z_h)/N.$$

Since F_N jumps at each distinct value, we have that

$$F_N(Z_h - 0) = (\# \text{ of values less than } Z_h)/N,$$

and

$$F_N(Z_h + 0) - F_N(Z_h - 0) = (\# \text{ of values equal to } Z_h)/N.$$

Thus, from (2.69), we have

$$r(Z_h) = NF_N(Z_h - 0) + (N/2)[F_N(Z_h + 0) - F_N(Z_h - 0)] + (1/2).$$

This is the same as

$$(1/N)2[r(Z_h) - (N+1)/2] = [F_N(Z_h + 0) + F_N(Z_h - 0)] - 1. \qquad (2.75)$$

This relation can also be expressed in terms of the survival function, $S_N = 1 - F_N$, as

$$(1/N)2[r(Z_h) - (N+1)/2] = 1 - [S_N(Z_h + 0) + S_N(Z_h - 0)]. \qquad (2.76)$$

The expressions on the right can be used as a generalization of ranks. This type of generalization has been used by Peto and Peto (1972) in dealing with data that contain some right-censored observations.

Now using the ranks defined by (2.69), we define the Wilcoxon rank sum statistic as

$$W_Y^* = sum \ of \ ranks \ of \ the \ y\text{-}observations. \qquad (2.77)$$

To be in conformity with the relation (2.66), a modified definition of the Mann-Whitney statistic is used. This statistic is denoted by U_{XY}^* and is defined as

$$U_{XY}^* = number \ of \ (X_i, Y_j) \ pairs \ for \ which \ X_i < Y_j$$
$$+ (1/2)(number \ of \ (X_i, Y_j) \ pairs \ for \ which \ X_i = Y_j). \qquad (2.78)$$

A detailed explanation about why this definition is a natural one is given in Appendix A2.

It is obvious that $U_{XY}^* = U_{XY}$, when there are no ties. Further, we have

$$U_{XY}^* = W_Y^* - \frac{n(n+1)}{2}. \qquad (2.79)$$

It is natural to replace W_Y by W_Y^* in the critical regions of Table 2.15 to obtain the modified critical regions. Since the null distribution is not tabulated, it is suggested that we use the critical regions defined by the standardized variable

$$Z_{W^*} = [W_Y^* - E(W_Y^*|H_0)]/\sqrt{var(W_Y^*|H_0)}, \qquad (2.80)$$

which is approximately distributed as the standard normal variable. It can be shown that

$$E(W_Y^*|H_0) = n(N+1)/2. \qquad (2.81)$$

Let S^{*2} be the variance of the ranks in the combined sample, that is,

$$S^{*2} = (1/(N-1))[sum \ of \ squares \ of \ the \ ranks - (N/4)(N+1)^2]$$
$$= \frac{1}{12}\{N(N+1)\} - \frac{1}{12(N-1)}\sum_{i=1}^{k} d_i(d_i^2 - 1),$$

where k is the number of distinct observations and d_i is the number of times the ith tied observation is repeated. Now the $var(W_Y^*|H_0)$ can be shown to be

$$var(W_Y^*|H_0) = (nm/N)S^{*2}. \qquad (2.82)$$

The derivation of the mean and the variance of W_Y^* is given in Appendix A2. The critical regions of approximate size α are given in Table 2.16. Before we illustrate this testing procedure, we give an expression to W_Y^* using the w-ranks. It is easy to see that

$$(1/2)\sum_{h=m+1}^{N} w(Z_h) = W_Y^* - E(W_Y^*|H_0).$$

Table 2.16 *Wilcoxon tests for tied data*

Alternative	Critical Region
H_{A1}	$Z_{W^*} > z_{1-\alpha}$
H_{A2}	$Z_{W^*} < z_{\alpha}$
H_{A3}	$|Z_{W^*}| > z_{(1-\alpha/2)}$

Table 2.17 *Results of a clinical trial*

			Response		
Treatment	1	2	3	4	5
A	1	13	16	15	7
B	5	21	14	9	3
Col. Total	6	34	30	24	10
Average	(7/2)	6+(35/2)	40+(31/2)	70+(25/2)	94+(11/2)
Ranks	=3.5	23.5	=55.5	=82.5	=99.5

Also

$$var(W_Y^*|H_0) = (nm/N)(1/4)\left(\sum_{h=1}^{N} w(Z_h)^2/(N-1)\right).$$

Thus

$$Z_{W^*} = \frac{\sum_{h=m+1}^{N} w(Z_h)}{\sqrt{(nm/N)\left(\sum_{h=1}^{N} w(Z_h)^2/(N-1)\right)}}. \tag{2.83}$$

We illustrate this testing procedure using ordered categorical data in the following example. A SAS program for carrying out this test is given in Appendix B2.

Example 2.8. An extreme case of tied observations occurs when the responses are categorical and the Wilcoxon procedure can be used if there is an ordering among the categories. DeJonge (1983) gave the results of a clinical trial, which is a comparative study involving two treatments, A and B. The response variable is the clinical change and it has been categorized as: worse (1), no change (2), slight improvement (3), moderate improvement (4) and marked improvement (5). The data and the average ranks are given in Table 2.17.

We use (2.69) to calculate the average ranks. For example,

rank for group 1 = $r(Z_1)$ = (no. in first category + 1)/2 = (6 + 1)/2 = 3.5.

Further,

rank for group 2 = (no. in first category) + (34 + 1)/2 = 6 + (35/2) = 23.5

The other ranks are calculated in a similar manner. Now the sum of the ranks for the observations on treatment A is

$$W_A^* = 1(3.5) + 13(23.5) + 16(55.5) + 15(82.5) + 7(99.5) = 3131.$$

Further,

$$E(W_A^*|H_0) = 52(105)/2 = 2730.$$

Here $k = 5, d_1 = 6, d_2 = 34, d_3 = 30, d_4 = 24$, and $d_5 = 10$. Using (2.82) we get the variance as

$$var(W_A^*|H_0) = 21951.068.$$

Thus the Z-statistic is

$$Z_{W^*} = (3131 - 2730)/[21951.068]^{1/2} = 2.71$$

Since we are interested in learning about the differences in the treatment effects, we take the two-sided alternatives. Since this value is greater than $z_{0.975} = 1.96$, the result is significant at level $\alpha = 0.05$. So we conclude that the two treatments differ in their effects.

Confidence interval for the shift parameter Δ

An important special case of a stochastic ordering between the two random variables X and Y is the so-called shift relationship introduced in (2.46), which is

$$G(x) = F(x - \Delta), \quad \text{for all } x,$$

where F is the cdf of X and G is the cdf of Y. This means that the distribution of $X + \Delta$ is the same as the distribution of Y. It is easy to verify that X is stochastically smaller than Y when $\Delta > 0$. However, X is stochastically larger than Y when $\Delta < 0$. It has already been mentioned that Δ can be interpreted as the difference between the medians, namely $med(Y) - med(X)$, in general. Further, if $E(X)$ and $E(Y)$ exist, then Δ can be seen to be the difference between the two means, namely, $E(Y) - E(X)$. The null hypothesis and the various alternatives can now be expressed in terms of the shift parameter Δ. The tests of Table 2.14. are restated in Table 2.18.

The main interest is the construction of a confidence interval using the statistic U_{XY}. We start with the problem of testing

$$H_0 : \Delta = \Delta_0 \quad against \quad H_A : \Delta \neq \Delta_0.$$

Table 2.18 *Mann-Whitney tests for shift parameter*

Alternative	Δ-Version	Critical Region
$H_{A1} : X st < Y$	$\Delta > 0$	$U_{XY} \geq c_1$
$H_{A2} : X st > Y$	$\Delta < 0$	$U_{XY} \leq c_2$
$H_{A3} : H_{A1} \text{ or } H_{A2}$	$\Delta \neq 0$	$U_{XY} \leq c_3 \text{ or } U_{XY} \geq c_4$

Under H_0, the samples $(X_1, X_2, \ldots, X_m)$ and $(Y_1 - \Delta_0, Y_2 - \Delta_0, \ldots, Y_n - \Delta_0) = (Y'_1, Y'_2, \ldots, Y'_n)$ are from the same distribution. So we can compute the statistic $U_{XY'}$, using the X-sample and Y'-sample and set up a test. It is easy to see that $U_{XY'}$ is the number of (X_i, Y_j) pairs for which $Y_j - X_i > \Delta_0$. The alternative for the current problem corresponds to H_{A3} in Table 2.18. The acceptance region of the relevant test is

$$c_3 + 1 \leq U_{XY'} \leq c_4 - 1. \tag{2.84}$$

Since $U_{XY'}$ has a symmetrical distribution, under H_0, the critical value c_4 is equal to $mn - c_3$. Now the confidence region is given by

$$\{\Delta_0 \mid c_3 + 1 \leq U_{XY'} \leq mn - c_3 - 1\}, \tag{2.85}$$

where c_3 is the $100(\alpha/2)$ percentile of U_{XY}. This region will turn out to be an interval. To obtain this interval we proceed as follows. Consider the differences $v_{ij} = Y_j - X_i$ and order them to obtain $v_{(1)} \leq v_{(2)} \leq \cdots \leq v_{(mn)}$. The confidence interval can be seen to be the interval $[v_{(c_3+1)}, v_{(mn-c_3)})$. The relevant explanation appears in Appendix A2. Using the normal approximation with the continuity correction, the integer c_3 can be approximated as

$$c_3 = \left\lfloor \frac{mn}{2} - 0.5 + z_{\alpha/2}\sqrt{mn(N+1)/12} \right\rfloor.$$

A computer program for calculating this interval is given in Appendix B2.

2.4.4 Analysis of proportional hazards model

This model is useful in the analysis of time to an event type response variable. The model prescribes a relationship between the hazard functions of the two distribution functions F and G, which are being compared. Let $\lambda_F(.)$ and $\lambda_G(.)$ be the two hazard functions. In other words,

$$\lambda_F(x) = \frac{f(x)}{1 - F(x)}; \quad \lambda_G(x) = \frac{g(x)}{1 - G(x)},$$

where f and g are probability densities. The corresponding cumulative hazard functions are

$$\Lambda_F(x) = \int_0^x \lambda_F(t)dt; \quad \Lambda_G(x) = \int_0^x \lambda_G(t)dt.$$

This model states that one hazard function is a scalar multiple of the other hazard function. In other words, the relationship between the two hazard functions is

$$\lambda_G(x) = (1 + \theta)\lambda_F(x),$$

where $1 + \theta$ is the proportionality constant. This relation implies a similar relation between the cumulative hazard functions $\Lambda_F(x)$ and $\Lambda_G(x)$. So

$$\Lambda_G(x) = (1 + \theta)\Lambda_F(x). \tag{2.86}$$

Using the fact that $\Lambda_F(x) = -log[1 - F(X)$ and $\Lambda_G(x) = -log[1 - G(X)$, from the relation (2.86), we obtain that

$$G(x) = 1 - [1 - F(x)]^{1+\theta}. \tag{2.87}$$

Thus the null hypothesis we want to test can be stated as $H_0 : \theta = 0$. Under this model there is a stochastic ordering between F and G, which is equivalent to a stochastic ordering between the random variables X and Y as indicated in Appendix A. So we can use either a median test or WMW test. However, we will discuss a different procedure that is related to the popular *logrank test*.

It is known that the variable $\Lambda_F(X) = -log[1 - F(X)]$ follows the exponential distribution with mean 1. By the same token $\Lambda_G(Y) = -log[1 - G(Y)]$ also follows the same exponential distribution. From the model we have $\Lambda_G(Y) = (1 + \theta)\Lambda_F(Y)$. Hence the variables $\Lambda_F(X)$ and $(1 + \theta)\Lambda_F(Y)$ follow an exponential distribution with mean 1. Testing the null hypothesis, $\theta = 0$, amounts to testing that both the variables $\Lambda_F(X)$ and $\Lambda_F(Y)$ follow the exponential distribution with mean 1. A suitable test criterion, based on the likelihood ratio method, in terms of the transformed variables is

$$T = \frac{\sum_i \Lambda_F(X_i)}{\sum_j \Lambda_F(Y_j)}.$$

Large or small values of T will lead to the rejection of H_0. The main limitation of this procedure is the fact that Λ_F is an unknown function. A natural way of overcoming this obstacle will be presented.

The exponential (or Savage) scores test

Since F is an unknown function, and F is the common distribution under the null hypothesis, we can estimate F using the combined sample. This estimate of F can give us an estimate of Λ_F, which can be used in T to get a test statistic. Alternatively, we can use an estimate of the cumulative hazard function Λ_F. This alternative scheme will be described, assuming that there are *no ties* in the combined sample. A popular estimate of Λ_F is the one proposed by Nelson (1972). It is a step function and it jumps at each sample value.

Let $(Z_1, \ldots, Z_m, Z_{m+1}, \ldots, Z_{m+n}) = (X_1, \ldots, X_m, Y_1, \ldots, Y_n)$ be the combined sample. The value of the estimate at Z_i is

$$\hat{\Lambda}(Z_i) = \frac{1}{N} + \frac{1}{N-1} + \cdots + \frac{1}{N - R_i + 1}, \tag{2.88}$$

where R_i is the rank of Z_i in the combined sample and $N = m + n$ is the size of the combined sample. It can be shown that

$$\hat{\Lambda}(Z_i) = E(V_{(R_i, N)}), \tag{2.89}$$

where $V_{(k,N)}$ is the kth order statistic in a sample of size N from the exponential distribution with mean 1. The quantities $E(V_{(k,N)})$ are called *exponential ordered scores*.

Now in the quantity T we replace the Λ_F function by the estimate $\hat{\Lambda}$ function to obtain T_1 and this statistic can be used as a test statistic. Since the sum of the numerator and the denominator of T_1 is N, we just use either the numerator or the denominator of T_1 as the test statistic. For definiteness, we use the numerator of T_1. Thus the resulting statistic is

$$T_{EX} = \sum_i E[V_{(R(X_i),N)}], \qquad (2.90)$$

where $R(X_i)$ is the rank of X_i in the combined sample. Savage (1956) proposed this statistic for the usual two-sample problem, when we suspect that the samples are from exponential distributions with different means. This statistic is related to the *logrank test statistic*. This relationship will be explained in a later section where methods for censored data are discussed.

The null distribution of this statistic can be obtained from the distribution of the ranks as in Example 2.5. The asymptotic normal distribution can be used to define the critical regions. The mean and the variance can be derived as in the case of the statistic W_Y. They are

$$E(T_{EX}|H_0) = m; \quad var(T_{EX}|H_0) = \frac{nm}{N}S^2, \qquad (2.91)$$

where

$$S^2 = \frac{1}{(N-1)} \sum_{k=1}^{N} [E(V_{(k,N)}) - 1]^2.$$

The test criterion is the standardized variable

$$Z_{EX} = \frac{T_{EX} - m}{\sqrt{var(T_{EX}|H_0)}}, \qquad (2.92)$$

which can be viewed as the standard normal variable and the critical regions of approximate size α tests are given in Table 2.19. A score test for this setting will be discussed in Appendix A2.

Remark 2.15. The estimate of Λ_F can be easily extended to the general case of ties. However, in that case it cannot be related to exponential scores.

Remark 2.16. Exact P-values can be obtained from SAS and StatXact packages among others. In SAS Savage scores are used. These are obtained by subtracting 1 from the exponential order scores.

Table 2.19 *Exponential ordered scores tests*

Alternative	θ-Version	Critical Region
$H_{A1} : X st < Y$	$\theta < 0$	$Z_{EX} \leq z_\alpha$
$H_{A2} : X st > Y$	$\theta > 0$	$Z_{EX} \geq z_{1-\alpha}$
$H_{A3} : H_{A1} or H_{A2}$	$\theta \neq 0$	$|Z_{EX}| \geq z_{(1-\alpha/2)}$

Example 2.9. Data on lifetimes (in hours) of certain types of electronic items produced under different conditions are collected and are given as follows.
Sample 1(X): 79   150   96   167   15   285   28   111
Sample 2(Y): 250   23   427   42   333   232   488   121   159
We want to assess the differences between the two lifetime distributions.

We will test the hypothesis of equality of the two distributions. A SAS program has been used to get the required statistics (see Appendix B2). From the computer output, the sum of the Savage scores for sample 1 is –3.1691 and thus $T_{EX} = -3.1691 + 8 = 4.8309$. Also $Z_{EX} = -1.6727$, with an approximate P-value of 0.0944 for two-sided alternatives. The exact P-value is 0.0931. Thus we do not reject the null hypothesis of equality of the two lifetime distributions at significance level $\alpha = 0.05$.

2.4.5 Smirnov test

In Chapter 1 we considered Kolmogorov's goodness-of-fit test. An extension of this procedure to the two-sample problem is the test proposed by Smirnov. In the goodness-of-fit test we compared the empirical (sample) distribution function with the hypothesized distribution function. A natural thing to do here is to compare the two empirical distributions and summarize the differences to obtain a test statistic.

Let $F_m(.)$ and $G_n(.)$ be the sample distribution functions of the X-sample and Y-sample, respectively. We consider three statistics for summarizing the differences between the two sample distributions. They are

$$D_{m,n}^{+} = sup_{(-\infty < x < \infty)}[F_m(x) - G_n(x)], \qquad (2.93)$$

$$D_{m,n}^{-} = sup_{(-\infty < x < \infty)}[G_n(x) - F_m(x)], \qquad (2.94)$$

$$D_{m,n} = sup_{(-\infty < x < \infty)}[|\, F_m(x) - G_n(x) \,|]. \qquad (2.95)$$

It will be shown later that

$$D_{m,n} = max[D_{m,n}^{+}, D_{m,n}^{-}]. \qquad (2.96)$$

Similar to previous sections we consider stochastic alternatives to the null hypothesis $H_0 : F = G$. These alternatives and the critical regions of the corresponding tests are given in Table 2.20.

To verify that the null distributions of these statistics do not depend on the common population distribution and to compute the statistics we derive equivalent expressions. Let $(Z_{(1)}, Z_{(2)}, \ldots, Z_{(N)})$ be the order statistics of the

Table 2.20 *Smirnov tests*

Alternative	Critical Region
$H_{A1} : X st < Y$	$D_{m,n}^{+} \geq c_1$
$H_{A2} : X st > Y$	$D_{m,n}^{-} \geq c_2$
$H_{A3} : H_{A1}\ or\ H_{A2}$	$D_{m,n} \geq c_3$

combined sample. Also let

$$\delta_j = \begin{cases} 1 & \text{if } Z_{(j)} \text{ is an } X, \\ 0 & \text{if } Z_{(j)} \text{ is an } Y, \end{cases}$$

for $j = 1, 2, \ldots, N$. Now the sample distribution functions can be seen to be

$$F_m(Z_{(j)}) = \frac{1}{m}[\delta_1 + \delta_2 + \cdots + \delta_j],$$

and

$$G_n(Z_{(j)}) = \frac{1}{n}[(1 - \delta_1) + (1 - \delta_2) + \cdots + (1 - \delta_j)]$$

$$= \frac{j}{n} - \frac{m}{n} F_m(Z_{(j)}).$$

Let $A_j = \sum_{\gamma=1}^{j} \delta_\gamma$, so that it stands for the number of X-values among the smallest j values of the combined sample. The difference between F_m and G_n can be seen to be

$$F_m(Z_{(j)}) - G_n(Z_{(j)}) = \frac{N}{mn}\left[A_j - \frac{mj}{N}\right] \equiv \frac{N}{mn} B_j. \qquad (2.97)$$

The order statistics $Z_{(i)}$s partition the real line into $(N+1)$ intervals, $I_0 = (-\infty, Z_{(1)})$, $I_j = [Z_{(j)}, Z_{(j+1)})$ for $j = 1, 2, \ldots, N-1$ and $I_N = [Z_{(N)}, \infty)$. Since in each of these intervals F_m and G_n are constant, we have

$$sup_{t \in I_j}[F_m(t) - G_n(t)] = [F_m(Z_{(j)}) - G_n(Z_{(j)})] = \frac{N}{mn} B_j,$$

for $j = 1, \ldots, N-1$. Also the difference $[F_m(t) - G_n(t)]$ is zero in the intervals I_0 and I_N. Further, $B_N = 0$. Thus

$$D_{m,n}^+ = sup_t[F_m(t) - G_n(t)] = max\left\{0, \frac{N}{mn} B_1, \ldots, \frac{N}{mn} B_{N-1}\right\}.$$

Hence we have

$$D_{m,n}^+ = max\left[\frac{N}{mn} B_1, \ldots, \frac{N}{mn} B_N\right] = \frac{N}{mn} max[B_1, \ldots, B_N]. \qquad (2.98)$$

The null distribution of the B's depends on the null distribution of the A's. However, the null distribution of the A's does not depend on the common distribution. Thus the statistic $D_{m,n}^+$ is a distribution-free statistic. If we suspect that $Xst < Y$, a large value of $D_{m,n}^+$ supports this alternative.

Arguing in a similar manner, we get

$$D_{m,n}^- = \frac{N}{mn} max[-B_1, \ldots, -B_N],$$

$$D_{m,n} = \frac{N}{mn} max[|B_1|, \ldots, |B_N|];$$

hence

$$D_{m,n} = max[D_{m,n}^+, D_{m,n}^-].$$

Since the null distributions of these statistics are independent of the common distribution, the tests based on these statistics are distribution-free tests.

It is important to note that the test for the alternatives H_{A3} can also be used for the *general class of alternatives* $H_A : F \neq G$. A class of alternatives known as *crossing hazards alternatives*, where F and G cross each other, is of special interest in the analysis of survival data. For these alternatives, this test is more powerful than the *exponential ordered scores test*, which is a popular test for the analysis of survival data. Fleming et al. (1980) considered a modified version of this statistic and used it for the case of crossing hazards and showed that the new test has better power. As these details are beyond the level of our discussion, they are omitted.

A table of critical values needed for the test of one-sided alternatives, H_{A1}, has been prepared by Gail and Green (1976). Tables of exact distribution of the statistic $D_{m,n}$ have been prepared by Kim and Jennrich (1973).

Remark 2.17. To handle the case of alternatives H_{A2}, just rename the X-sample and the Y-sample as Y^*-sample and X^*-sample, respectively. In relation to the new variables X^* and Y^*, the problem is the same as the problem of testing for the alternatives H_{A1}.

Exact null distributions for equal sample sizes

When the sample sizes are equal, i.e., $m = n$, the null distribution simplifies considerably. It has been shown that for $k = 1, 2, \ldots, n$

$$P[D_{nn}^+ \geq k/n | H_0] = \binom{2n}{n-k} \div \binom{2n}{n}, \tag{2.99}$$

$$P[D_{nn} \geq k/n | H_0] = 2 \sum_{i=1}^{j} (-1)^{i+1} \binom{2n}{n-ik} \div \binom{2n}{n}, \tag{2.100}$$

where $j = \lfloor (n/k) \rfloor$ is the largest integer not exceeding (n/k). Derivation of these results is available in Pratt and Gibbons (1981). This can be used for computing the P-values. Computer programs for computing the P-values for the equal sample size case are given in Appendix B2.

Example 2.10. Consider the following two samples:
X-sample: 69, 15, 80, 89, 58, 38, 23, 43, 83, 87;
Y-sample: 35, 51, 76, 77, 62, 66, 32, 48, 50, 20.
The relevant calculations are given in Table 2.21. The graph of F_m against G_n is given in Figure 2.1.

Now the test statistics are

$$D_{10,10}^+ = 0.10; \ D_{10,10}^- = 0.40, \ D_{10,10} = 0.40.$$

The exact P-value for the two-sided alternatives is 0.4175. So we do not reject the null hypothesis.

Table 2.21 *Sample distribution functions*

j	$Z_{(j)}$	Sample ID	$F_m(Z_{(j)})$	$G_n(Z_{(j)})$	DV_j	$\mid DV_j \mid$
1	15	X	0.10	0.00	0.10	0.10
2	20	Y	0.10	0.10	0.00	0.00
3	23	X	0.20	0.10	0.10	0.10
4	32	Y	0.20	0.20	0.00	0.00
5	35	Y	0.20	0.30	−0.10	0.10
6	38	X	0.30	0.30	0.00	0.00
7	43	X	0.40	0.30	0.10	0.10
8	48	Y	0.40	0.40	0.00	0.00
9	50	Y	0.40	0.50	−0.10	0.10
10	51	Y	0.40	0.60	−0.20	0.20
11	58	X	0.50	0.60	−0.10	0.10
12	62	Y	0.50	0.70	−0.20	0.20
13	66	Y	0.50	0.80	−0.30	0.30
14	69	X	0.60	0.80	−0.20	0.20
15	76	Y	0.60	0.90	−0.30	0.30
16	77	Y	0.60	1.00	−0.40	0.40
17	80	X	0.70	1.00	−0.30	0.30
18	83	X	0.80	1.00	−0.20	0.20
19	87	X	0.90	1.00	−0.10	0.10
20	89	X	1.00	1.00	0.00	0.00

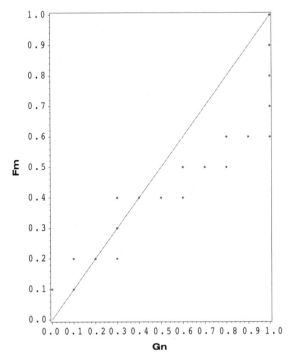

Figure 2.1 *P-P plot for the two-sample problem.*

2.4.6 P-P plot for the two-sample problem

Here we describe an interesting plotting procedure (i.e., graphical procedure) to test the null hypothesis $H_0 : F = G$. This procedure was suggested by Wilk and Gnanadesikan (1968). Here are the steps for setting up this plot.

For the sake of simplicity we assume that there are no ties in the combined sample. The order statistics of the combined sample are $Z_{(1)} < Z_{(2)} < \cdots < Z_{(N)}$. Now we plot $F_m(Z_{(j)})$ against $G_n(Z_{(j)})$ for $j = 1, \ldots, N$. Under the null hypothesis this plot should be close to the 45° line through the origin. At times it is visually difficult to assess the closeness of the plot to the 45° objective straight line. So it is suggested that we consider a measure of closeness of the plot to the objective straight line and if this measure is large, we would tend to reject the null hypothesis.

One possible measure is a function of the vertical distances of the points from the straight line. The differences are

$$DV_j = [F_m(Z_{(j)}) - G_n(Z_{(j)})].$$

It is easy to see that

$$max_j DV_j = \frac{mn}{N} D_{m,n}^+.$$

A large value for this measure signifies the rejection of H_0, which is the Smirnov test.

Another function of these differences that can be used to assess the closeness of the curve to the diagonal is the sum. This sum is related to the sum of the ranks of the Y-sample values. It is shown that (see Appendix A2)

$$\sum_j^N DV_j = \frac{N}{mn}[W_Y - E(W_Y|H_0)]. \qquad (2.101)$$

Thus using the sum as a test criterion is the same as conducting the Wilcoxon rank sum test.

So a visual examination of the plot to assess its closeness to the diagonal can be supplemented by the Smirnov test based on $D_{m,n}^+$ or WMW test if we are expecting differences defined by H_{A1}. If H_{A2} alternatives are appropriate, we consider the horizontal differences, namely $DH_j = -DV_j = G_n(Z_{(j)}) - F_m(Z_{(j)})$ and summarize them by the statistic D_{mn}^- or by the sum which is related to W_X, the sum of the ranks of the X-sample values. In essence it seems appropriate to supplement the graphical procedure by a statistical test. Thus we can view this plotting procedure as a precursor to the WMW test or the Smirnov test. An example illustrates the details for setting up the plot as well as the calculation of the test statistics.

Example 2.10 (cont'd.). The graph of F_m against G_n is given in Figure 2.1. These quantities are taken from Table 2.21.

We find

$$\sum DV_j = -2.2, \quad \sum -DV_j = 2.2.$$

To obtain the rank sum statistic W_Y, we need to find the mean

$$E(W_Y|H_0) = 10(21)/2 = 105,$$

and using (2.101) we get

$$W_Y = [(mn)/N](-2.2) + 105 = 94.$$

The other rank sum statistic is

$$W_X = [N(N+1)/2] - W_Y = 210 - 94 = 116.$$

The rank sums coincide with the values given by the computer program, which is included in Appendix B. If we are interested in the alternatives H_{A3}, the Smirnov statistic $D_{10,10}$ will be used. A table of the critical values of $D_{m,n}$, prepared by Kim and Jennrich (1973), is available in *Selected Tables in Mathematical Statistics*, Vol. 1. From this table we find that the 0.05-level critical value is 0.70. So we do not reject the null hypothesis.

A table of critical values for the test based on $D_{m,n}^+$ has been prepared by Gail and Green (1976).

2.4.7 Confidence interval for the difference between medians without shift assumption

In relation to the median tests and the WMW test, we introduced the shift model and derived a confidence interval for the shift parameter, which can be interpreted as the difference between the medians. In some applications this difference in medians is of interest and the shift assumption may not be appropriate. In this section, using the large sample theory, we construct a confidence interval using the difference between the sample medians.

This problem was considered by Albers and Löhnberg (1984) and a modification was suggested by Bristol (1990). These papers considered the problem in relation to the difference between quantiles. However, here we give the details for the difference between medians. Let $\xi(F)$ and $\xi(G)$ be the medians of the distributions F and G, respectively. Using independent random samples from F and G, we want to construct a confidence interval for the difference $\Delta = \xi(F) - \xi(G)$. Let $\hat{\xi}(F)$ and $\hat{\xi}(G)$ be usual estimates of the population medians. Then an estimate of the difference is $\hat{\Delta} = \hat{\xi}(F) - \hat{\xi}(G)$ and the asymptotic distribution of this estimate is a normal distribution with mean Δ and variance $var(\hat{\Delta})$ (see Appendix A2), where

$$var(\hat{\Delta}) = (1/4)\left\{ \frac{1}{mf^2(\xi(F))} + \frac{1}{ng^2(\xi(G))} \right\}.$$

To estimate this variance one may proceed as follows. We choose integers

$$k_m^{\pm} = m\left(0.5 \pm z_{\alpha/2} \frac{1}{\sqrt{4m}}\right)$$

and

$$k_n^{\pm} = n\left(0.5 \pm z_{\alpha/2}\frac{1}{\sqrt{4n}}\right).$$

Then we compute

$$Q_X = \{X_{(k_m^+)} - X_{(k_m^-)}\}^2$$

and

$$Q_Y = \{Y_{(k_n^+)} - Y_{(k_n^-)}\}^2.$$

The quantities $X_{(r)}$ and $Y_{(s)}$ are the order statistics of order r and s in the respective samples. Finally the variance estimate is $v^2 = (Q_X + Q_Y)$. Using this estimate, a confidence interval is

$$(\hat{\Delta} - 0.5v, \hat{\Delta} + 0.5v). \tag{2.102}$$

Bristol (1990) pointed out that for some sample sizes the above procedure may not produce an interval. He suggested a modified choice of the integers

$$kb_m^{\pm} = (m+1)\left[0.5 \pm C\frac{1}{\sqrt{4m}}\right].$$

and a similar definition for $kb_n^{\pm}$. Then the modified Qs are calculated and used to get the variance estimate. This modified variance estimate is $v_b^2 = (QB_X + QB_Y)/(4C^2)$. Finally, the confidence interval is

$$(\hat{\Delta} - z_{1-\alpha/2}v_b, \hat{\Delta} + z_{1-\alpha/2}v_b). \tag{2.103}$$

The constant C has to be chosen to satisfy the condition

$$C \le \frac{[(n-1)/2]\sqrt{4n}}{(n+1)}$$

and

$$C \le \frac{[(m-1)/2]\sqrt{4m}}{(m+1)}.$$

Bristol gives a table of maximum C-values and these will be useful for applying this method. In practice the maximum value of C is recommended for use.

Chakraborti and Desu (1986) generalized the method of Albers and Löhnberg for the case where the data include some right-censored observations.

Use of this interval for testing

Suppose we want to test the null hypothesis that the two medians are equal. The confidence interval for the difference can be used for this testing purpose. If the interval does not include zero we can reject the null hypothesis. This method of testing is called the *standard method* (see Schenker and Gentleman, (2001)). In this reference the authors compare the standard method with the *overlap method* and recommend the standard method for applications.

2.5 Linear rank statistics for the two-sample problem

The Wilcoxon statistic as well as the Mood statistic depend on the ranks. Both of them also can be represented as a linear function of Y-ranks, namely

$$T = \sum_{i=1}^{n} a(r_i),$$

where $r_1 < \cdots < r_n$ are the ranks of the ordered Y observations in the combined sample, and $a(.)$ is a known function. These were introduced for testing the hypothesis $H_0 : F(x) = G(x)$, using two independent samples from F and G. Here we explore the problem of finding optimal *linear rank statistics* for the two-sample problem.

The advantage with rank statistics is that the null distribution of the rank vector is completely known. However, to obtain a powerful test we need to examine the distribution of the rank vector under the alternatives. Hoeffding (1951) explored the problem of optimal nonparametric tests. From his work we can get the distribution of the Y-ranks. This result is given now.

We assume that F and G have densities f and g, and the support of F is same as the support of G. Let $R_1 < R_2 < \cdots < R_n$ be the ranks of Y's, in the combined sample.

Distribution of the Y-ranks

For integers $1 \leq r_1 < r_2 < \cdots < r_n \leq N$,

$$P(R_1 = r_1, R_2 = r_2, \ldots, R_n = r_n) = \binom{m+n}{m}^{-1} E\left[\prod_{j=1}^{n} \frac{g(F^{-1}(U_{(r_j)}))}{f(F^{-1}(U_{(r_j)}))} \right],$$

where $U_{(1)} < U_{(2)} < \cdots < U_{(N)}$ are the order statistics of a sample of size N from the uniform distribution on $(0, 1)$.

For a proof of this result refer to Hettmansperger (1984, p. 142).

When the nonparametric tests were proposed, researchers were concerned about their performance under parametric models. The question of optimal choice of the scores $a(i)$ in relation to a parametric family also needs to be resolved. In the current discussion, answers to some of these problems will be provided.

It is proposed to find locally the most powerful rank tests for the two-sample problem. In this discussion we are concerned with the alternatives that are close to the null hypothesis. Later this notion of closeness will be made clearer. Assume that there is a relationship between the two distributions F and G and this relationship is indexed by a scalar parameter θ. In other words, there exists a density function $f_*(x, \theta)$, such that

$$g(x) = f_*(x, \theta), \quad and \quad f(x) = f_*(x, 0),$$

so that for $\theta = 0$, $g(x) = f(x)$. We have seen that the shift model is a special case of this setup. We denote the probability function of the rank vector **R** by $p(\mathbf{r}|\theta)$. Let us consider the problem of testing

$$H_0 : \theta = 0, \quad versus \quad H_{A1} : \theta > 0. \tag{2.104}$$

We are concerned with *small values* of θ. This is what we mean by close alternatives. One can use the likelihood ratio method to obtain a test. The likelihood based on the Y-ranks is $p(\mathbf{r}|\theta)$. Appealing to Taylor's theorem, we have the following approximation:

$$log\, p(\mathbf{r}|\theta) \approx log\, p(\mathbf{r}|0) + \theta \left. \frac{\partial log\, p(\mathbf{r}|\theta)}{\partial \theta} \right|_{\theta=0} .$$

Thus the log of the likelihood ratio can be approximated as

$$log\, [p(\mathbf{r}|\theta)/p(\mathbf{r}|0)] \approx \theta \left. \frac{\partial log\, p(\mathbf{r}|\theta)}{\partial \theta} \right|_{\theta=0} \equiv \theta T(0). \tag{2.105}$$

The statistic $T(0)$ is called the (*efficient*) *score statistic*. Further, the likelihood ratio test is to

$$reject\ H_0\ in\ favor\ of\ H_{A1}\ if\ T(0) > c. \tag{2.106}$$

If the alternative is $H_{A2} : \theta < 0$, then the critical region is $T(0) < c$. These tests are the *locally most powerful* tests. Further details about this method are available in Cox and Hinkley (1974, p. 106). For a general discussion about the score tests see Rao (1973, p. 415). For various special relationships, the score statistic will take the form indicated at the beginning of this section.

Before we take up the calculation of T for various special cases, it will be advantageous to obtain a convenient expression for $T(\theta)$. First we note that

$$T(\theta) = \frac{\partial log\, p(\mathbf{r}|\theta)}{\partial \theta} = \frac{1}{p(\mathbf{r}|\theta)} \frac{\partial p(\mathbf{r}|\theta)}{\partial \theta}.$$

Assuming that the differentiation can be taken inside the expectation, we have

$$\frac{\partial p(\mathbf{r}|\theta)}{\partial \theta} = \binom{m+n}{m}^{-1} E\left[\frac{\partial}{\partial \theta} \prod_{j=1}^{n} \frac{g(F^{-1}(U_{(r_{(j)})}))}{f(F^{-1}(U_{(r_j)}))} \right] \equiv [p(\mathbf{r}|0)]S_\theta.$$

The statistic $T(0)$ is equal to S_0, which can be expressed as a sum, namely,

$$T_0 = \sum_{j=1}^{n} a_{OE}(r_j, f_*), \tag{2.107}$$

where for $i = 1, \ldots, N$,

$$a_{OE}(i, f_*) = E\left[\frac{\left. \frac{\partial f_*(F_*^{-1}(U_{(i)}), \theta)}{\partial \theta} \right|_{\theta=0}}{f_*(F_*^{-1}(U_{(i)}), 0)} \right]. \tag{2.108}$$

It is important to note that these scores, $a(i)$, *do depend* on the density f_*, and these are referred to as *optimal expected value scores*. In some cases, the computation of these expectations is somewhat complicated, and it has been proposed to replace them by the following *function scores*. These are

$$a_{OF}(i, f_*) = \left[\frac{\frac{\partial f_*(F_*^{-1}(E(U_{(i)})),\theta)}{\partial \theta}\Big|_{\theta=0}}{f_*(F_*^{-1}(E(U_{(i)}),0)} \right], \tag{2.109}$$

where $E(U_{(i)}) = (i/(N+1))$.

Calculation of the mean and the variance of T_0, under the null hypothesis, is given in Appendix A2. Using them and using the fact that the asymptotic distribution of T_0 is a normal distribution, we can construct critical regions of approximate size α. This asymptotic normality result is the famous Chernoff and Savage theorem (1958). Using these facts an approximation for the P-value can be obtained. For certain special models and special distributions f_*, we get the test statistics that were proposed for the testing problem.

We consider three models for detailed analysis. They are (1) location model or shift model, (2) proportional hazards model, and (3) scale model. In each case, the scores are derived for various distributions of interest.

2.5.1 Location model (Shift model)

Here we assume that

$$g(x) = f_*(x - \theta); \quad f(x) = f_*(x),$$

where f_* is a known density. Now the derivative

$$\frac{\partial}{\partial \theta} \left[\frac{g(x)}{f(x)} \right] = \frac{1}{f_*(x)} \frac{\partial}{\partial \theta} f_*(x - \theta)|_{\theta=0} = \frac{-f_*'(x)}{f_*(x)},$$

where $f_*'(x)$ is the derivative of $f_*(x)$. Hence the optimal expected value scores for the location model are

$$a_{LE}(i, f_*) = E\left(\frac{-f_*'(F_*^{-1}(U_{(i)}))}{f_*(F_*^{-1}(U_{(i)}))} \right), \tag{2.110}$$

and the optimal function scores are

$$a_{LF}(i, f_*) = \frac{-f_*'(F_*^{-1}(i/(N+1)))}{f_*(F_*^{-1}(i/(N+1)))}. \tag{2.111}$$

In the above expressions F_* is the cdf of the distribution defined by the density f_*. We now specialize these results to specific distributions.

Normal distribution (location)

Here the density $f_*(x) = \phi(x)$, the density of the standard normal distribution and hence

$$\frac{-f_*'(x)}{f_*(x)} = x.$$

Thus the *normal expected value scores* are

$$a_{LE}(i, NOR) = E(\Phi^{-1}(U_{(i)})), \tag{2.112}$$

where $U_{(1)} < \cdots < U_{(N)}$ are the order statistics of a random sample of size N from the uniform distribution on the interval $(0, 1)$. These scores were suggested by Fisher. A tabulation is available in Fisher and Yates (1936) and the test based on these scores is called *Fisher and Yates normal scores* test.

The function scores have been proposed by van der Waerden (1952, 1953). In other words, the *van der Waerden scores* are

$$a_{LF}(i, NOR) = \Phi^{-1}\left(\frac{i}{N+1}\right).$$

We can use the NPAR1WAY procedure of SAS to implement the linear rank test with these function scores.

Logistic distribution
Here the density f_* and the cdf F_* are

$$f_*(x) = \frac{e^{-x}}{(1 + e^{-x})^2},$$

and

$$F_*(x) = \frac{1}{(1 + e^{-x})}, \quad for \; -\infty < x < \infty,$$

so that

$$\frac{-f_*'(x)}{f_*(x)} = 2F_*(x) - 1.$$

Thus the optimal expected value scores are

$$a_{LE}(i, LO) = 2E(U_{(i)}) - 1 = 2\frac{i}{N+1} - 1 = \frac{2}{N+1}\left(i - \frac{N+1}{2}\right). \quad (2.113)$$

It may be noted that these scores are multiples of w-ranks. An equivalent choice is $a(i) = i$, which are the ranks proposed by Wilcoxon. In other words, the expected value scores statistic is equivalent to the Wilcoxon rank sum statistic.

Extreme value distribution
Here the density f_* is

$$f_*(x) = exp(x - e^x), \quad for \; -\infty < x < \infty,$$

and the cdf F_* is

$$F_*(x) = 1 - exp(-e^x), \quad for \; -\infty < x < \infty,$$

Hence

$$\frac{-f_*'(x)}{f_*(x)} = e^x - 1.$$

Thus the optimal expected value scores are

$$a_{LE}(i, EXT) = E(-ln(1 - U_{(i)})) - 1 = E(V_{(i)}) - 1, \quad (2.114)$$

where the V's are order statistics from the exponential distribution with mean 1. These scores are called *Savage scores*. The test with these scores can be implemented using the NPAR1WAY procedure of the SAS package. These optimal expected value scores are equivalent to the exponential ordered scores. The sum of the Savage scores is zero.

Laplace distribution (double exponential distribution)

Here the density f_* is

$$f_*(x) = \frac{1}{2}e^{-|x|}, \quad for \ -\infty < x < \infty.$$

It can be seen that

$$\frac{-f_*'(x)}{f_*(x)} = \begin{cases} -1 & \text{if } x < 0, \\ 1 & \text{if } x > 0. \end{cases}$$

In other words,

$$\frac{-f_*'(x)}{f_*(x)} = sgn(x).$$

Thus the expected value scores are

$$a_{LE}(i; DEXP) = E[sgn(F_*^{-1}(U_{(i)}))]. \tag{2.115}$$

The optimal function scores in this case are called *median scores*, which are related to Mood's median scores. This connection will be demonstated now. First we note that the function scores are

$$a_{LF}(i; DEXP) = sgn\left[F_*^{-1}\left(\frac{i}{N+1}\right)\right]. \tag{2.116}$$

The right-hand side function can be simplified as follows. We recall that the cdf $F_*(.)$ is

$$F_*(x) = \begin{cases} \frac{1}{2}e^x & \text{if } x \le 0, \\ 1 - \frac{1}{2}e^{-x} & \text{if } x > 0. \end{cases}$$

Thus

$$F_*^{-1}\left(\frac{i}{(N+1)}\right) = \begin{cases} log\left(\frac{2i}{(N+1)}\right) & \text{if } \frac{2i}{N+1} < 1, \\ -log2\left(1 - \frac{i}{(N+1)}\right) & \text{if } \frac{2i}{N+1} > 1. \end{cases}$$

So the function scores are

$$a_{LF}(i; DEXP) = sgn\left[F_*^{-1}\left(\frac{i}{N+1}\right)\right]$$

$$= sgn\left[\frac{2i}{N+1} - 1\right] = sgn[2i - (N+1)].$$

The score is taken as zero for $i = (N+1)/2$. We refer to these scores as the *median scores* and these are linear functions of Mood's median scores. Thus this test is equivalent to Mood's median test, which can be implemented in the SAS system through the NPAR1WAY procedure.

2.5.2 Proportional hazards model

Here

$$1 - G(x) = [1 - F(x)]^{(1+\theta)},$$

so that the densities are related as

$$g(x) = (1 + \theta)[1 - F(x)]^{\theta} f(x).$$

In the notation of this section, we have

$$f_*(x, \theta) = (1 + \theta)[1 - F_*(x, 0)]^{\theta} f_*(x, 0).$$

Hence

$$\left[\frac{\left. \frac{\partial f_*(x, \theta)}{\partial \theta} \right|_{\theta=0}}{f_*(x, 0)} \right] = 1 + log[1 - F_*(x, 0)].$$

Thus the optimal expected scores are

$$a_{PHE}(i, f_*) = E[1 + log(1 - U_{(i)})] = 1 - E(V_{(i)}), \qquad (2.117)$$

where the $U_{(i)}$'s are uniform order statistics and the $V_{(i)}$'s are order statistics from the exponential distribution with mean 1. Also, here the scores *do not* depend on the density f_*. They are easy to calculate since

$$E(V_{(i)}) = \sum_{j=1}^{i} \frac{1}{N - j + 1}.$$

A proof of this result is given in Appendix A2. These scores are equivalent to Savage scores. So the test with these scores can be implemented using the NPAR1WAY procedure of the SAS package as indicated earlier.

Even though these optimal expected ordered scores are easy to calculate, we examine the function scores, which are

$$a_{PHF}(i) = 1 + log[1 - E(U_{(i)})] = 1 + log\left[1 - \frac{i}{N+1}\right].$$

In other words,

$$a_{PHF}(i) = 1 - log(N + 1) + log(N + 1 - i).$$

We note that $(N+1-i)$ is the reverse rank of a value that received the rank i. Recall that the value with $(N+1-i)$ as the reverse rank is the $(N+1-i)$th highest value in the data. Thus these scores are equivalent to log (reverse) ranks.

2.5.3 Scale model

Under this model,

$$G(x) = F_*(x(1 + \theta)); \quad F(x) = F_*(x).$$

Hence

$$g(x) = (1 + \theta) f_*(x(1 + \theta)).$$

Thus

$$\left[\frac{\frac{\partial f_*(x,\theta)}{\partial \theta}\big|_{\theta=0}}{f_*(x,0)} \right] = 1 + x \frac{f_*'(x)}{f_*(x)}.$$

Hence the optimal expected scores are

$$a_{SE}(i, f_*) = E \left[1 + F_*^{-1}(U_{(i)}) \frac{f_*'(F_*^{-1}(U_{(i)}))}{f_*(F_*^{-1}(U_{(i)}))} \right]. \tag{2.118}$$

We now consider two specific distributions.

Exponential distribution (scale)
‾‾‾‾‾‾‾‾‾‾‾‾‾‾‾‾‾‾‾‾‾
Here

$$f_*(x) = e^{-x}, \quad for \ x > 0.$$

So

$$1 + x \frac{f_*'(x)}{f_*(x)} = 1 - x;$$

and hence the optimal expected scores are

$$a_{SE}(i, EXP) = 1 - E(F_*^{-1}(U_{(i)})) = 1 - E(V_{(i)}), \tag{2.119}$$

where $V_{(i)}$ are the order statistics from the exponential distribution with mean 1. In other words, the optimal expected scores are equivalent to the exponential ordered scores.

Here we can work with transformed variables, $\log(X)$ and $\log(Y)$. The distributions of the transformed variables satisfy the shift model and f_* is the density of the extreme value distribution considered earlier. The scores we got there are the same as the ones we obtained here.

Normal distribution (scale)
‾‾‾‾‾‾‾‾‾‾‾‾‾‾‾‾‾‾
Here

$$f_*(x) = \phi(x).$$

It can be verified that the expected scores are

$$a_{SE}(i, NOR) = 1 - E(\Phi^{-1}(U_{(i)})^2). \tag{2.120}$$

The test based on these scores was derived by Capon (1961).

For this case the function scores are suggested by Klotz (1963) and they are

$$a_{SF}(i, NOR) = 1 - \left[\Phi^{-1} \left(\frac{i}{N+1} \right) \right]^2.$$

2.6 Analysis of censored data

A generalization of the Wilcoxon-Mann-Whitney test to censored data was proposed independently by Gilbert (1962) and Gehan (1965). Mantel (1966) showed how to adopt the Mantel-Haenszel procedure to survival data. The same statistic has been proposed by Peto and Peto (1972) and Cox (1972). This statistic is a generalization of the exponential ordered scores test of Savage. Peto and Peto (1972) and Prentice (1978) proposed other generalizations of the Wilcoxon test. We describe the Gehan's Wilcoxon test and the logrank test of Peto and Peto, followed by a general class of test statistics, proposed by Tarone and Ware (1977). This class includes Gehan's statistic, the logrank statistic, and the Prentice Wilcoxon.

2.6.1 Gehan's Wilcoxon test

In the Wilcoxon procedure we combined the two samples and ranked the observations, and then used the ranks to construct the test statistic. We have seen that instead of ranks we could use the w-scores. Here the w-scores can be computed using the following scoring function:

$$sgn(a, b) = \begin{cases} -1, & \text{if } a < b \text{ or } a \leq b^+; \\ 1, & \text{if } a > b \text{ or } a^+ \geq b; \\ 0, & \text{otherwise.} \end{cases}$$

In the definition, it is understood that a and/or b may be censored, in which case a plus $(+)$ sign is used to denote this fact. It is easy to interpret the above function as

$$sgn(a, b) = \begin{cases} -1, & \text{if } a \text{ is decidedly below } b; \\ 1, & \text{if } a \text{ is decidedly above } b; \\ 0, & \text{otherwise.} \end{cases} \tag{2.121}$$

Let us rename the combined sample as

$$(X_1, \ldots, X_m, Y_1, \ldots, Y_n) = (Z_1, \ldots, Z_m, \ldots, Z_{m+n}).$$

Let $N = m + n$; now we define the w-scores

$$w(Z_h) = \sum_{k=1}^{N} sgn(Z_k, Z_h).$$

In words,

$$w(Z_h) = \# \text{ of } Z_k \text{ that are decidedly} > Z_h$$
$$-\# \text{ of } Z_k \text{ that are decidedly} < Z_h. \tag{2.122}$$

This quantity is the score attached to Z_h. Gehan's test can be stated in terms of the statistic GW_Y, which is the sum of scores attached to the Y-sample and is

$$GW_Y = \sum_{h=m+1}^{m+n} w(Z_h). \tag{2.123}$$

Table 2.22 *Permutation tests based on Gehan's Wilcoxon scores*

Alternative	Critical Region
$H_{A1} : X st < Y$	$Z_{GW} \leq z_\alpha$
$H_{A2} : X st > Y$	$Z_{GW} \geq z_{1-\alpha}$
$H_{A3} = H_{A1} \cup H_{A2}$	$\mid Z_{GW} \mid \geq z_{(1-\alpha/2)}$

Since the sum of the w-scores is zero, we have $GW_X = -GW_Y$. In developing the tests, we assume that the *two censoring distributions are equal to each other*. The test statistic is the standardized version of GW_Y. To define this statistic, we need to find the mean and the variance of GW_Y under H_0, which states that the survival distributions are equal to each other. Since we are assuming the equality of the censoring distributions we get the mean and the variance from the permutation distribution. It can be shown that the mean and the variance under the permutation distribution are

$$E_P(GW_Y) = 0; \quad var_P(GW_Y) = \frac{mn}{N(N-1)} \sum_{h=1}^{N} w^2(Z_h). \tag{2.124}$$

Thus, the standardized test statistic is

$$Z_{GW} = \frac{GW_Y}{\sqrt{Var_P(GW_Y)}}, \tag{2.125}$$

and its distribution can be approximated by the standard normal distribution. The critical regions of approximate α-level tests are given in Table 2.22. These tests are called the *permutation tests*.

Remark 2.17. When the data are complete, the statistic GW_Y becomes $-2[W_Y^* - n(N+1)/2]$, and thus this test is a generalization of the Wilcoxon rank sum test. In a later section a modification of this test, without the assumption of equality of the censoring distributions, will be presented. This modification is based on the work of Breslow (1970) and the details are given in Subsection 2.6.3.

Remark 2.18. In some discussions the quantities $w(z)/(N+1)$ are referred to as *Wilcoxon scores* (Kalbfleisch and Prentice, 1980, p. 147). In the case of complete data these scores are the optimal expected value scores for the logistic distribution (see (2.113)).

Example 2.11. Consider the samples $\{X_1, X_2, \ldots, X_6\} = \{6, 6^+, 7, 9^+, 10, 13\}$ and $\{Y_1, Y_2, \ldots, Y_5\} = \{1, 8^+, 11, 12^+, 14^+\}$. Thus the combined sample is

$$\{Z_1, Z_2, \ldots, Z_{11}\} = \{6, 6^+, 7, 9^+, 10, 13, 1, 8^+, 11, 12^+, 14^+\}$$

Now we calculate the w-scores for all observations.

$$
\begin{aligned}
w(Z_1) &= 0+1+1+1+1+1-1+1+1+1+1 = 8 \\
w(Z_2) &= -1+0+0+0+0+0-1+0+0+0+0 = -2 \\
w(Z_3) &= -1+0+0+1+1+1-1+1+1+1+1 = 5 \\
w(Z_4) &= -1+0-1+0+0+0-1+0+0+0+0 = -3 \\
w(Z_5) &= -1+0-1+0+0+1-1+0+1+1+1 = 1 \\
w(Z_6) &= -1+0-1+0-1+0-1+0--1+0+1 = -4 \\
w(Z_7) &= 1+1+1+1+1+1+0+1+1+1+1 = 10 \\
w(Z_8) &= -1+0-1+0+0+0-1+0+0+0 = -3 \\
w(Z_9) &= -1+0-1+0-1+1-1+0+0+1+1 = -1 \\
w(Z_{10}) &= -1+0-1+0-1+0-1+0-1+0+0 = -5 \\
w(Z_{11}) &= -1+0-1+0-1-1-1+0-1+0+0 = -6
\end{aligned}
$$

The sum of these scores is zero. The statistic $GW_Y = 10 - 3 - 1 - 5 - 6 = -5$. To find the variance of this statistic under H_0, we need to get the sum of squares of the w-scores. This sum is 290. Hence the permutation variance is

$$
var_P(GW_Y) = \frac{(6 \times 5)}{11} \frac{290}{10} = 79.0909.
$$

Thus the value of the standardized statistic Z_{GW} is 0.56. In relation to the two-sided alternatives, the approximate P-value is 0.5754. Thus this statistic is not significant at level 0.05.

2.6.2 Logrank test

In Gehan's Wilcoxon procedure, we essentially attached scores to the observations, and the sum of the scores of one of the samples is used as the test statistic. A slightly different set of scores will be considered for obtaining the logrank test statistic. In the case of complete data, the scores are the estimates of the cumulative hazard function based on the combined sample. In the censored data case the scores are related to the estimated cumulative hazard function, $\hat{\Lambda}(t)$.

To define the function $\hat{\Lambda}(t)$, we need to introduce some notation. Let $t_{(1)} < \cdots < t_{(k)}$ be the distinct uncensored observations in the combined sample. Let m_i be the number of uncensored values equal to $t_{(i)}$. Let r_i be the number of observations, complete or censored, which are at least $t_{(i)}$. Then, for $t \in [t_{(i)}, t_{(i+1)})(i = 1, 2, \ldots, k - 1)$, the estimate is

$$
\hat{\Lambda}(t) = \sum_{j \le i} \frac{m_j}{r_j}. \tag{2.126}
$$

The score attached to $t_{(i)}$ is $1 - \hat{\Lambda}(t_{(i)})$ and the score for an observation censored at t is $-\hat{\Lambda}(t)$. We call these scores the *logrank scores*. It may be noted that the larger the uncensored observation is, the smaller the score it receives. Censored observations receive negative scores. The sum of all scores is zero. As in the case of Wilcoxon's test, the logrank test is based on the test statistic S_Y, which is the sum of the logrank scores of the Y-sample

Table 2.23 *Permutation tests using logrank scores*

Alternative	Critical Region
$H_{A1} : X st < Y$	$Z_S \leq z_\alpha$
$H_{A2} : X st > Y$	$Z_S \geq z_{1-\alpha}$
$H_{A3} : H_{A1} \cup H_{A2}$	$\mid Z_S \mid > z_{(1-\alpha/2)}$

observations. Denoting the logrank score for Z_j by L_j, we have

$$S_Y = \sum_{j=m+1}^{m+n} L_j. \tag{2.127}$$

To construct the critical regions for various alternatives, the large sample distribution of the test statistic is used. Here again, it is assumed that the two censoring distributions are equal. So we find the mean and variance of the permutation distribution. It can be shown that

$$E_P(S_Y) = 0; \quad var_P(S_Y) = \frac{mn}{N(N-1)} \sum_{j=1}^{N} L_j^2. \tag{2.128}$$

Now the asymptotic distribution of $Z_S = S_Y / \sqrt{var_P(S_Y)}$ is the standard normal distribution. This result is used to define the critical regions, which are listed in Table 2.23.

Remark 2.19. The estimate $\hat{\Lambda}(t)$ reduces to the Nelson's estimate, when there are no censored observations and no ties in the combined sample. So the logrank scores are related to the exponential scores considered earlier.

Remark 2.20. The logrank test statistic S_Y is same as the test statistic proposed by Cox (1972). However, a different variance estimate is used by Cox to define the test. The proposal by Cox does not assume the equality of the censoring distributions. We will discuss this point further in Subsection 2.6.3.

Remark 2.21. As the sum of all scores is zero, as pointed out earlier, we have $S_Y = -S_X$, where S_X is the sum of scores for the X sample. Thus we can also express the critical regions in terms of S_X.

Example 2.12. Here we illustrate the calculation of the logrank statistic using the data of Example 2.11. The relevant calculations are given in Table 2.24. The sum of the scores is 0.0001. This should be zero and the discrepancy is due to rounding errors. The statistic S_Y is -1.2045 and the sum of squares of the scores is 4.7341. Thus the variance is

$$var_P(S_Y) = \frac{(6 \times 5)}{(11 \times 10)} 4.7341 = 1.29911.$$

Table 2.24 *Calculation of logrank test statistic*

Obs.	Sample ID	$r(t)$	$m(t)$	$m(t)/r(t)$	$\hat{\Lambda}(t)$	L	L^2
1	Y	11	1	$(1/11)$	0.0909	0.9091	0.8265
6	X	10	1	$(1/10)$	0.1909	0.8091	0.6546
6+	X	–	–	–	0.1909	−0.1909	0.0364
7	X	8	1	$(1/8)$	0.3159	0.6841	0.4680
8+	Y	–	–	–	0.3159	−0.3159	0.0998
9+	X	–	–	–	0.3159	−0.3159	0.0998
10	X	5	1	$(1/5)$	0.5159	0.4841	0.2344
11	Y	4	1	$(1/4)$	0.7659	0.2341	0.0548
12+	Y	–	–	–	0.7659	−0.7659	0.5866
13	X	2	1	$(1/2)$	1.2659	−0.2659	0.0707
14+	Y	–	–	–	1.2659	−1.2659	1.6025

Hence the test statistic Z_S is -1.0601, which is not significant at the 5% level in relation to the two-sided alternatives. The approximate two-sided P-value is 0.2892.

2.6.3 Tarone and Ware test

Tarone and Ware (1977) considered the two-sample problem from a contingency table perspective. They reinterpreted the Gehan's statistic and the logrank statistic as follows. Corresponding to each uncensored value $t_{(j)}$ of the combined sample, let $m_{j1}(m_{j2})$ be the multiplicity in the X-sample (Y-sample) and let $r_{j1}(r_{j2})$ be the number of observations in the X-sample (Y-sample) that are greater than or equal to t_j. Tarone and Ware (1977) showed that

$$GW_Y = \sum_{j=1}^{k} r_j \left[m_{j2} - r_{j2} \frac{m_j}{r_j} \right],$$

and

$$S_Y = \sum_{j=1}^{k} \left[m_{j2} - r_{j2} \frac{m_j}{r_j} \right],$$

where $m_j = m_{j1} + m_{j2}$, and $r_j = r_{j1} + r_{j2}$.

Based on the works of Radhakrishna (1965) and Mantel and Haenszel (1959), they proposed the test statistic TW_Y, which is

$$TW_Y = \sum_{j=1}^{k} \sqrt{r_j} \left[m_{j2} - r_{j2} \frac{m_j}{r_j} \right]. \tag{2.129}$$

It is easy to see that the three statistics GW_Y, S_Y, and TW_Y are members of the following class of statistics,

$$U_w = \sum_{j=1}^{k} w(r_j) \left[m_{j2} - r_{j2} \frac{m_j}{r_j} \right], \tag{2.130}$$

where $w(.)$ is a weight function. It may be noted that

$$w(r_j) = \begin{cases} r_j, & \text{for } GW_Y, \\ 1, & \text{for } S_Y, \\ \sqrt{r_j}, & \text{for } TW_Y. \end{cases}$$

The test criterion is the standardized statistic

$$Z_w = \frac{U_w}{v(w)}, \tag{2.131}$$

where $v^2(w)$ is an estimate of the asymptotic null variance of the statistic U_w, and it is

$$v^2(w) = \sum_{j=1}^{k} w^2(r_j) \frac{(r_{j1}r_{j2})}{r_j^2(r_j-1)} m_j(r_j - m_j) \equiv \sum_{j=1}^{k} v^2(j). \tag{2.132}$$

In deriving this variance expression, the equality of the censoring distributions is not assumed. Here we should take $v^2(j) = 0$, whenever $r_j = 1$.

The critical regions are constructed using the asymptotic normality of Z_w. The SAS package uses this variance for Gehan's Wilcoxon and logrank tests. These modified tests are called the *Wilcoxon and logrank tests*. In other words, the Wilcoxon test and the logrank test of the SAS package are not the same as Gehan's Wilcoxon test and the logrank test discussed in Subsections 2.6.1 and 2.6.2. The logrank test of SAS is the same as the test proposed by Cox (1972), which was also discussed by Mantel (1966). Using the data considered in Example 2.11, we will illustrate the calculation of the three test statistics. Now the Gehan's Wilcoxon statistic and its null variance are

$$GW_Y = \sum_{j=1}^{6} GW_Y(j) = -5, \quad v_{GW}^2 = \sum_{j=1}^{6} v_{GW}^2(j) = 80.$$

Also, the logrank test statistic and its null variance are

$$S_Y = \sum_{j=1}^{6} S_Y(j) = -1.2046, \quad v_S^2 = \sum_{j=1}^{6} v_S^2(j) = 1.4154.$$

The Tarone and Ware statistic and its null variance are

$$TW_Y = \sum_{j=1}^{6} TW_Y(j) = -2.4187, \quad v_{TW}^2 = \sum_{j=1}^{6} v_{TW}^2(j) = 9.5773.$$

The test criterion for two-sided alternatives is $X^2 = U_w^2/v_w^2$, which is approximately a chi-square variable with 1 degree of freedom. In Table 2.25, we give various statistics and the corresponding P-values. These results are from the output of a computer program given in Appendix B2. All three tests do not reject the null hypothesis at the 0.5 level.

Table 2.25 *Various test results*

Name	Statistic X^2	P-Value
Wilcoxon	0.3125	0.5762
Logrank	1.0125	0.3113
TW	0.6108	0.4345

j	1	2	3	4	5	6
t_j	1	6	7	10	11	13
m_{j1}	0	1	1	1	0	1
m_{j2}	1	0	0	0	1	0
m_j	1	1	1	1	1	1
r_{j1}	6	6	4	2	1	1
r_{j2}	5	4	4	3	3	1
r_j	11	10	8	5	4	2
$GW_Y(j)$	6	−4	−4	−3	1	−1
$S_Y(j)$	0.5454	−0.4	−0.5	−0.6	0.25	−0.5
$TW_Y(j)$	1.8091	−1.2649	−1.4142	−1.3416	0.5	−0.7071
$v^2_{GW}(j)$	30	24	16	6	3	1
$v^2_S(j)$	0.2479	0.2400	0.2500	0.2400	0.1875	0.2500
$v^2_{TW}(j)$	2.7273	2.4000	2.0000	1.2000	0.7500	0.5000

2.6.4 Testing for equivalence with censored data

Here we are interested in testing the equivalence hypothesis of two survival distributions. The equivalence hypothesis is defined in terms of the relative risk. Let X and Y be the survival times under two different treatments and let the corresponding hazard functions be $\lambda_X(.)$ and $\lambda_Y(.)$. The relative risk is defined as the ratio of the hazard functions, which is assumed to be a constant. In other words, the two survival distributions satisfy the proportional hazards assumption. Let

$$r(t) = \frac{\lambda_X(t)}{\lambda_Y(t)}$$

be the relative risk. We assume that $r(t) = \rho$. The problem of equivalence is usually formulated as testing

$$H_0 : \rho = \rho_0 \quad against \quad H_A : \rho < \rho_0, \tag{2.133}$$

where $\rho_0 > 1$. This is a slightly different from the formulation proposed by Com-Nougue, Rodary, and Patte (1993). Here rejecting the null hypothesis is interpreted as the equivalence of the two survival distributions under the two treatments. We propose the same test as the one proposed by Com-Nougue et al., which is a generalization of the logrank test.

Table 2.26 *Pattern of observed deaths at time* $t_{(i)}$

Sample	# of Events (deaths)	# of Non-Events (# survived)	Total (# at risk)
X	m_{iX}	$r_{iX} - m_{iX}$	r_{iX}
Y	m_{iY}	$r_{iY} - m_{iY}$	r_{iY}
Total	m_i	$r_i - m_i$	r_i

We will now describe the test. Let n_X be the size of the X-sample and n_Y be the size of the Y-sample. Each sample may contain some right-censored observations. Combine the two samples and let $t_{(1)} < t_{(2)} < \cdots < t_{(k)}$ be the ordered uncensored observations in the combined sample. Now the pattern of deaths can be described by a series of 2×2 tables. A typical one that describes the information relative to time $t_{(i)}$ is given in Table 2.26. Let p_{iX} and p_{iY} be the probabilities for an event at $t_{(i)}$ for the X- and Y-samples. Under H_0, we have that $p_{iX}/p_{iY} = \rho_0$. Further, ν_i is the probability that an event is from the X-sample. Given that m_i events occurred at $t_{(i)}$, the variable m_{iX} has a binomial distribution with parameters m_i and ν_i. The unconditional distribution of m_{iX} is the binomial distribution with parameters r_{iX} and p_{iX}. So we have

$$
\begin{aligned}
r_{iX} p_{iX} = E(m_{iX}) &= EE(m_{iX}|m_i) \\
&= \nu_i E(m_i) = \nu_i E(m_{iX} + m_{iY}) \\
&= \nu_i (r_{iX} p_{iX} + r_{iY} p_{iY}).
\end{aligned}
\tag{2.134}
$$

Hence solving for ν_i, we get

$$
\nu_i = \frac{r_{iX} p_{iX}}{(r_{iX} p_{iX} + r_{iY} p_{iY})} = \frac{r_{iX} \rho_0}{(r_{iX} \rho_0 + r_{iY})}.
\tag{2.135}
$$

Using this value we get

$$
E_0(m_{iX}|m_i) = m_i \cdot \nu_i, \quad var_0(m_{iX}|m_i) = m_i \cdot \nu_i(1 - \nu_i).
$$

We use the test statistic

$$
T = \frac{\sum_{i=1}^{k}(m_{iX} - E_0(m_{iX}|m_i))}{\sqrt{\sum_{i=1}^{k} var_0(m_{iX}|m_i)}},
\tag{2.136}
$$

which is asymptotically a $N(0,1)$ variable under the null hypothesis. Further, small values of T indicate the rejection of H_0 and declare equivalence. Hence the approximate P-value is

$$
P\text{-value} \approx P(N(0,1) < T(obs)),
\tag{2.137}
$$

where $T(obs)$ is the observed value of T.

Remark 2.22. When $\rho_0 = 1$, this test is the same as the logrank test with asymptotic variance.

2.7 Asymptotic relative efficiency (Pitman efficiency)

Whenever we have more than one procedure for answering a statistical question, we may want to make a choice. To help make this choice a comparison of the performance of one procedure with that of another procedure is needed. We restrict our discussion to the comparison of tests of hypotheses. For comparison purposes, we need to choose a performance criterion. The commonly used criterion is the power for a set of alternatives. Conceptually, to get better power we may need a larger sample size. If n_1 and n_2 are the sample sizes required for tests 1 and 2, of the same level, to achieve the same power, then the ratio (n_2/n_1) could be used to indicate the relative efficiency of test 1 relative to test 2.

To make the notion of relative efficiency of test T_1 relative to test T_2 explicit and come up with an expression we proceed as follows. It is appropriate to compare two tests for a particular testing problem when they are of the same level. To make headway we examine a very specific testing problem. Suppose we want to test the null hypothesis

$$H_0 : \theta = \theta_0 \quad against \quad H_{A1} : \theta > \theta_0. \tag{2.138}$$

We consider tests based on a single sample. This discussion is mainly heuristic and not a rigorous mathematical derivation.

Consider a test defined in terms of the statistic T_n, where n is the sample size. Suppose that the asymptotic distribution is a normal distribution with mean $\mu_n(\theta)$ and variance $\sigma_n^2(\theta)$. It is not necessary that these μ and σ^2 are the mean and variance of T_n; it is sufficient that the distribution of $(T_n - \mu_n(\theta))/\sigma_n(\theta)$ is asymptotically standard normal. The critical region of an approximate size α test is

$$T_n > \mu_n(\theta_0) + z_{1-\alpha} \cdot \sigma_n(\theta_0). \tag{2.139}$$

We now consider a sequence of local alternatives $\theta_n = \theta_0 + \delta/\sqrt{n}$, where $\delta > 0$. We examine the power of the test for these alternatives. The power at θ_n is

$$\Pi_n(\theta_n) = P\left(\frac{T_n - \mu_n(\theta_n)}{\sigma_n(\theta_n)} > u_n\right),$$

where

$$u_n = \frac{\mu_n(\theta_0) - \mu_n(\theta_n)}{\sigma_n(\theta_n)} + \frac{z_{(1-\alpha)}\sigma_n(\theta_0)}{\sigma_n(\theta_n)}.$$

Now we get an approximation to u_n. From an appropriate Taylor's expansion for $\mu_n(\theta_n)$, we have

$$\mu_n(\theta_n) \approx \mu_n(\theta_0) + (\theta_n - \theta_0)\mu_n'(\theta_0)$$

$$= \mu_n(\theta_0) + \frac{\delta}{\sqrt{n}}\mu_n'(\theta_0).$$

We assume that $\sigma_n(\theta_n) \approx \sigma_n(\theta_0)$. So we have

$$u_n \approx -\frac{\delta}{\sqrt{n}}\frac{\mu'_n(\theta_0)}{\sigma_n(\theta_0)} + z_{(1-\alpha)}.$$

Hence the asymptotic power is

$$\Pi^*_n(\theta_n) = 1 - \Phi\left(-\frac{\delta}{\sqrt{n}}\frac{\mu'_n(\theta_0)}{\sigma_n(\theta_0)} + z_{1-\alpha}\right)$$

$$= \Phi\left(\frac{\delta}{\sqrt{n}}\frac{\mu'_n(\theta_0)}{\sigma_n(\theta_0)} - z_{1-\alpha}\right). \tag{2.140}$$

It is easy to see that, for given δ and α, this asymptotic power depends on $e_n(T)$, where

$$e_n(T) = \frac{\mu'_n(\theta_0)}{\sqrt{n}\sigma_n(\theta_0)}.$$

We assume that this quantity has a limit as $n \to \infty$, which is denoted by $e(T)$, and this limit is called the *efficacy of test T*.

Efficacy of test T
The efficacy of test T is defined as

$$e(T) = \lim_{n\to\infty}\frac{\mu'_n(\theta_0)}{\sqrt{n}\sigma_n(\theta_0)}. \tag{2.141}$$

In terms of this quantity, the limiting power is

$$\Pi^*_T(\theta) = \Phi\left(\delta e(T) - z_{1-\alpha}\right). \tag{2.142}$$

Relative efficiency (Pitman efficiency) of T_1 with respect to T_2
We now look into the comparision of two test procedures T_1 and T_2, which are defined in relation to sample sizes n_1 and n_2, respectively. The two tests are of size α and all the conditions we assumed for deriving the efficacy hold. Let us consider the two sequences of alternatives

$$\theta_{1,n_1} = \theta_0 + \delta_1/\sqrt{n_1}; \quad \theta_{2,n_2} = \theta_0 + \delta_2/\sqrt{n_2}.$$

Using (2.137), we see that the limiting powers are

$$\Pi^*_{T_1}(\theta) = \Phi(\delta_1 e(T_1) - z_{1-\alpha}),$$

and

$$\Pi^*_{T_2}(\theta) = \Phi(\delta_2 e(T_2) - z_{1-\alpha}),$$

where $e(T_1)$ and $e(T_2)$ are the efficacies of T_1 and T_2, respectively. These powers will be equal if and only if

$$\delta_1 e(T_1) = \delta_2 e(T_2),$$

and the alternatives are equal whenever

$$\delta_1/\sqrt{n_1} = \delta_2/\sqrt{n_2}.$$

Thus the asymptotic relative efficiency of T_1 with respect to T_2 is

$$ARE(T_1, T_2) = lim_{n1,n_2 \to \infty}(n_2/n_1)$$

$$= (\delta_2/\delta_1)^2 = \frac{e^2(T_1)}{e^2(T_2)}. \qquad (2.143)$$

It can be easily verified that

1. $0 \le ARE(T_1, T_2) \le \infty$.

2. $ARE(T_1, T_2) = \frac{1}{ARE(T_2, T_1)}$.

3. $ARE(T_1, T_3) = ARE(T_1, T_2)ARE(T_2, T_3)$.

Uses of ARE

The ARE determined in (2.143) has two practical uses:

1. If T_1 and T_2 are two α-level tests, and if n_1 is the sample size required for test T_1 to achieve power Π_1 at an alternative θ_1, then the sample size required for test T_2 to achieve the same power Π_1 at the same alternative θ_1 is

$$n_2 = n_1.ARE(T_1, T_2). \qquad (2.144)$$

2. Suppose that T_1 and T_2 are two α-level tests based on the same sample size n, the powers of T_1 and T_2 are the same at the alternatives $\theta_0 + \frac{\delta_1}{\sqrt{n}}$, and $\theta_0 + \frac{\delta_2}{\sqrt{n}}$, where

$$\delta_2 = \delta_1 \cdot \sqrt{ARE(T_1, T_2)}. \qquad (2.145)$$

These results for the two-sample tests will be discussed in this section. Some one-sample results connected with paired samples will be given in Chapter 3.

ARE calculations for two-sample tests

After indicating the needed modifications for the two-sample case, we will illustrate the relative efficiency calculations. Clearly, in the case of tests for the two-sample problems, the quantity $e_n(T)$ will depend on both sample sizes m and n and the limiting process to consider is that the total sample size $N = m + n$ tends to infinity such that $(m/N) \to \lambda$, where $0 < \lambda < 1$.

In the following we consider the two-sample problem under the shift model, that is,

$$G(x) = F(x - \Delta),$$

and we are interested in testing

$$H_0 : \Delta = 0 \quad against \quad H_{A1} : \Delta > 0.$$

When F is a normal distribution, in the parametric analysis we would use the two-sample t-test. In this chapter we discussed three other tests, namely, Mathisen's test, Mood's median test, and the WMW test. Now we derive relative efficiencies of the WMW test and Mood's median test relative to the t-test. This calculation involves the calculation of efficacies of these tests.

Efficacy of two-sample t-test
The t-test statistic is defined as

$$T = (\bar{Y} - \bar{X})/\sqrt{s_p^2(1/m + 1/n)}$$

$$= [(\bar{Y} - \bar{X})/\sigma]/\sqrt{(s_p^2/\sigma^2)(1/m + 1/n)}. \qquad (2.146)$$

Here s_p^2 is the pooled sample variance. As $N \to \infty$, s_p^2 tends to σ^2 and we can take

$$\mu_T(\Delta) = \frac{\Delta/\sigma}{\sqrt{1/m + 1/n}},$$

which leads to

$$\mu_T'(0) = \frac{1}{\sigma \cdot \sqrt{1/m + 1/n}}.$$

Since $\sigma_T^2(0)$ can be taken as 1, we have

$$e(T) = \lim_{N \to \infty} \left(\frac{1}{\sigma \cdot \sqrt{1/m + 1/n}\sqrt{N}} \right)$$

$$= \frac{\sqrt{\lambda(1 - \lambda)}}{\sigma}. \qquad (2.147)$$

Now we will calculate the efficacy of the WMW test.

Efficacy of WMW test
It is convenient to consider the Mann-Whitney statistic U_{XY}. The mean of this statistic is

$$\mu_U(\Delta) = m \cdot nP(X < Y)$$

$$= m \cdot n \int_{-\infty}^{\infty} F(y)g(y)dy$$

$$= m \cdot n \int_{-\infty}^{\infty} F(y)f(y - \Delta)dy$$

$$= m \cdot n \int_{-\infty}^{\infty} F(x + \Delta)f(x)dx. \qquad (2.148)$$

Hence we have

$$\mu_U'(0) = m \cdot n \int_{-\infty}^{\infty} f^2(x)dx. \qquad (2.149)$$

Recalling that

$$\sigma_U^2(0) = m \cdot n(m+n+1)/12,$$

we have the efficacy of U is

$$e(U) = \lim_{N \to \infty} \left(\frac{m \cdot n \int_{-\infty}^{\infty} f^2(x)dx}{\sqrt{N}\sqrt{\frac{mn(m+n+1)}{12}}} \right)$$

$$= \sqrt{12}\sqrt{\lambda(1-\lambda)} \int_{-\infty}^{\infty} f^2(x)dx. \tag{2.150}$$

ARE (WMW, T)

Finally, the relative efficiency of the WMW test relative to the two-sample t-test is

$$ARE(W,T) = ARE(U,T) = 12\sigma^2 \left[\int_{-\infty}^{\infty} f^2(x)dx \right]^2. \tag{2.151}$$

When $f(x)$ is the density of a standard normal distribution, (2.151) reduces to

$$ARE_N(W,T) = 3/\pi.$$

Sample size for WMW test

Consider the case where we have samples of the same size ($m = n$) from two normal distributions, with the same variance σ^2. The common sample size n_T for an α-level two-sample t-test with a power of at least $1 - \beta$ at the shift alternative $\delta = \delta_1$, using a normal approximation for the power, is obtained. It is

$$n_T = \frac{2\sigma^2(z_\alpha + z_\beta)^2}{\delta_1^2}.$$

Hence, from (2.151), the required sample size for the WMW test is

$$n_w = \frac{n_T}{ARE_N(W,T)} = \frac{2\sigma^2(z_\alpha + z_\beta)^2 \pi}{3\delta_1^2}. \tag{2.152}$$

This formula is also given by Lehmann (1998, p. 74). This sample size determination problem has also been considered by Noether (1987).

Efficacy of Mood's median test

As before, we consider the problem of testing $H_0 : \Delta = 0$ against the alternative $H_A : \Delta > 0$. Mood's test is based on the statistic V, the number of X-observations exceeding the median M of the combined sample. However, the critical region should be of the form $T > C$, so we have to take $T_M = m - V$. This statistic follows a hypergeometric distribution under the null hypothesis.

For large samples, M is approximately equal to ν, where ν is given by

$$mF(\nu) + nG(\nu) = \frac{m+n}{2}. \tag{2.153}$$

Now we compute the mean of V under the alternative. It is

$$\mu_{T_M} = m - \mu_V(\Delta) = m - m(1 - F(\nu)) = mF(\nu);$$

so the derivative is

$$\mu'_{T_M}(\Delta) = mf(\nu)\frac{d\nu}{d\Delta}. \tag{2.154}$$

Replacing $G(x)$ by $F(x - \Delta)$ in (2.153) and differentiating with respect to Δ we get

$$\frac{dC\nu}{d\Delta} = \frac{nf(\nu - \Delta)}{mf(\nu) + nf(\nu - \Delta)}.$$

Hence

$$\frac{d\nu}{d\Delta}|_{\Delta=0} = \frac{n}{n+m}. \tag{2.155}$$

Therefore

$$\mu'_{T_M}(0) = \frac{mn}{n+m}f(\nu). \tag{2.156}$$

Under H_0, from (2.153), we note that $\nu = F^{-1}(1/2)$. Hence (2.156) becomes

$$\mu'_{T_M}(0) = \frac{mn}{n+m}f(F^{-1}(1/2)). \tag{2.157}$$

The null variance is

$$\sigma^2_{T_M}(0) = \sigma^2_V(0) = \frac{mn}{4(m+n)}.$$

Thus the efficacy of T_M, with $m = \lambda N$, is

$$e(T_M) = \lim_{N \to \infty} \left[\frac{\frac{mn}{m+n}f(F^{-1}(1/2))}{(1/2)\sqrt{mn}} \right]$$

$$= 2\sqrt{\lambda(1-\lambda)}f(F^{-1}(1/2)). \tag{2.158}$$

Finally, from (2.158) and (2.147), we get

$$ARE(V,T) = ARE(T_M,T) = 4\sigma^2(f(F^{-1}(1/2)))^2. \tag{2.159}$$

When F is the standard normal distribution, the above ARE simplifies to

$$ARE_N(Mood, T) = 2/\pi.$$

Further, we have

$$ARE_N(V, U_{m,n}) = \frac{2/\pi}{3/\pi} = 2/3. \tag{2.160}$$

Table 2.27 *Estimates of ARE's relative to F-test*

	Cox's Test	Logrank Test	Gehan's Wilcoxon Test
No Censoring	0.88	0.88	0.72
47% Censoring	0.98	0.98	0.77

Graphical estimation of ARE

From (2.142), we observe that the plot of $\Phi^{-1}(power)$ against δ is a straight line. Thus an estimate of the ARE of two tests can be obtained from the slopes of the normal probability plots of the power curves.

We generate the power curves using simulations for fixed n and varying δ. Using the normal probability plots of the power curves we can estimate the slopes. Then ARE estimates the square of the ratio of the slopes of the simulated power curves (see Cox and Stuart, 1955).

This idea was used by Lee, Desu, and Gehan (1975) to estimate the ARE's of the Cox test, the logrank test, and Gehan's Wilcoxon test relative to the F-test for uncensored data and for censored data from exponential distributions. The results are given in Table 2.27.

2.8 Appendix A2: Mathematical supplement

A2.1 Derivation of the conditional distribution of A given $T = t$

We have

$$P(A = a, B = b|\theta_1, \theta_2) = \binom{m}{a}\theta_1^a(1 - \theta_1)^{m-a}\binom{n}{b}\theta_2^b(1 - \theta_2)^{n-b},$$

which under the null hypothesis $H_0 : \theta_1 = \theta_2 = \Theta$, simplifies to

$$P(A = a, B = b|\Theta) = \binom{N}{t}\Theta^t(1 - \Theta)^{N-t}\frac{\binom{m}{a}\binom{n}{b}}{\binom{N}{t}},$$

where $N = m + n$ and $t = a + b$. We observe that under H_0, $T(= A + B)$ has the binomial distribution with parameters N and Θ. So

$$P(T = t|H_0) = \binom{N}{t}\Theta^t(1 - \Theta)^{N-t}.$$

Now

$$
\begin{aligned}
P(A = a|T = t, H_0) &= P(A = a, T = t|H_0)/P(T = t|H_0) \\
&= P(A = a, B = t - a|H_0)/P(T = t|H_0) \\
&= \binom{m}{a}\binom{n}{t-a}\Big/\binom{N}{t} \\
&= \binom{t}{a}\binom{N-t}{m-a}\Big/\binom{N}{m}.
\end{aligned}
$$

The last equality follows by expressing the bionomial coefficients in terms of factorials and combining them in a different manner.

It is of some interest to make a note of the mean and the variance of this conditional distribution. Using the results for a hypergeometric distribution, we have

$$E(A|T = t, H_0) = \frac{mt}{N},$$

and

$$var(A|T = t, H_0) = \frac{m(t/N)(1 - (t/N))(N - m)}{(N - 1)}$$

$$= E(A|T = t, H_0)\frac{(N - t)(N - m)}{(N - 1)N}.$$

Clearly

$$E(A|T = t, H_0) = \hat{E}(A|H_0).$$

Noting

$$var(A|T = t, H_0) = var\left(A - m\frac{T}{N}\Big|T, H_0\right),$$

and

$$var\left(A - m\frac{T}{N}\Big|H_0\right) = \frac{mn}{N}\Theta(1 - \Theta),$$

we have

$$\hat{v}ar\left(A - m\frac{T}{N}\Big|H_0\right) = \frac{mn}{N}\hat{\Theta}(1 - \hat{\Theta}) = \frac{N - 1}{N}var(A|T, H_0).$$

Hence

$$var(A|T, H_0) \approx \hat{v}ar(A|H_0).$$

It follows that

$$Z = \frac{A - E(A|T = t, H_0)}{\sqrt{var(A|T = t, H_0)}}$$

has zero mean and unit variance. This statistic is approximately equal to the statistic Z of (2.10). Thus a test based on the approximate normal distribution of this Z is similar to the test considered in Section 2.1.

A2.2 Maximum likelihood estimation in the case of clinical equivalence

The likelihood of the two samples is

$$L = \left[\binom{m}{A}\theta_1^A(1 - \theta_1)^{m-A}\right] \times \left[\binom{n}{B}\theta_2^B(1 - \theta_2)^{n-B}\right].$$

Here $\theta_2 = \theta_1 - \Delta_0$. Thus the log likelihood is

$$l = lnL = ln\binom{m}{A} + Aln\,\theta_1 + (m-A)ln(1-\theta_1)$$

$$+ ln\binom{n}{B} + Bln(\theta_1 - \Delta_0) + (n-B)ln(1-\theta_1+\Delta_0)$$

so that the derivative of l is

$$\frac{dl}{d\theta_1} = \frac{A}{\theta_1} - \frac{m-A}{1-\theta_1} + \frac{B}{\theta_1-\Delta_0} - \frac{n-B}{1-\theta_1+\Delta_0}.$$

This derivative is the same as

$$\frac{dl}{d\theta_1} = \frac{A-m\theta_1}{\theta_1(1-\theta_1)} + \frac{B-n(\theta_1-\Delta_0)}{(\theta_1-\Delta_0)(1-\theta_1+\Delta_0)}.$$

Setting this derivative to zero and replacing θ_1 by z, for convenience, we get the maximum likelihood equation

$$(A-mz)[(z-\Delta_0)(1-z+\Delta_0)] + z(1-z)[B-n(z-\Delta_0)] = 0.$$

We need to solve this equation for z. This is a cubic equation in z. Multiplying by $1/m$, we rewrite it as

$$a^* z^3 + b^* z^2 + c^* z + d^* = 0,$$

where

$$a^* = (1+\eta); \quad b^* = -[\hat{\theta}_1 + \eta\hat{\theta}_2 + (1+\eta) + \Delta_0(\eta+2)];$$

and

$$c^* = \Delta_0^2 + (2\hat{\theta}_1 + 1 + \eta)\Delta_0 + \hat{\theta}_1 + \eta\hat{\theta}_2; \quad d^* = -\Delta_0(1+\Delta_0)\hat{\theta}_1,$$

η being the ratio (n/m), $\hat{\theta}_1$ being A/m, and $\hat{\theta}_2$ being B/n. To get the required solution we compute

$$v = (b^*/3a^*)^3 - b^* c^*/(6(a^*)^2) + d^*/(2a^*),$$

$$u = sgn(v)[\{b^*/(3a^*)\}^2 - c^*/(3a^*)]^{1/2}.$$

Using u and v, we compute

$$w = (1/3)[\pi + cos^{-1}(v/u^3)].$$

Now the solution is given by

$$z = 2ucos(w) - (b^*/3a^*).$$

A2.3 Koopman's interval for the ratio of two binomial θ's

Here the parameter of interest is the ratio $\psi = \theta_1/\theta_2$. We want to construct a confidence interval for ψ. The statistics A and B follow binomial distributions and they are independent. To derive a confidence interval a test of $H_0 : \psi = \psi_0$ against $H_1 : \psi \neq \psi_0$ will be considered. Pearson's goodness-of-fit test is based on U_{ψ_0}, which is given by

$$U_{\psi_0} = \frac{(A - m\hat{\theta}_1)^2}{m\hat{\theta}_1(1 - \hat{\theta}_1)} + \frac{(B - n\hat{\theta}_2)^2}{n\hat{\theta}_2(1 - \hat{\theta}_2)},$$

where $\hat{\theta}_1$ and $\hat{\theta}_2$ are the maximum likelihood estimates under the null hypothesis. It may be noted that $\hat{\theta}_2 = \hat{\theta}_1/\psi_0$.

Now let us derive the maximum likelihood estimate $\hat{\theta}_1$ under the null hypothesis. The log likelihood function, under H_0, is

$$l = const + A \cdot ln\theta_1 + (m - A) \cdot ln(1 - \theta_1)$$

$$+ B \cdot ln\left(\frac{\theta_1}{\psi_0}\right) + (n - B) \cdot ln\left(1 - \frac{\theta_1}{\psi_0}\right).$$

Thus the derivative of l with respect to θ_1 is

$$\frac{dl}{d\theta_1} = \frac{A}{\theta_1} - \frac{(m - A)}{1 - \theta_1} + \frac{B}{\theta_1} - \frac{(n - B)}{\psi_0 - \theta_1}.$$

This derivative can be simplified as

$$\frac{dl}{d\theta_1} = \frac{A - m\theta_1}{\theta_1(1 - \theta_1)} + \frac{B\psi_0 - n\theta_1}{\theta_1(\psi_0 - \theta_1)}.$$

Equating this derivative to zero, we get the likelihood equation

$$\frac{A - m\theta_1}{\theta_1(1 - \theta_1)} = -\frac{B\psi_0 - n\theta_1}{\theta_1(\psi_0 - \theta_1)}.$$

This equation is the same as

$$(m + n)\theta_1^2 - [\psi_0(m + B) + A + n]\theta_1 + (A + B)\psi_0 = 0.$$

Solving this quadratic equation, we get the ML estimator of θ_1 as

$$\hat{\theta}_1 = \frac{1}{2(m + n)}[\psi_0(m + B) + (A + n)$$

$$- \left(\{\psi_0(m + B) + A + n\}^2 - 4\psi_0(m + n)(A + B)\right)^{1/2}].$$

Also $\hat{\theta}_2 = \hat{\theta}_1/\psi_0$. Using the relationship between A , B and θ_1, the statistic U can be computed as

$$U_{\psi_0} = \frac{(A - m\hat{\theta}_1)^2}{m\hat{\theta}_1(1 - \hat{\theta}_1)}\left[1 + \frac{m(\psi_0 - \hat{\theta}_1)}{n(1 - \hat{\theta}_1)}\right].$$

The asymptotic distribution of U is the chi-square distribution with 1 degree of freedom. Hence an approximate $1 - \alpha$ confidence region is

$$\{\psi : U_\psi < \chi_1^2(1 - \alpha)\}.$$

This region will reduce to an interval. The end points ψ_l and ψ_r are the two solutions of the quadratic equation $U_\psi = \chi_1^2(1 - \alpha)$. We need a computer program to find these solutions.

A2.4 Calculation of exact P-values for the problem of Section 2.3:
Extension of Fisher's exact test

To calculate the exact P-values we need to derive a conditional distribution, as in the case of Fisher's exact test of Subsection 2.1.2. We argue as follows.

The random variables A_i follow a multinomial distribution and the variables B_i also follow a multinomial distribution. The two samples are independent. So the joint probability function of A_i's and B_i's is

$$P(A_1 = a_1, \ldots, A_k = a_k; B_1 = b_1, \ldots, B_k = b_k) = m! \prod_{i=1}^{k} \frac{\pi_{xi}^{a_i}}{a_i!} \times n! \prod_{i=1}^{k} \frac{\pi_{yi}^{b_i}}{b_i!}.$$

Under H_0, $\pi_{xi} = \pi_{yi} = \pi_i$, so we have

$$P(A_1 = a_1, \ldots, A_k = a_k; B_1 = b_1, \ldots, B_k = b_k | H_0) = m!n! \prod_{i=1}^{k} \frac{\pi_i^{c_i}}{a_i!b_i!},$$

where $c_i = a_i + b_i$. The P-value is the sum of the probabilities of all frequency configurations that are more extreme than the one observed. Since π_i are unknown, we cannot use the above probability function to compute the P-value. However, we can rewrite the above function as the probability function of $C_i = A_i + B_i$ under H_0 times another function. This is the idea behind Fisher's exact test. Now

$$P(A_1 = a_1, \ldots, A_k = a_k; B_1 = b_1, \ldots, B_k = b_k | H_0)$$

$$= N! \left(\prod_{i=1}^{k} \frac{\pi_i^{c_i}}{c_i!} \right) \frac{m!n!}{N!} \prod_{i=1}^{k} \binom{c_i}{a_i}.$$

The second part is the probability function of the conditional distribution of A_i's and B_i's, given $C_i = c_i$. In other words

$$P(A_1 = a_1, \ldots, A_k = a_k; B_1 = b_1, \ldots, B_k = b_k | C_i = c_i, H_0) = \frac{m!n!}{N!} \prod_{i=1}^{k} \binom{c_i}{a_i},$$

which is a multivariate hypergeometric distribution. Usually this function is used to find the exact P-value.

In general, we could use more than one statistic for testing the null hypothesis. In Section 2.3 we considered the Pearson's goodness-of-fit statistic T. The P-value is the sum of probabilities, according to the conditional distribution, of all frequency configurations for which the value of the statistic T is greater than $T(obs)$, the observed value of T. In other words, the P-value is

$$P\text{-}value(T) = \sum_{\epsilon} P(\mathbf{a}, \mathbf{b} | \mathbf{c}^*, H_0),$$

where c_i^* is the observed value of $C_i = A_i + B_i$ and the set e is

$$\epsilon = \{(\mathbf{a}, \mathbf{b}) | T > T(obs)\}.$$

We could use the likelihood ratio statistic and then the exact P-value would be computed in a manner similar to the one just described.

Extension of Fisher's test.

Here we calculate the sum of probabilities of all configurations that are less likely than the observed configuration, according to the conditional distribution. In other words

$$P\text{-}value(\mathit{Fisher}) = \sum_{\Gamma} P(\mathbf{a}, \mathbf{b}|\mathbf{c}^*, H_0),$$

where c_i^* is the observed value of $C_i = A_i + B_i$ and the set Γ is

$$\Gamma = \{(\mathbf{a}, \mathbf{b})|P(\mathbf{a}, \mathbf{b}|\mathbf{c}^*, H_0) \leq P(\mathbf{a}^*, \mathbf{b}^*|\mathbf{c}^*, H_0), \}$$

where $\mathbf{a}^*$ and $\mathbf{b}^*$ are the observed values of $\mathbf{a}$ and $\mathbf{b}$, respectively.

This calculation can be done using a program in the SAS package.

A2.5 Some models that induce stochastic ordering

In Section 2.4 we considered the assumption that the two distribution functions F and G satisfy an ordering. Sometimes this relationship is described in explicit terms through a semi-parametric model, such as

$$G(x) = A(F(x), \Delta),$$

where Δ is a scalar parameter and A is a completely specified function. It is also assumed that for a particular value Δ_0 of Δ, $G(x) = F(x)$. In other words, $G(x) = A(F(x), \Delta_0)$. First we note two general results and then examine some special cases, which have been considered frequently.

Result 1.

Let X and Y be two continuous random variables with cdf's F and G, respectively. Then

$$
\begin{aligned}
med\ X = med\ Y \quad & if\ F(x) = G(x), \\
< med\ Y \quad & if\ F(x) \geq G(x), i.e., X st < Y, \\
> med\ Y \quad & if\ F(x) \leq G(x), i.e., X st > Y.
\end{aligned}
$$

Result 2.

Let X and Y be continuous random variables with cdf's F and G, respectively. Assume that $E(X)$ and $E(Y)$ exist. Then

$$
\begin{aligned}
E(X) = E(Y) \quad & if\ F(x) = G(x), \\
< E(Y) \quad & if\ F(x) \geq G(x), i.e., X st < Y, \\
> E(Y) \quad & if\ F(x) \leq G(x), i.e., X st > Y.
\end{aligned}
$$

To prove this result, we need the following lemma.

Lemma.

Let X be a continuous random variable with finite expectation and let $F(.)$ be the cdf and $f(.)$ be the pdf of X. Then

$$E(X) = \int_0^\infty [1 - F(x)]dx - \int_{-\infty}^0 F(x)dx.$$

Proof. First we observe that

$$E(X) = \int_{-\infty}^\infty xf(x)dx$$

$$= \int_{-\infty}^0 xf(x)dx + \int_0^\infty xf(x)dx$$

$$= I_1 + I_2,$$

and we evaluate the two integrals. Now, using integration by parts, we have

$$I_{1a} = \int_a^0 xf(x)dx = -aF(a) - \int_a^0 F(x)dx.$$

Taking the limit as $a \to -\infty$, we get

$$I_1 = lt I_{1a} = 0 - \int_{-\infty}^0 F(x)dx.$$

Let

$$I_{2b} = \int_0^b xf(x)dx.$$

First we observe that

$$I_{2b} = \int_0^b xf(x)dx = bF(b) - \int_0^b F(x)dx$$

$$= -b[1 - F(b)] + \int_0^b [1 - F(x)]dx.$$

Now taking limits as $b \to \infty$, we get

$$I_2 = lt I_{2b} = 0 + \int_0^\infty [1 - F(x)]dx.$$

In view of these representations of I_1 and I_2, we get the final result.

Remark. For positive random variables, we have

$$E(X) = \int_0^\infty [1 - F(x)]dx = \int_0^\infty S_F(x)dx,$$

where $S_F(.)$ is the survival function.

We proceed to prove result 2.

Proof of result 2. When $F(x) \geq G(x)$, we have

$$-F(x) \leq -G(x); \quad and \quad 1 - F(x) \leq 1 - G(x).$$

Using this fact and the expression for $E(X)$, given by the lemma, we have

$$E(X) \leq \int_0^\infty [1 - G(x)]dx - \int_{-\infty}^0 G(x)dx,$$

which means that

$$E(X) \leq E(Y).$$

A similar proof can be given for the other case, $F(x) \leq G(x)$.

Shift model (location model)
    We have already discussed the shift model, which states

$$G(x) = F(x - \Delta).$$

Under this model there is a stochastic ordering between F and G, and $F = G$, when $\Delta = 0$. The distribution of $X + \Delta$ is the same as the distribution of Y. Thus

$$med\, Y = med\, X + \Delta, \quad and \quad E(Y) = E(X) + \Delta.$$

Example 1. Suppose that X has $N(0, 1)$ distribution and Y has $N(\Delta, 1)$ distribution. Then the distributions of X and Y are related through a shift model.
    Under this model the relationship between the probability densities is

$$g(x) = f(x - \Delta).$$

Scale model
    Here we *assume* that $F(0) = 0$ and $G(0) = 0$, so that the associated random variables X and Y are *positive-valued* random variables. Further,

$$G(x) = F\left(\frac{x}{\Delta}\right),$$

where Δ is a positive scalar parameter. Now it is easy to see that for

$$\Delta = 1, \; F(x) = G(x),$$
$$< 1, \; F(x) \leq G(x), \; or \; X \; st > Y,$$
$$> 1, \; F(x) \geq G(x), \; or \; X \; st < Y.$$

In general, the distribution of ΔX is the same as the distribution of Y. Hence

$$med\, Y = \Delta med\, X, \quad E(Y) = \Delta E(X).$$

Further, the relationship between the probability densities is

$$g(x) = \frac{1}{\Delta} f\left(\frac{x}{\Delta}\right).$$

    For making an inference about Δ, it is convenient to consider the transformed variables

$$X^* = log(X) \quad and \quad Y^* = log(Y).$$

Now the distributions of the transformed variables X^* and Y^* are related through a shift model with $log(\Delta)$ as the shift parameter. Thus any problem of interest under the scale model can be answered using the shift model for the transformed variables.

Example 2. Suppose that X follows an exponential distribution with mean 1 and Y follows an exponential distribution with mean Δ. Then the distributions F and G satisfy the scale model.

When the scale model holds, the relationship between the survival functions is

$$S_G(x) = S_F\left(\frac{x}{\Delta}\right).$$

We want to remark that in some discussions the parameter Δ is taken as $1+\theta$, so that $G(x) = F(x)$, for $\theta = 0$.

Proportional hazards model

Lehmann (1953) proposed the model

$$G(x) = [F(x)]^\Delta$$

to define a set of alternatives to the homogeneity hypothesis and discussed certain properties of the power of some rank tests for these alternatives. The proportional hazards (PH) model has been used extensively in reliability studies and survival data analysis. Under the PH model, the ratio of the hazard functions of the distributions F and G is a constant. In other words,

$$\lambda_G(x) = (1 + \theta)\lambda_F(x).$$

This relation implies that the cumulative hazard function of the distribution G is a multiple of the cumulative hazard function of the distribution F. In other words,

$$\Lambda_G(x) = (1 + \theta)\Lambda_F(x),$$

since $\Lambda_F(x) = \int_0^x \lambda_F(u)du$.

Now we can establish a relationship between the survival functions. We have

$$S_G(x) = exp\{-\Lambda_G(x)\}$$
$$= exp\{-(1 + \theta)\Lambda_F(x)\}, \text{ under } PH \text{ model}$$
$$= [exp\{-\Lambda_F(x)\}]^{(1+\theta)} \equiv [S_F(x)]^{(1+\theta)}.$$

From this relationship, it immediately follows that

$$G(x) = 1 - [1 - F(x)]^{(1+\theta)}.$$

Now it is easy to see that for

$$\theta = 0, \quad G(x) = F(x),$$
$$> 0, \quad G(x) \geq F(x), i.e., X st > Y,$$
$$< 0, \quad G(x) \leq F(x), i.e., X st < Y.$$

Example 2 also serves as an example for this model.

A2.6 The null distribution of T_a

This derivation uses the fact that the conditional distribution of T_a, given $X_{(a)}$ is the binomial distribution with parameters n and $G(X_{(a)})$ and the probability density function of $X_{(a)}$ is

$$h(x) = a \binom{m}{a} [F(x)]^{a-1} [1 - F(x)]^{m-a} f(x).$$

Thus

$$P(T_a = j) = \int_{-\infty}^{\infty} P(T_a = j | X_{(a)} = x) h(x) dx$$

$$= \int_{-\infty}^{\infty} \binom{n}{j} [G(x)]^j [1 - G(x)]^{n-j} h(x) dx.$$

This result implies

$$P(T_a = j | H_0) = \int_{-\infty}^{\infty} \binom{n}{j} [F(x)]^j [1 - F(x)]^{n-j} h(x) dx.$$

Substituting for $h(x)$ and evaluating the integral we get

$$P(T_a = j | H_0) = \binom{n}{j} a \binom{m}{a} Beta(j + a, n - j + m - a + 1)$$

$$= \binom{n}{j} \frac{m!}{(a-1)!(m-a)!} \frac{(a+j-1)!(m+n-a-j)!}{(m+n)!}$$

$$= \frac{m}{m+n} \frac{\binom{n}{j}\binom{m-1}{a-1}}{\binom{m+n-1}{a+j-1}} = \frac{\binom{a+j-1}{j}\binom{N-a-j}{n-j}}{\binom{N}{n}}.$$

To derive an expression for the variance of T_a, we calculate the second moment,

$$E(T_a^2 | H_0) = E[E(T_a^2 | X_{(a)}, H_0)]$$

$$= E[nF(X_{(a)})\{1 - F(X_{(a)})\} + \{nF(X_{(a)})\}^2]$$

$$= \frac{na}{(m+1)} + (n^2 - n) \left[\frac{a(a+1)}{(m+2)(m+1)} \right].$$

Using this expression the required variance is

$$var(T_a|H_0) = E(T_a^2|H_0) - [E(T_a|H_0)]^2$$

$$= \frac{na}{(m+1)} + (n^2 - n)\left[\frac{a(a+1)}{(m+2)(m+1)}\right] - \frac{n^2a^2}{(m+1)^2}$$

$$= \frac{na[(m+1)(N-a+1) - na]}{(m+1)^2(m+2)}$$

$$= \frac{na(m+1-a)(N+1)}{(m+1)^2(m+2)}$$

$$= \frac{n\lambda(1-\lambda)(N+1)}{(m+2)},$$

where $\lambda = a/(m+1)$.

This statistic is also used in Chapter 4, where we consider an extension of Mathisen's median test to the k-sample problem. We call this statistic *Mathisen's control percentile statistic*, when the X-sample is from a control population and the Y-sample is from an experimental population. In terms of λ the expectation of T_a is $n\lambda$.

Relationship to the hypergeometric distribution

Let us consider a random variable Z. Then, for $z = a + j$, the distribution of Z is defined as

$$P(Z = z) = P(T_a = j|H_0) = \frac{\binom{m}{a-1}\binom{N-m}{z-a}}{\binom{N}{z-1}} \cdot \frac{m-a+1}{N-z+1}$$

$$= \frac{\binom{z-1}{a-1}\binom{N-z}{m-a}}{\binom{m+n}{n}}.$$

It can be seen that Z represents the number of draws required to accumulate a red balls from a collection of m red balls and n green balls. Balls are drawn at random without replacement. Let U denote the number of red balls in a sample of $a + t - 1$ balls drawn without replacement from the same collection. Now

$$P(T_a \geq t|H_0) = P(Z \geq t+a)$$

$$= 1 - P(Z \leq t+a-1)$$

$$= 1 - P(U \geq a),$$

because stopping in at most $t + a - 1$ draws implies that we could get at least a red balls in $t + a - 1$ draws. So

$$P(T_a \geq t|H_0) = P(U \leq a-1)$$

$$= \sum_{u=0}^{a-1} \frac{\binom{m}{u}\binom{n}{a-1+t-u}}{\binom{m+n}{a-1+t}}.$$

Thus the P-values can be computed using the cdf of a hypergeometric distribution, which is a SAS function.

A2.7 Confidence interval for Δ from Mathisen's test

We consider the problem of testing

$$H_0 : \Delta = \Delta_0 \quad against \quad H_{A3} : \Delta \neq \Delta_0.$$

In order to use the tests developed earlier, we have to use the transformed data $(X_1, X_2, \ldots, X_m)$ and $(Y_1', Y_2', \ldots, Y_n')$, where $Y_j' = Y_j - \Delta_0$. We use the two-sided test on this transformed data. We note that the null hypothesis is not rejected whenever

$$Y'_{(c_3+1)} \leq X_{(s)} \quad or \quad Y'_{(c_4)} > X_{(s)}$$

where $s = \lfloor (m/2) \rfloor + 1$. In other words, all Δ_0 values that satisfy the condition

$$Y_{(c_3+1)} - \Delta_0 \leq X_{(s)} < Y_{c_4} - \Delta_0$$

are not rejected. Hence Δ values that belong to the interval $(Y_{(c_3+1)} - X_{(s)}, Y_{(c_4)} - X_{(s)})$ will not be rejected by a test of size α. In other words, the above interval is a confidence interval for Δ with a confidence coefficient $1 - \alpha$.

Large sample approximations for the constants c_3 and c_4 are obtained by appealing to the large sample normal distribution of the statistic T_M. For example, we need to choose c_3 so that

$$\alpha/2 \geq P(X_{(s)} < Y_{(c_3+1)}|H_0) = P(T_s \leq c_3|H_0).$$

Considering the standardized version of T_s, with approximate mean and approximate variance and using the continuity correction, we get

$$c_3 + 0.5 - (n/2) = z_{\alpha/2} \left(\frac{nN}{4m} \right)^{1/2}.$$

In other words, we take c_3 as

$$c_3 = \left\lfloor (n/2) - 0.5 - z_{(1-\alpha/2)} \left(\frac{nN}{4m} \right)^{1/2} \right\rfloor.$$

The other constant is determined in a similar manner.

A2.8 A class of distribution-free statistics

A class of statistics that are useful in the general two-sample problem will be indicated. The Mood's statistic V is a member of this class. Let F_m and G_n be the sample distribution functions of the X-sample and the Y-sample, respectively. Let $(Z_{(1)}, Z_{(2)}, \ldots, Z_{(N)})$ be the ordered values in the combined sample. For $t = 1, 2, \ldots, N$, let

$$A_s = number\ of\ X\text{-}values\ less\ than\ or\ equal\ to\ Z_{(s)}.$$

It is easy to see that the statistic $A_a = m - V$, where V is the Mood's statistic. We can relate A_s and the X-sample distribution function, and this relationship is

$$A_s = m F_m(Z_{(s)}).$$

In the combined sample we have mX's and nY's. There are $\binom{N}{m}$ distinguishable ways of arranging the $N(= m + n)$ values of the combined sample in a row. These possible permutations are equally likely under the null hypothesis. To calculate the probability of the event $\{A_s = k\}$, we need to count the number of arrangements that are favorable to this event. These favorable arrangements are those in which there are exactly k X's among the bottom s values, and exactly $(m - k)$ X's among the top $N - s$ values. Clearly, the number of such arrangements is $\binom{s}{k}\binom{N-s}{m-k}$. Thus we have

$$P[A_s = k | H_0] = \frac{\binom{s}{k}\binom{N-s}{m-k}}{\binom{N}{m}},$$

for integers k between $max(s - n, 0)$ and $min(m, s)$. As this distribution is a hypergeometric distribution, we have

$$E(A_s | H_0) = m(s/N) = s(m/N),$$

and

$$var(A_s | H_0) = \frac{mns(N - s)}{N^2(N - 1)}.$$

Connection to Smirnov test statistic

In the context of the Smirnov test for the two-sample problem we consider the differences $D_s^+ = F_m(Z_{(s)}) - G_n(Z_{(s)})$. This difference is a linear function of A_s. In fact

$$D_s^+ = \frac{1}{m} A_s - \frac{1}{n}(s - A_s)$$

$$= \left(\frac{1}{m} + \frac{1}{n}\right) A_s - (s/n)$$

$$= \frac{N}{mn} A_s - (s/n)$$

$$= \frac{N}{mn}[A_s - E(A_s | H_0)].$$

The joint distribution of the entire set $\{D_t^+, t = 1, \ldots, N\}$ is independent of the common distribution, under the null hypothesis. So under the null hypothesis the statistic $D_{m,n}^+$ is a distribution-free statistic. The same is true for the statistic $D_{m,n}^-$. Hence the statistic $D_{m,n}$ is a distribution-free statistic under the null hypothesis.

A2.9 The null distribution of V

This can be obtained from the results of Section A8. However, a direct derivation is of interest and this derivation is given below.

In the combined sample we have mX's and nY's. There are $\binom{N}{m}$ distinguishable ways of arranging the $N(= m + n)$ values of the combined sample in a
row. These possible permutations are equally likely under the null hypothesis. Now to calculate the probability of the event $\{V = v\}$, we need to count
the number of arrangements that are favorable to this event. These favorable
arrangements are those in which there are exactly v X's among the top $N - a$
values and exactly $(m - v)$ X's among the bottom a values. The number of
such arrangements is $\binom{a}{m-v}\binom{N-a}{v}$. Thus we have

$$P(V = v|H_0) = \frac{\binom{a}{m-v}\binom{N-a}{v}}{\binom{N}{m}}.$$

By expressing the binomial coefficients as factorials and by regrouping we
get

$$P(V = v|H_0) = \frac{\binom{m}{v}\binom{n}{N-a-v}}{\binom{N}{N-a}},$$

for $v = max(m - a, 0), 1, \ldots, min(m, N - a)$.

A2.10 Confidence interval for Δ from Mood's median test

Consider the problem of testing

$$H_0 : \Delta = 0 \quad against \quad H_{A3} : \Delta \neq 0.$$

Recall that the critical region of the test for this two-sided alternative is

$$V \leq v_3 \quad or \quad V \geq v_4,$$

which is the same as

$$A_s \leq a_3 \quad or \quad A_s \geq a_4,$$

where $a_4 = m - v_3$ and $a_3 = m - v_4$. As per this test, we would "accept" a
Δ value if there are at least $(a_3 + 1)$, X-values and at most $(a_4 - 1)$, X-values
among the s smallest of the combined sample $(X_1, \ldots, X_m, Y_1 - \Delta, \ldots, Y_n - \Delta)$
and we would reject a Δ value otherwise. Let $Z_{(s)}$ be the sth order statistic of
this combined sample. Let us analyze the set $A_s \leq a_3$. This set corresponds
to the event that the s smallest values include at most a_3, X-values and hence
at least $s - a_3, Y - \Delta$ values. In other words, in this set, we have

$$Z_{(s)} < X_{(a_3+1)} \quad and \quad Z_{(s)} \geq Y_{(s-a_3)} - \Delta.$$

This is the same as

$$X_{(a_3+1)} > Y_{(s-a_3)} - \Delta.$$

A similar argument results that the set, $A_s \geq a_4$, corresponds to the event

$$X_{(a_4)} < Y_{(s-a_4+1)} - \Delta.$$

Thus we reject a Δ value whenever

$$X_{(a_4)} < Y_{(s-a_4+1)} - \Delta \quad or \quad X_{(a_3+1)} > Y_{(s-a_3)} - \Delta.$$

In other words a Δ value would not be rejected if

$$Y_{(s-a_4+1)} - X_{(a_4)} \leq \Delta \leq Y_{(s-a_3)} - X_{(a_3+1)}.$$

Replacing a's by v's, the required confidence interval is

$$(Y_{(s-m+v_3+1)} - X_{(m-v_3)}, Y_{(s-m+v_4)} - X_{(m-v_4+1)}).$$

A2.11 Null distribution of the rank vector

In the two-sample problem, we obtain the ranks of the observations in the combined sample and use them for constructing a test criterion. Let us denote the combined sample as

$$(Z_1, \ldots, Z_m, Z_{m+1}, \ldots, Z_{m+n}) = (X_1, \ldots, X_m, Y_1, \ldots, Y_n).$$

Under the null hypothesis, this $\mathbf{Z}$ vector is a set of independent and identically distributed continuous random variables. Let $\mathbf{S}$ be the rank vector, where the ith component S_i is the rank of Z_i in the combined sample. Clearly the set of possible values for $\mathbf{S}$ is the set of $N!$ permutations of $(1, 2, \ldots, N)$.

To understand the details about the derivation we consider a simple case where $m = 1$, and $n = 2$. We can enumerate the possible orderings of the Z variables that determine the values of $\mathbf{S}$. This information appears in the following table.

Distribution of the rank vector

Order of Z's	(S_1, S_2, S_3)	Prob.
$Z_1 < Z_2 < Z_3$	$(1, 2, 3)$	$1/6$
$Z_1 < Z_3 < Z_2$	$(1, 3, 2)$	$1/6$
$Z_2 < Z_1 < Z_3$	$(2, 1, 3)$	$1/6$
$Z_2 < Z_3 < Z_1$	$(2, 3, 1)$	$1/6$
$Z_3 < Z_1 < Z_2$	$(3, 1, 2)$	$1/6$
$Z_3 < Z_2 < Z_1$	$(3, 2, 1)$	$1/6$

All these orderings are equally likely, since Z's are i.i.d. continuous random variables. Thus each ordering has a probability $(1/3!) = (1/6)$. Hence all values of the rank vector are equally likely. From this distribution we can obtain the distribution of (R_1, R_2), the ranks of ordered Y-values. For example, corresponding to the first two values of $\mathbf{S}$, the value of (R_1, R_2) is $(2, 3)$; so this value carries a probability of $(2/3!)$. Similar results hold for the other possible values of (R_1, R_2). This means that all possible values for (R_1, R_2) are equally likely.

We can generalize this argument. In the general case, the rank vector can take $N!$ values, which are the permutations of $(1, 2, \ldots, N)$. Each value for the rank vector corresponds to an ordering among the Z variables. All orderings of the Z values are equally likely. Hence the rank vector has a uniform distribution over the set of permutations of $(1, 2, \ldots, N)$. So each value is assumed with a probability $1/N!$.

First we will make some observations about the marginal distribution of S_i, and the marginal distribution of (S_i, S_j). We will then discuss the marginal distribution of Y-ranks.

Marginal distribution of S_i

Clearly the possible values for S_i are $(1, 2, \ldots, N)$. Further, for $k = 1, \ldots, N$,

$$P(S_i = k) = \frac{(N-1)!}{N!} = \frac{1}{N}.$$

This distribution does not depend on i. In other words, the variables S_i follow the same distribution.

It is easy to see that

$$E(S_i) = \sum_{k=1}^{N}(k/N) = \frac{(N+1)}{2}.$$

It can also be seen that

$$var(S_i) = \frac{(N+1)(N-1)}{12}.$$

Marginal distribution of a pair (S_i, S_j)

Now for $i \neq j$ and $k \neq l$, it can be seen that

$$P(S_i = k, S_j = l) = \frac{(N-2)!}{N!} = \frac{1}{N(N-1)}.$$

The range of this distribution is the set of ordered distinct pairs (k, l), chosen from the set $(1, 2, \ldots, N)$. Clearly there are $N(N-1)$ such pairs. The distribution is uniform over all possible values.

Here again all pairs (S_i, S_j) follow the same distribution. It can be shown that

$$cov(S_i, S_j) = \frac{-(N+1)}{12}.$$

Using these results we compute the mean and the variance of the Wilcoxon statistic.

Mean of W_Y under H_0

First we note that

$$W_Y = \sum_{i=m+1}^{m+n} S_i.$$

Hence

$$E(W_Y|H_0) = nE(S_{m+1}|H_0) = \frac{n(N+1)}{2}.$$

Variance of W_Y under H_0

First we note that

$$\operatorname{var}(W_Y|H_0) = \sum_{i=m+1}^{m+n} \operatorname{var}(S_i|H_0) + \sum_{k \neq l} \operatorname{cov}(S_k, S_l|H_0).$$

This expression simplifies to

$$\operatorname{var}(W_Y|H_0) = n \operatorname{var}(S_{m+1}|H_0) + n(n-1)\operatorname{cov}(S_{m+1}, S_{m+2}|H_0).$$

Using the values for the variance and the covariance and simplifying we get

$$\operatorname{var}(W_Y|H_0) = \frac{mn(N+1)}{12}.$$

It should be noted that the mean and the variance of W_Y can also be obtained from the marginal distribution of Y-ranks, $(S_{m+1}, \ldots, S_{m+n})$.

Distribution of the Y-ranks

We will make some observations about the marginal distribution of the Y-ranks. First, we can verify that the possible values are ordered subsets of size n that can be chosen from the set $(1, \ldots, N)$. The number of such subsets is the same as the number of permutations of N distinct objects taken n at a time, which is $[N!/(N-n)!]$. Since all of these values are equally likely, each value has a probability $(N-n)!/N! = m!/N!$.

In view of the fact that the possible values are the ordered subsets of a finite population and all these values are equally likely, one can identify the distribution of Y-ranks with the distribution of values in a random sample of size n, chosen at random without replacement from the finite population consisting of values $(1, \ldots, N)$. So the distribution of W_Y is the same as the distribution of a sample sum, where the sample is a simple random sample of size n, drawn without replacement from a finite population. This observation can be used to find the mean and the variance from the results about simple random samples from a finite population. The mean of W_Y is

$$E(W_Y|H_0) = E(sample\ sum) = n \cdot E(sample\ mean)$$
$$= n(pop.\ mean) = n\frac{(N+1)}{2}.$$

Recalling that the variance of our finite population is

$$\sigma_F^2 = \frac{1}{N-1} \sum_{i=1}^{N} [i - (N+1)/2]^2 = \frac{N(N+1)}{12},$$

we have

$$var(W_Y | H_0) = n^2 \, Var(sample \ mean) = \frac{n(N-n)\sigma_F^2}{N} = \frac{mn(N+1)}{12}.$$

Remark. The distribution of the ranks of the ordered Y-values can be obtained from the distribution of the Y-ranks. Each value for the ordered ranks correspond to $n!$ values for the (unordered) ranks. Thus each value has a probability $n!(m!/N!) = [1/\binom{N}{n}]$. In a worked-out example this fact has been assumed and the distribution of W_Y has been obtained.

The reason we gave the derivation of the mean and the variance of W_Y, without using the distribution of the ranks, is that a similar derivation can be used in the discussion on linear rank statistics.

Symmetry of the null distribution of W_Y

Let us rank the combined sample in the reverse order and let W_Y' be the sum of the reverse ranks of the Y-sample. Noting that the observation having rank i gets a reverse rank $N - i + 1$, we have

$$W_Y' = \sum_{j=1}^{n} \{N + 1 - r(Y_j)\} = n(N+1) - W_Y.$$

Hence

$$W_Y' - \frac{n(N+1)}{2} = \frac{n(N+1)}{2} - W_Y.$$

It is easy to see that the distribution of W_Y is the same as the distribution of W_Y'. Now consider

$$P_0 \left(W_Y \leq \frac{n(N+1)}{2} - w \right) = P_0 \left(W_Y' \leq \frac{n(N+1)}{2} - w \right)$$

$$= P_0 \left(W_Y' - \frac{n(N+1)}{2} \leq -w \right)$$

$$= P_0 \left(\frac{n(N+1)}{2} - W_Y \leq -w \right)$$

$$= P_0 \left(W_Y \geq \frac{n(N+1)}{2} - w \right).$$

Thus the null distribution of W_Y is symmetric about $[n(N+1)/2]$.

A2.12 Mean and variance of linear rank statistics

Locally, the most powerful rank tests for the two-sample problem depends on linear rank statistics, such as

$$T = \sum_{j=1}^{n} a(S_{m+j}),$$

where $a(i)$ are specified constants. For example, if we set $a(S_i) = S_i$ the statistic T reduces to the Wilcoxon rank sum W_Y. In large samples the null distribution of T can be approximated by a normal distribution. In this connection we need to find the mean and the variance of T under the null hypothesis.

Mean of T under H_0

Finding the mean requires the calculation of the mean of $a(S_i)$. Now

$$E(a(S_i)) = \sum_{k=1}^{N} \frac{a(k)}{N} \equiv \bar{a},$$

where $\bar{a}$ is the average of $(a(1), \ldots, a(N))$. Noting that this quantity does not depend on i, we have

$$E(T) = n\bar{a}.$$

Variance of T under H_0

We need to find $var(a(S_i))$ and the covariance, $cov(a(S_k), a(S_l))$. Now

$$var(a(S_i)) = (1/N) \sum_{j=1}^{N} [a(j)]^2 - [\bar{a}]^2 = \frac{1}{N} \sum_{j=1}^{N} [a(j) - \bar{a}]^2.$$

It can be shown that

$$cov(a(S_k), a(S_l)) = -\frac{1}{N-1} var(a(S_i)).$$

Finally, we have

$$var(T) = \sum_{j=1}^{n} var(a(S_{m+j})) + 2 \sum_{j<j'} cov(a(S_{m+j}), a(S_{m+j'}))$$

$$= n \; var(a(S_{m+1})) + n(n-1) cov(a(S_{m+1}), a(S_{m+2}))$$

$$= n \left(1 - \frac{(n-1)}{N-1} \right) var(a(S_{m+1}))$$

$$= \frac{nm}{N(N-1)} \sum_{j=1}^{N} [a(j) - \bar{a}]^2.$$

<u>Remark.</u> The expressions for $E(T)$ and $var(T)$ can also be obtained from finite sampling results as mentioned in relation to W_Y.

A2.13 Motivation for the definition of U_{XY}^* as in (2.78)

In the case of untied data, we started with the statistic U_{XY} and related it to the rank sum statistic W_Y. In a tied case, we start with the rank sum statistic W_Y^* and examine the same using the algebraic expressions for the ranks.

First we observe that

$$W_Y^* = \sum_{i=1}^{n} r(Z_{m+i}),$$

where $(Z_1, \ldots, Z_m, Z_{m+1}, \ldots, Z_N) = (X_1, \ldots, X_m, Y_1, \ldots, Y_n)$ is the combined sample and $r(Z_h)$ is the rank of Z_h in the combined sample. Let us examine the sum of ranks by rewriting each rank as a sum as per the definition (2.70). Thus

$$\sum_{i=1}^{n} r(Z_{m+i}) = \sum_{i=1}^{n} \sum_{k=1}^{m} c(X_k, Y_i) + \sum_{i=1}^{n} \sum_{j=1}^{n} c(Y_j, Y_i) + (n/2).$$

The first sum S_1 can be seen to be equal to

$S_1 = $ *number of pairs* (X_k, Y_i) *for which* $X_k < Y_i$
$\quad + (1/2)$ *number of pairs* (X_k, Y_i) *for which* $X_k = Y_i$.

The second sum S_2 can be related to the ranks, r^*, of Y's in the Y-sample. This relation from (2.71) is

$$S_2 = \sum_{j=1}^{n} [r^*(Y_j) - (1/2)] = \sum_{j=1}^{n} j - \frac{n}{2}$$
$$= [n(n+1)/2] - (n/2).$$

Thus we have

$$W_Y^* = S_1 + [n(n+1)/2] - (n/2) + (n/2),$$

which leads to the result

$$W_Y^* - [n(n+1)/2] = S_1.$$

Thus one can use S_1 as a generalization of U_{XY}.

A2.14 Two properties of midranks

First we find the sum of the midranks and then we derive an expression for the sum of squares of the midranks.

When there are no ties, the sum of the ranks of N values is $(1+2+\cdots+N) = N(N+1)/2$. Here we verify the fact that the sum of midranks is also equal to $N(N+1)/2$ so that the average of the ranks, in general, is $(N+1)/2$ as claimed in connection with the w-ranks.

First we recall that

$$W_Y^* - [n(n+1)/2] = U_{XY}^*.$$

We can define U_{YX}^*, a generalization of U_{YX}, as

$$U_{YX}^* = \text{number of pairs } (X_k, Y_j) \text{ for which } X_k > Y_j$$
$$+ (1/2) \text{ number of pairs } (X_k, Y_j) \text{ for which } X_k = Y_j.$$

Now arguing as in Appendix A2.13, we see that W_X^*, the sum of X-ranks, is given by

$$W_X^* = [m(m+1)/2] + U_{YX}^*.$$

Thus the sum S of all N ranks is

$$S = W_Y^* + W_X^*$$
$$= U_{XY}^* + [n(n+1)/2] + U_{YX}^* + [m(m+1)/2]$$
$$= mn + [(m^2 + n^2 + n + m)/2]$$
$$= (N^2 + N)/2 = N(N+1)/2.$$

Thus in general the sum of the ranks of N observations is $N(N+1)/2$, which is equal to the sum $\sum_{j=1}^{N} j$.

Sum of squares of midranks

When there are ties, we find the distinct values and their frequencies. Let the distinct ordered values be $Z_{(1)}^* < Z_{(2)}^* < \cdots < Z_{(k)}^*$. Further let m_j be the number of values tied at $Z_{(j)}^*$. This set of values is referred to as the jth group. Let

$$M_j = m_1 + m_2 + \cdots + m_j, \quad for \; j = 1, 2, \ldots, k.$$

Each of the values tied at $Z_{(j)}^*$ is assigned a rank that is the average of $(M_{j-1} + 1, M_{j-1} + 2, \ldots, M_{j-1} + m_j)$, which is

$$R_j^* = M_{j-1} + \frac{m_j + 1}{2}.$$

Let ς_j be the sum of squares of the ranks for the jth group, when all values were distinct, which is

$$\varsigma_j = (M_{j-1} + 1)^2 + (M_{j-1} + 2)^2 + \cdots + (M_{j-1} + m_j)^2.$$

Now expanding the squares on the right side and combining them we have

$$\varsigma_j = m_j \left[M_{j-1} + \frac{m_j + 1}{2} \right]^2 + \frac{m_j (m_j^2 - 1)}{12}.$$

Thus the sum of squares of the midranks is

$$\sum_{j=1}^{k} m_j (R_j^*)^2 = \sum_{j=1}^{k} m_j \left[M_{j-1} + \frac{m_j + 1}{2} \right]^2$$

$$= \sum_{j=1}^{k} \varsigma_j - \frac{1}{12} \sum_{j=1}^{k} m_j (m_j^2 - 1).$$

It may be noted that the first sum is the sum of squares of the ranks when there are no ties, and the second sum is zero when there are no ties. So the first sum is $\sum_{i=1}^{N} i^2 = N(N+1)(2N+1)/6$. Finally, the sum of squares of the midranks is

$$\sum_{j=1}^{k} m_j (R_j^*)^2 = \frac{N(N+1)(2N+1)}{6} - \frac{1}{12} \sum_{j=1}^{k} m_j (m_j^2 - 1).$$

Mean and variance of W_Y^*

When there are ties, the distribution of W_Y^* is conditional on the observed pattern of ties. Under this condition (P), the collection of midranks is fixed and the distribution of W_Y^* under the null hypothesis is the same as the sample sum, where the sample is a random sample from the finite population of midranks. The sample is drawn without replacement. Using the results for random samples from the finite population, we have

$$E_0(W_Y^*|P) = n \cdot (sample\ mean) = n(N+1)/2.$$

Further,

$$var_0(W_Y^*|P) = n^2 \cdot var(sample\ mean) = \frac{nm}{N}(Pop.\ variance).$$

Here, using the earlier result of the sum of squares of the midranks, we can express the population variance in terms of the variance when there are no ties.

A2.15 Confidence interval for Δ from the WMW test

We argue as in the case of Mathisen's test. Let us consider the problem of testing

$$H_0 : \Delta = \Delta_0 \quad against \quad H_A : \Delta \neq \Delta_0.$$

We apply the two-sided MW test to the transformed data $(X_1, \ldots, X_m)$ and $(Y_1', \ldots, Y_n')$, where $Y_j' = Y_j - \Delta_0$. The test is to

$$reject\ H_0\ if\ U_{XY'} \leq c_3\ or\ U_{XY^*} \geq c_4.$$

Thus the values of Δ_0 that will not be rejected are those for which

$$\{\Delta_0 | U_{XY^*} > c_3\ or\ U_{XY'} < c_4\}.$$

We can interpret this region as an interval.

To do so we argue as follows. $U_{XY'}$ represents the number of pairs $(X_i, Y_j - \Delta_0)$, for which $X_i < Y_j - \Delta_0$. Equivalently, when $U_{XY'} = u$, we have exactly u differences $v_{ij} = Y_j - X_i$ above Δ_0. This means that Δ_0 is between the $(mn - u)$th and $(mn - u + 1)$th order statistics of the differences. So the acceptance region is

$$\bigcup_{c_3+1}^{c_4-1} \{U_{XY^*} = u\} = \bigcup_{c_3+1}^{c_4-1} \{v_{ij}(mn - u) \leq \Delta_0 < v_{ij}(mn - u + 1)\}.$$

In other words, the acceptance region is the half open interval $[v_{ij}(mn-c_4+1), v_{ij}(mn-c_3)]$. Here $v_{ij}(u)$ is the uth order statistic of the set of differences v_{ij}. Since $c_3 = mn - c_4$, the interval can be restated as $[v_{ij}(c_3 + 1), v_{ij}(mn - c_3))$.

A2.16 Score test statistic for the PH model

In Section 2.4 we considered the two-sample problem of testing the null hypothesis, $H_0 : F(x) = G(x)$, assuming that $G(x) = 1 - [1 - F(x)]^{1+\theta}$. Here we derive a test statistic using the score method. This statistic is a linear function of the exponential ordered scores statistic mentioned there.

The likelihood of the two samples is

$$L = \left[\prod f(x_i) \right] \left[\prod g(y_j) \right].$$

We replace the density functions in terms of the hazard functions. Then the likelihood is

$$L = \prod [\lambda_F(x_i) exp(-\Lambda_F(x_i))] \prod [\lambda_G(y_j) exp(-\Lambda_G(y_j))],$$

so that the log of the likelihood is

$$l = logL = \sum_i log[\lambda_F(x_i)] - \sum_i \Lambda_F(x_i) + \sum_j log[\lambda_G(y_j)] - \sum_j \Lambda_G(y_j).$$

Using the PH model, this log likelihood can be rewritten as

$$l = \sum_i log[\lambda_F(x_i)] - \sum_i \Lambda_F(x_i) + \sum_j log[(1 + \theta)\lambda_F(y_j)] - (1 + \theta) \sum_j \Lambda_F(y_j).$$

Now the partial derivative of the log likelihood with respect to θ is

$$\frac{\partial l}{\partial \theta} = \frac{n}{1 + \theta} - \sum_j \Lambda_F(y_j).$$

Under the null hypothesis, $\theta = 0$. $\Lambda_F(.)$ is unknown. The usual score method applied to parametric problems suggests that we should set the parameter values to the hypothesized values and the unspecified parameters should be replaced by their estimators. Here we can set $\theta = 0$ and the unknown parameter is a function. Extending the idea behind the score method we can use an estimate of this function. Thus we replace $\Lambda_F(.)$ by an estimate based on the combined sample and the resulting statistic is taken as the test statistic. We have two choices for the estimator of $\Lambda_F(.)$. They are (1) Nelson's estimate $\hat{\Lambda}_N(.)$, and (2) $-log\hat{S}(.)$, where $\hat{S}(.)$ is the estimate of the survival function. Both these estimates have to be constructed using the combined sample. In large samples these two estimates do not differ very much. Nelson's estimate is usually used. Here this gives the test statistic

$$T_{PH}(Y) = \sum_j [1 - \hat{\Lambda}_N(y_j)].$$

Instead we could use

$$T_{PH}(X) = \sum_i [1 - \hat{\Lambda}_N(x_i)].$$

It can be shown that the sum of these two statistics is zero. Thus the test proposed in Section 2.4 is equivalent to the score test.

Using this procedure we obtain the estimator $\hat{\Lambda}_N(.)$ and score the observations, Z's, the scores being $1 - \hat{\Lambda}_N(Z)$. Finally, the sum of the scores for one sample is the test statistic. These scores are called the *logrank scores*.

In the SAS system the scores, $\hat{\Lambda}_N(Z_i) - 1$, are called the Savage scores, and the corresponding test is called the Savage test.

A2.17 Expectation of $V_{(i,N)}$, of Section 2.5

We have a random sample of size N from the exponential distribution with pdf

$$f_X(x) = \begin{cases} e^{-x}, & \text{for } x > 0; \\ 0 & \text{otherwise.} \end{cases}$$

The order statistics of the sample, for convenience, will be denoted as $V_{(1)} < V_{(2)} < \cdots < V_{(N)}$. The joint density function of the order statistics is

$$f_V(v_1, v_2, \ldots, v_N) = \begin{cases} N! \prod_i f_X(v_i) & \text{for } v_1 < v_2 < \cdots < v_N, \\ 0 & \text{otherwise.} \end{cases}$$

We consider a set of variable W's defined as

$$W_1 = V_{(1)}, W_2 = V_{(2)} - V_{(1)}, \ldots, W_N = V_{(N)} - V_{(N-1)}.$$

The joint pdf of W's can be seen to be

$$f_W(w_1, w_2, \ldots, w_N) = \begin{cases} N! exp\left[-\sum_{j=1}^N (N - j + 1)w_j\right] & \text{for } w_j > 0, \\ 0 & \text{otherwise.} \end{cases}$$

This implies that the variable W's are independent, and the pdf of W_j is

$$f_{W_j}(w_j) = \begin{cases} (N - j + 1)exp[-(N - j + 1)w_j] & \text{for } w_j > 0, \\ 0 & \text{otherwise.} \end{cases}$$

Hence

$$E(W_j) = \frac{1}{(N - j + 1)}.$$

Finally, noting that $V_{(i)} = \sum_{j=1}^i W_j$, we have

$$E(V_{(i)}) = \sum_{j=1}^i E(W_j) = \sum_{j=1}^i \frac{1}{(N - j + 1)}.$$

Sum of these expectations
The sum of these expectations is N. This result will be established now:

$$\sum_{i=1}^{N} E(V_{(i)}) = \sum_{i=1}^{N}\sum_{j=1}^{i} \frac{1}{N-j+1} = \sum_{j=1}^{N} \frac{1}{N-j+1} \sum_{i=j}^{N} 1$$

$$= \sum_{j=1}^{N} \frac{1}{N-j+1}(N-j+1) = N.$$

A2.18 Asymptotic distribution of $X_{(k)}$ the kth order statistic of a random sample of size n

The following is a sketch of the proof that the limiting distribution of $X_{(k)}$ is a normal distribution.

Let p be a positive fraction and let k be the integer $\lfloor np \rfloor + 1$. Let ξ_p be the quantile $F^{-1}(p)$. The order statistic $X_{(k)}$ is an estimator of ξ_p.

Step I:
Consider the distribution function of $X_{(k)}$. For real x, we have

$$P(X_{(k)} \le x) = P[F(X_{(k)}) \le F(x)] = P[Beta(k, n-k+1) \le F(x)],$$

since the distribution of $F(X_{(k)})$ is a beta distribution. Now recall that

$$E(F(X_{(k)})) = E[Beta(k, n-k+1)] = k/(n+1) \approx p;$$

and

$$var(F(X_{(k)})) = var[Beta(k, n-k+1)]$$

$$= \frac{k(n-k+1)}{(n+1)^2(n+2)} \approx \{p(1-p)/n\}.$$

Using the approximate mean and the approximate variance, we have

$$P[F(X_{(k)}) \le F(x)] = P\left[Z \le \frac{F(x)-p}{\sqrt{p(1-p)/n}}\right],$$

where $Z = [F(X_{(k)}) - p]/\sqrt{p(1-p)/n}$. However, the distribution of Z can be approximated by the $N(0,1)$ distribution. In other words, as $n \to \infty$

$$distribution\ of\ \frac{[F(X_{(k)}) - p]}{\sqrt{p(1-p)/n}} \to N(0,1).$$

Step II:
Let us assume that $F(x)$ is twice differentiable in a neighborhood of ξ_p, $F'(\xi_p) \ne 0$ and $F''(\xi_p) \ne 0$. Consider the Taylor expansion approximation used in the delta method. This enables us to approximate $F(X_{(k)})$, as

$$F(X_{(k)}) \approx F(\xi_p) + f(\xi_p)(X_{(k)} - \xi_p),$$

where $f(x) = F'(x)$. Thus the variables $\frac{[F(X_{(k)})-p]}{\sqrt{p(1-p)/n}}$ and $\frac{f(\xi_p)(X_{(k)}-\xi_p)}{\sqrt{p(1-p)/n}}$ follow the same limiting normal distribution. So, as $n \to \infty$

$$\text{distribution of } \frac{f(\xi_p)(X_{(k)} - \xi_p)}{\sqrt{p(1-p)/n}} \to N(0,1).$$

This result is usually restated as

$$X_{(k)} \sim N\left(\xi_p, \frac{p(1-p)}{n[f(\xi_p)]^2}\right),$$

where $k = \lfloor np \rfloor + 1$. Since $(k/n) \approx p$, $\xi_p = F^{-1}(k/n)$ and the limiting result can be restated as

$$X_{(k)} \sim N\left(F^{-1}(k/n), \frac{(k/n)(1-(k/n))}{nf^2[F^{-1}(k/n)]}\right).$$

In the case of a sample median, $p = (1/2)$ and we take $k = \lfloor (n/2) \rfloor + 1$. The limiting distribution of $X_{(k)}$ is $N(\xi_{(1/2)}, \frac{1}{4nf^2[\xi_{(1/2)}]})$.

A2.19 Proof of (2.101)

In Subsection 2.4.6, it was asserted that $\sum_{j=1}^{N} DV_j$ is a linear function of the Wilcoxon rank sum statistic. We derive that relationship.

First we note that

$$\frac{mn}{N} \sum_{j=1}^{N} DV_j = \sum_{j=1}^{N}\left[A_j - \frac{mj}{N}\right] = \sum_{j=1}^{N} A_j - \frac{m(N+1)}{2}.$$

Now we will simplify the sum of A_j's.

$$\sum_{j=1}^{N} A_j = m \sum_{j=1}^{N} F_m(Z_{(j)})$$

$$= m \sum_{i=1}^{m} F_m(X_{(i)}) + m \sum_{j=1}^{n} F_m(Y_{(j)})$$

$$= \sum_{i=1}^{m} i + \sum_{j=1}^{n} U_{XY_{(j)}} \equiv \frac{m(m+1)}{2} + U_{XY}.$$

In the above, $U_{XY_{(j)}}$ stands for the number of X-values that are less than $Y_{(j)}$. Now using this expression, we get

$$\frac{mn}{N} \sum_{j=1}^{N} DV_j = \frac{m(m+1)}{2} + U_{XY} - \frac{m(N+1)}{2}$$

$$= U_{XY} - \frac{mn}{2}$$

$$= W_Y - \frac{n(n+1)}{2} - \frac{mn}{2} = W_Y - \frac{n(N+1)}{2}.$$

In other words,

$$\sum_{j=1}^{N} DV_j = \frac{N}{mn}\left[W_Y - \frac{n(N+1)}{2}\right].$$

2.9 Appendix B2: Computer programs

B2.1 Fisher's test for a 2×2 table

Here we are comparing two samples with binary data, using Fisher's test. We use the data from Example 2.2.

```
data test;
input plant$ class$ count @@;
lines;
1 1 7 1 0 8 2 1 5 2 0 10
;
proc freq order=data;
weight count;
table plant*class/norow nocol nopercent exact ;
title 'Computations for 2 X 2 table';
run;
```

Output

```
                 Computations for 2 X 2 table
                      The FREQ Procedure
                   Table of plant by class


        plant       class

        Frequency|1          |0          |  Total
        ---------+---------+---------+
        1        |     7 |      8 |      15
        ---------+---------+---------+
        2        |     5 |     10 |      15
        ---------+---------+---------+
        Total         12         18         30



        Statistics for Table of plant by class

        Statistic               DF     Value        Prob
        -------------------------------------------------
        Chi-Square               1     0.5556      0.4561
        Likelihood Ratio
```

```
Chi-Square                    1    0.5576    0.4552
Continuity Adj.
Chi-Square                    1    0.1389    0.7094
Mantel-Haenszel
Chi-Square                    1    0.5370    0.4637
```

Fisher's Exact Test

```
-----------------------------------
Cell (1,1) Frequency (F)         7
Left-sided Pr <= F          0.8682
Right-sided Pr >= F         0.3552

Table Probability (P)       0.2234
Two-sided Pr <= P           0.7104
```

B2.2 Testing for clinical equivalence

Here we calculate the test for one-sided equivalence with binary data. This method uses maximum likelihood estimation for estimating the variance of the test statistic. We need to solve a third-degree equation. Details are taken from Farrington and Manning (1990). We use the data from Example 2.3. We compare a new treatment with a standard treatment. So θ_1 is the effect rate of chemotherapy and θ_2 is the effect rate of radiation treatment. The rupture rates are compared.

```
data one;
n2=100;
n1=200;
r2=57;
r1=120;
p1=r1/n1;
p2=r2/n2;
d=0.1;
delta0=d;
rho=n2/n1;
a=1+rho;
b = -(1 + rho +p1 +rho*p2+d*(rho+2));
c= d*(d+(2*p1)+rho+1)+p1+rho*p2;
f=-p1*d*(1+d);
rba=b/(3*a);
v= rba**3 - b*c/(6*a*a) + f/(2*a);
u= sign(v)*sqrt(rba*rba - c/(3*a));
w= ((22/7)+arcos(v/(u*u*u)))/3;
p1d=2*u*cos(w) - rba;
p2d=p1d-d;
```

```
zn=p1-p2-d;
v2=(p1d*(1-p1d)+(p2d*(1-p2d)/rho))/n1;
v=sqrt(v2);
zd=zn/v;
alpha = 0.05;
zc=probit(alpha);
if zd le zc then result=1;
else result=0;
pv=probnorm(zd);
diff=p1-p2;
var=v2;
se=v;
p11=round(p1,.0001);
p21=round(p2,.0001);
diff1=p11-p21;
var1=round(var,.0001);
vD=round(se,.0001);
pv1=round(pv,.0001);
theta1d=round(p1d,.0001);
zd1=round(zd,.0001);
proc print;
title 'data and specifications';
var r1 n1 r2 n2 delta0 alpha ;
proc print ;
title 'sample proportions,difference,std.error';
var p11 p21 diff1 var1 vD;
proc print;
title 'ML estimate and test result';
var theta1d  zd1 result pv1;
run;
```

Output

data and specifications

r1	n1	r2	n2	delta0	alpha
120	200	57	100	0.15	0.05

sample proportions,difference ,std.error

p11	p21	diff1	vD
0.6	0.57	0.03	0.0604

ML estimate and test result

theta1d	zd1	result	pv1
0.6377	-1.9852	1	.0236

B2.3 Sample size for one-sided test

This is a program for computing the sample size for a two-sample binomial testing problem. The null hypothesis is $p_1 = p_2$ against the alternative $p_1 > p_2$. The requirements are that

$$P(type I\ error) \le \alpha,$$

and the power at (p_{11}, p_{21}) is greater than or equal to $1 - \beta$. First, we get the simple solution using normal approximation, which has been improved by Casagrande et al. (1978). Recently Levin and Chen (1999) gave a careful derivation of this approximation. This result is close to the sample size needed when we use Fisher's test. The exact calculation is quite tedious. The critical region of the exact test is $X \ge x_u$, where

$$P(X \ge x_u | X + Y = m, H_0) \le \alpha.$$

We are assuming that we have two samples each of size n.

```
data one;
p2=0.25;
diff=0.05;
p1=p2+diff;
alpha=0.05;
power=0.90;
beta=1-power;
z11=probit(1-alpha);
z22=probit(beta);
pav=(p1+p2)/2;
v0=2*pav*(1-pav);
sd0=sqrt(v0);
v1=p1*(1-p1)+p2*(1-p2);
sd1=sqrt(v1);
del=(p1-p2)**2;
nnum=(z11*sd0-z22*sd1)**2;
nnor=nnum/del;
n=int(nnor)+ 1;
a=nnum;
b=1+4*((p1-p2)/a);
b1=sqrt(b);
ncf=((1+b1)/2)**2;
ncc=nnor*ncf;
nlc=int(ncc)+1;
proc print;
title 'specifications';
var p1 p2 alpha power ;
proc print;
title 'Sample size using ordinary normal approximation';
```

```
     var p1 p2 alpha power n ;
     proc print;
     title ' Casagrande ,Pike, and Smith sample size';
     var p1 p2 alpha power nlc ;
     run;
```

Output

specifications

Obs	p1	p2	alpha	power
1	0.3	0.25	0.05	0.9

Sample size using ordinary normal approximation

Obs	p1	p2	alpha	power	n
1	0.3	0.25	0.05	0.9	1365

Casagrande, Pike, and Smith sample size

Obs	p1	p2	alpha	power	nlc
1	0.3	0.25	0.05	0.9	1404

B2.4 Analysis of a 2 × 3 table

Here we are comparing two groups. The response can belong to one of three categories. We want to test the homogeneity hypothesis.

```
data college;
input sex@;
do pref= 1 to 3;
   input num@;
   output;
   end;
lines;
1 20 45 35
2 15 40 20
;
proc print;
run;
proc freq;
weight num;
table sex*pref/nocol norow nopercent cellchi2 chisq exact;
run;
```

Output

```
              Obs    sex    pref   num

               1      1      1     20
               2      1      2     45
               3      1      3     35
               4      2      1     15
               5      2      2     40
               6      2      3     20
```

```
                 The FREQ Procedure
                 Table of sex by pref
```

```
      sex                pref

      Frequency      |
      Cell Chi-Square|     1|        2|       3|Total
      ---------------+--------+--------+--------+
            1 |     20 |     45 |     35 |100
              |      0 | 0.2626 | 0.4058 |
      ---------------+--------+--------+--------+
            2 |     15 |     40 |     20 |75
              |      0 | 0.3501 | 0.5411 |
      ---------------+--------+--------+--------+
      Total          35       85       55   175
```

```
      Statistics for Table of sex by pref

      Statistic        DF     Value      Prob
      ------------------------------------------
      Chi-Square        2     1.5597    0.4585
      Likelihood Ratio
      Chi-Square        2     1.5705    0.4560
      Mantel-Haenszel
      Chi-Square        1     0.5904    0.4423
```

```
              Fisher's Exact Test
      ------------------------------------
      Table Probability (P)         0.0097
      Pr <= P                       0.4402
```

B2.5 Wilcoxon procedure for complete data

This is an illustration of comparing two samples using the Wilcoxon procedure. Exact P-values are computed using the exact option. We use the data from Example 2.7.

```
data one;
  input group response@@;
  lines;
   1 69 1 61 1 76
   2 73   2 85 2 79 2 72
   ;
proc npar1way wilcoxon;
class group;
var response;
exact;
run;
```

Output

The NPAR1WAY Procedure
Wilcoxon Scores (Rank Sums) for Variable response
Classified by Variable group

group	N	Sum of Scores	Expected Under H0	Std Dev Under H0
1	3	8.0	12.0	2.828427
2	4	20.0	16.0	2.828427

Wilcoxon Two-Sample Test

Statistic (S)	8.0000
Normal Approximation	
Z	-1.2374
One-Sided Pr < Z	0.1080
Two-Sided Pr > |Z|	0.2159
t Approximation	
One-Sided Pr < Z	0.1311
Two-Sided Pr > |Z|	0.2622
Exact Test	
One-Sided Pr <= S	0.1143
Two-Sided Pr >= |S - Mean|	0.2286

Z includes a continuity correction of 0.5.

B2.6 Wilcoxon test for ordered categorical data

This is an illustration of calculating the Wilcoxon statistic for comparing two samples with ordered categorical data. We use the data from Example 2.8. The first step is to expand the data into individual observations.

```
data one;
input treat$ resp freq@@;
lines;
 A 1 1 A 2 13 A 3 16 A 4 15 A 5 7
 B 1 5 B 2 21 B 3 14 B 4 9 B 5 3
 ;
 proc print;
```

Before we can calculate the test statistic, we need to create a data set using the frequencies

```
data two;
set one;
do i=1 to freq;
output two;
end;
run;
```

We are ready to calculate the Wilxocon test statistic

```
proc npar1way Wilcoxon data=two;
class treat;
var resp;
title 'Wilcoxon test for ordered categorical data';
run;
```

Output

Obs	treat	resp	freq
1	A	1	1
2	A	2	13
3	A	3	16
4	A	4	15
5	A	5	7
6	B	1	5
7	B	2	21
8	B	3	14
9	B	4	9
10	B	5	3

```
                Wilcoxon test for ordered categorical data
                          The NPAR1WAY Procedure
                Wilcoxon Scores (Rank Sums) for Variable resp
                          Classified by Variable treat
                          Sum of      Expected        Std Dev
         treat      N     Scores      Under H0        Under H0
         ----------------------------------------------------------
           A        52    3131.0       2730.0        148.158928
           B        52    2329.0       2730.0        148.158928


                      Average scores were used for ties.



                         Wilcoxon Two-Sample Test

              Statistic                        3131.0000

              Normal Approximation
              Z                                   2.7032
              One-Sided Pr >  Z                   0.0034
              Two-Sided Pr >  |Z|                 0.0069

              t Approximation
              One-Sided Pr >  Z                   0.0040
              Two-Sided Pr >  |Z|                 0.0080


              Z includes a continuity correction of 0.5.
```

B2.7 Confidence interval for Δ from the WMW test

This is a program for computing a distribution-free confidence interval for the shift parameter. We use the data from Example 2.10. The expression for c3 is given in (2.85) and cc is the confidence coefficient. We calculate below integers ll and ul. First we create $D = Y - X$ by creating an array of X-values and an array of Y-values. We pick the the order statistics $D_{(ll)}$ and $D_{(ul)}$, which are the end points of the confidence interval.

```
data pairs;
nx=10;
ny=10;
nd=nx*ny;
input x1-x10 y1-y10;
array X(I) x1-x10;
array Y(J) y1-y10;
do I = 1 to nx;
do J= 1 to ny;
```

```
D=Y-X;output;
end;
end;
lines;
69 15 80 89 58 38 23 43 83 87 35 51 76 77 62 66 32
48 50 20
;
```

We calculate the integers ll and ul.

```
data one;
nx=10;
ny=10;
cc=0.90;
halpha=(1-cc)/2;
nd=nx*ny;
mean=0.5*(nx*ny-1);
var=nd*(nx+ny+1)/12;
sd=sqrt(var);
l=mean +sd*probit(halpha);
c3=floor(l);
ll=c3+1;
ul=nd-c3;
tcc=probnorm((ll-mean)/sd);
acc=1-2*tcc;
call symput('ll',ll);
call symput('ul',ul);
proc print;
title 'sample sizes and confidence coefficient';
var nx ny cc;
run;
proc print;
title 'required order statistics & actual cc';
var ll ul acc;
run;
data three;
proc sort data=pairs(keep=d) out=two;
by d;
run;
data four;
set two;
if(_n_=symget('ll') or _n_=symget('ul'));
proc transpose data=four out=output(rename=(col1=lower
                                   col2=upper)drop=_name_);
proc print data=output;
title 'end points of confidence interval';
run;
```

Output

```
                sample sizes and confidence coefficient
                         nx    ny    cc

                         10    10   0.9
                required order statistics, & actual cc

                    ll     ul    p1 p2  acc

                    28     73    28 73 0.89589

                   end points of confidence interval

                         lower  upper

                          -29    12
```

B2.8 Savage test

This is an illustration of comparing two samples using the Savage test. The exact *P*-value is computed using the "exact" option. We use the data from Example 2.9.

```
title 'SAVAGE TEST';
 data fail;
 input group hours @@;
 liness;
  1 79 1 150 1 96 1 167 1 15 1 285 1 28 1 111
  2 250 2 23 2 427 2 42 2 333 2 232 2 488 2 121 2 159
  ;
 proc npar1way savage;
 class group;
 var hours;
 exact;
 run;
```

The edited output follows.

```
                        SAVAGE TEST
                  The NPAR1WAY Procedure
         Savage Scores (Exponential) for Variable hours
                Classified by Variable group
                     Sum of      Expected     Std Dev
      group    N    Scores       Under H0     Under H0
      ------------------------------------------------------
        1      8   -3.169118       0.0        1.894606
        2      9    3.169118       0.0        1.894606
```

Savage Two-Sample Test

Statistic (S)   -3.1691

Normal Approximation
Z   -1.6727
One-Sided Pr < Z   0.0472
Two-Sided Pr > |Z| 0.0944

Exact Test
One-Sided Pr <= S   0.0447
Two-Sided Pr >= |S - Mean| 0.0931

Savage One-Way Analysis
Chi-Square   2.7979
DF   1
Pr > Chi-Square   0.0944

B2.9 Smirnov test

Here we calculate the Smirnov two-sample test statistics. We use the data
from Example 2.5. The exact P-values are appropriate for equal sample sizes.
The asymptotic P-values are given.

```
Data dn;
input group$ score@@;
lines;
1 69   1 15   1 80   1 89   1 58   1 38   1 23 1 43   1 83 1 87
2 35   2 51   2 76   2 77   2 62   2 66   2 32 2 48   2 50 2 20


;
Proc sort;
by score;
Data two;
set dn;
If group=1 then DEL=1;
else DEL1=0;
DEL2=1-DEL1;
proc print;
Title 'COMBINED ORDERED SAMPLE';'
proc iml;
use two;
M=10;
N=10;
```

```
read all variables{score} into M1;
read all variables{DEL1} into M2;
read all variables{DEL2} into M3;
M4=(1/m)*CUSUM(M2);
M5=(1/n)*CUSUM(M3);
M6=M4-M5;
DP=MAX(M6);
M7=-M6;
M8=ABS(M6);
DM=max(M7);
D=max(DP,DM);
D=ROUND(D,.0001);
DP=ROUND(DP,.0001);
DV=sum(M6);
print M1 M4 M5 M6 M8;
print 'various statistics';
print  DP DM D DV;
K=DP*N;
START BIN(X,Y);
 SUM=GAMMA(X+1)/(GAMMA(Y+1)*GAMMA(X-Y+1));
 RETURN(SUM);
FINISH BIN;
NU=BIN(2*N,N-K);
DI=BIN(2*N,N);
P11=NU/DI;
P1=round(P11,.0001);
K1=D*N;
J=INT(N/K1);
SUM=0;
DO I=1 TO J BY 1;
 TERMI=(-1)**(I+1)*BIN(2*N,N-I*K1)/BIN(2*N,N);
 SUM=SUM+TERMI;
END;
P22=2*SUM;
P2=ROUND(P22,.0001);
PRINT 'EXACT ONE-SIDED P-VALUE(DP)' P1;
PRINT 'EXACT P-VALUE(D)' P2;
K2=DP*DP*(M*N)/(M+N);
AP1=exp(-2*k2);
K3=D*D*(M*N)/(M+N);
AP2=2*exp(-2*k3);
Print 'ASYM. ONE-SIDED P-Value(DP)' AP1;
Print 'ASYM. P-Value(D)' AP2;
EXIT;
RUN;
```

Output

COMBINED ORDERED SAMPLE

GROUP	SCORE	DEL1	DEL2
1	15	1	0
2	20	0	1
1	23	1	0
2	32	0	1
2	35	0	1
1	38	1	0
1	43	1	0
2	48	0	1
2	50	0	1
2	51	0	1
1	58	1	0
2	62	0	1
2	66	0	1
1	69	1	0
2	76	0	1
2	77	0	1
1	80	1	0
1	83	1	0
1	87	1	0
1	89	1	0

M1 is the order statistics of the combined sample. M4 is the sample cdf $F_m(.)$.
M5 is the sample cdf $G_n(.)$. M6 is DV_j. M8 is $|DV_j|$.

M1	M4	M5	M6	M8
15	0.1	0	0.1	0.1
20	0.1	0.1	0	0
23	0.2	0.1	0.1	0.1
32	0.2	0.2	0	0
35	0.2	0.3	-0.1	0.1
38	0.3	0.3	0	0
43	0.4	0.3	0.1	0.1
48	0.4	0.4	0	0
50	0.4	0.5	-0.1	0.1
51	0.4	0.6	-0.2	0.2
58	0.5	0.6	-0.1	0.1
62	0.5	0.7	-0.2	0.2
66	0.5	0.8	-0.3	0.3

69	0.6	0.8	-0.2	0.2
76	0.6	0.9	-0.3	0.3
77	0.6	1	-0.4	0.4
80	0.7	1	-0.3	0.3
83	0.8	1	-0.2	0.2
87	0.9	1	-0.1	0.1
89	1	1	0	0

DP is D^+, DM is D^-, D is $max(D^+, D^-)$, and DV is $\sum DV_j$.

	DP	DM	D	DV
various statistics	0.1	0.4	0.4	-2.2

```
                                              P1
            EXACT ONE-SIDED P-VALUE(DP)   0.9091


                                        P2
                EXACT P-VALUE(D)     0.4175


                                          AP1
        ASYM. ONE-SIDED P-Value(DP) 0.9048374
                                          AP2
            ASYM. P-Value(D)    0.403793
```

B2.10 Wilcoxon and logrank tests for censored data

Here we compare two samples with censored data. We use the data from
Example 2.11.

```
data days;
input group T censor@@;
lines;
1 6 0 1 6 1 1 7 0 1 9 1 1 10 0 1 13 0
2 1 0 2 8 1 2 11 0 2 12 1 2 14 1
  ;
proc lifetest notable;
time T*censor(1);
strata group;
run;
```

Output

The LIFETEST Procedure

Summary of the Number of Censored and
Uncensored Values

Stratum	GROUP	Total	Failed	Censored	Percent Censored
1	1	6	4	2	33.33
2	2	5	2	3	60.00
Total		11	6	5	45.45

Testing Homogeneity of Survival Curves
for T over Strata

Rank Statistics

GROUP	Log-Rank	Wilcoxon
1	1.2045	5.0000
2	-1.2045	-5.0000

Covariance Matrix for the Log-Rank Statistics

GROUP	1	2
1	1.41543	-1.41543
2	-1.41543	1.41543

Covariance Matrix for the Wilcoxon Statistics

GROUP	1	2
1	80.0000	-80.0000
2	-80.0000	80.0000

Test of Equality over Strata

Test	Chi-Square	DF	Pr > Chi-Square
Log-Rank	1.0251	1	0.3113
Wilcoxon	0.3125	1	0.5762

2.10 Problems

1. A study to evaluate a diagnostic test was conducted. A group of 60 persons with a disease were tested and 54 persons had a positive test. When 100 healthy persons were tested only 20 had a positive test. The ratio of the fraction of true positives to the fraction of false positives characterizes the ability of the test to detect the disease. Find a 95% confidence interval for the ratio of interest. This problem is based on an example from Koopman (1984).

2. A prospective randomized study for comparing conventional chemotherapy and high-dose radiation therapy for cancer was conducted and the results (artificial data) are given below:

Response Type	Low Dose	High Dose
Complete	6	25
Very good partial	11	20
Partial	40	41
Minimal	16	6
Progressive disease	22	10

Can we conclude that the two therapies are equally effective?

3. Consider the two independent samples $(X_1, \ldots, X_m)$ and $(Y_1, \ldots, Y_n)$. Let $F_m(.)$ and $G_n(.)$ be the corresponding sample distribution functions. Show that the Mann-Whitney statistics are given by

$$U_{XY} = m \sum_{j=1}^{n} F_m(Y_j) \quad and \quad U_{YX} = n \sum_{i=1}^{m} G_n(X_i).$$

4. Verify the result that

$$U_{YX} = W_X - m(m+1)/2,$$

assuming that there are no tied values in the combined sample.

5. In Appendix A2.12, it is claimed that

$$cov(a(S_k), a(S_l)) = -var(a(S_k))/(N-1).$$

Prove this claim.

6. In order to evaluate undesirable effects of ozone an experiment was conducted at California Primate Research Center, University of California, Davis. One group of 22 rats were kept in an ozone environment for 7 days and their weight gains were noted. A second group of 23 similar rats were kept in

an ozone-free environment for 7 days and their weight gains were noted. The resulting weight data were reported in Doksum and Sievers (1976). The data are given below.

Control (ozone-free) group: 41.0, 38.4, 24.4, 25.9, 21.9, 18.3, 13.1, 27.3, 28.5, −16.9, 26.0, 17.4, 21.8, 15.4, 27.4, 19.2, 22.4, 17.7, 26.0, 29.4, 21.4, 26.6, 22.7.
Ozone group: 10.1, 6.1, 20.4, 7.3, 14.3, 15.5, −9.9, 6.8, 28.2, 17.9, −9.0, −12.9, 14.0, 6.6, 12.1, 15.7, 39.9, −15.9, 54.6, −14.7, 44.1, −9.0.
The researchers felt that the presence of ozone may retard the growth. Do these data provide enough support for the researchers' hypothesis?

7. Hoel (1972) presented mortality data gathered in a laboratory experiment with two groups of mice that received a radiation dose of 300r at an age of 5–6 weeks. One group was in a germ-free environment, whereas the other group lived in a usual laboratory environment. An autopsy was performed to determine the cause of death. The following data refer to survival times for the mice that died because of thymic lymphoma.

Group 1 (conventional): 159, 189, 191, 198, 200, 207, 220, 235, 245, 250, 256, 261, 265, 266, 280, 343, 356, 383, 403, 414, 428, 432.
Group 2 (germ free): 158, 192, 193, 193, 194, 202, 212, 215, 229, 230, 237, 240, 244, 247, 259, 300, 301, 321, 337, 415, 434, 444, 485, 496, 529, 537, 624, 707, 800.
Is there adequate evidence to infer that the survival distributions are different?

8. The following data on remission times of leukemia patients appeared in Lee and Desu (1972). Censored values are indicated by a '+' sign.
Treatment 1: 8.0, 10.0, 10.0, 12.0, 14.0, 20.0, 48.0, 70.0, 75.0, 99.0, 103.0, 162.0, 169.0, 195.0, 222.0, 161.0+, 199.0+, 217.0+, 245.0+.
Treatment 2: 8.0, 10.0, 11.0, 23.0, 25.0, 25.0, 28.0, 31.0, 31.0, 40.0, 48.0, 89.0, 124.0, 143.0, 12.0+, 159.0+, 190.0+, 196.0+, 197.0+, 205.0+, 219.0+.
Can we conclude that the remission time distributions are equal?

9. The following data on survival times (in months) of Hodgkin's disease patients appeared in Bartolucci and Dickey (1977).
Group A: 1.25, 1.41, 4.98, 5.25, 5.38, 6.92, 8.89, 10.98, 11.18, 13.11, 13.21, 16.33, 19.77, 21.08, 22.07, 42.92, 21.84+, 31.38+, 32.62+, 37.18+.
Group B: 1.05, 2.92, 3.61, 4.20, 4.49, 6.72, 7.31, 9.08, 9.11, 16.85, 14.49+, 18.82+, 26.59+, 30.26+, 41.34+.
Is there enough evidence to conclude that the two groups are different?

10. Pike (1966) reported the results of a two-group experiment on vaginal cancer in female rats insulted with the carcinogen DMBA. The variable under study is the time to cancer mortality in days. The two groups are distinguished by pretreatment regimen. The data are given below.

Group 1: 143, 164, 188, 188, 190, 192, 206, 209, 213, 216, 220, 227, 230, 234, 246, 265, 304, 216+, 244+.

Group 2: 142, 156, 163, 198, 205, 232, 232, 233, 233, 233, 233, 239, 240, 261, 280, 280, 296, 296, 323, 204+, 344+.

Obtain the approximate P-value of the logrank test for testing the hypothesis of no difference between the effects of the pretreatment regimen on the survival.

2.11 References

Albers, W. and Löhnberg, P. (1984). An approximate confidence interval for the difference between quantiles in a biomedical problem. *Statist. Neer.*, **38**, 20–22.

Bartolucci, A.A. and Dickey, J.M. (1977). Comparative Bayesian and traditional inference for gamma-modeled survival data. *Biometrics*, **33**, 343–354.

Breslow, N. (1970). A generalized Kruskal-Wallis test for comparing k samples subject to unequal pattern of censorship. *Biometrika*, **57**, 579–594.

Bristol, D.R. (1990). Distribution-free confidence intervals for the difference between quantiles. *Statist. Neer.*, **44**, 87–90.

Brookmeyer, R. and Crowley, J. (1982). A k-sample median test for censored data. *J. Amer. Statist. Assoc.*, **77**, 433–440.

Brown, G.W. and Mood, A.M. (1948). Homogeneity of several samples. *Amer. Statist.*, **2(3)**, 22.

Capon, J. (1961). Asymptotic efficiency of certain locally most powerful rank tests. *Ann. Math. Statist.*, **32**, 88–100.

Casagrande, J.T., Pike, M.C., and Smith, P.G. (1978). An improved approximate formula for calculating sample sizes for comparing two binomial proportions. *Biometrics*, **34**, 483–486.

Chakraborti, S. (1984). A generalization of the control median test. Unpublished Ph.D. dissertation, State University of New York at Buffalo.

Chakraborti, S. and Desu, M.M. (1986). A distribution-free confidence interval for the difference between quantiles with censored data. *Statist. Neer.*, **40**, 93–98.

Chernoff, H. and Savage, I.R. (1958). Asymptotic normality and efficiency of certain nonparametric test statistics. *Ann. Math. Statist.*, **29**, 972–994.

Cochran, W.G. and Cox, G.M. (1957). *Experimental Designs*, 2nd edition, John Wiley & Sons, New York.

Com-Nougues, C., Rodary, C., and Patte, C. (1993). How to establish equivalence when data are censored: a randomized trial of treatments for B non-Hodgkin lymphoma. *Statist. Med.*, **12**, 1353–1364.

Cox, D.R. (1972): Regression models and life tables. *J. Royal Statist. Soc., Series B*, **34**, 187–220.

Cox, D.R. and Hinkley, D.V. (1974). *Theoretical Statistics*, Chapman and Hall Ltd., London.

DeJonge, H. (1983). Deficiencies in clinical reports for registration of drugs. *Statist. Med.*, **2**, 155–166.

Desu, M.M. and Raghavarao, D. (1990). *Sample Size Methodology*, Academic Press, Boston.

Dunnett, C.W. and Gent, M. (1977). Significance testing to establish equivalence between treatments, with special reference to data in the form of 2 × 2 table. *Biometrics*, **33**, 593–602.

Epstein, B. (1954). Tables for the distribution of the number of exceedances. *Ann. Math. Statist.*, **25**, 762–768.

Doksum, K.A. and Sievers, G.L. (1976). Plotting with confidence: graphical comparisons of two populations. *Biometrika*, **63**, 421–434.

Eilbott, J. and Nadler, J. (1965). On precedence life testing. *Technometrics*, **7**, 359–377.

Epstein, B. (1954). Tables for the distribution of the number of exceedances. *Ann. Math. Statist.*, **25**, 762–768.

Farrington, C.P. and Manning, G. (1990). Test statistics and sample size formulae for comparative binomial trials with null hypothesis of non-zero risk difference or non-unity relative risk. *Statist. Med.*, **9**, 1447–1454.

Finney, D.J., Latscha, R., Bennett, B.M., and Hsu, P. (1963). *Tables for Testing Significance in a* 2 × 2 *Contigency Table*, Cambridge University Press, Cambridge, U.K.

Fisher, R.A. and Yates, F. (1963). *Statistical Tables*, Hafner, New York.

Fleming, T.R., O'Fallon, J.R., O'Brian, P.C., and Harrington, D.P. (1980). Modified Kolmogorov-Smirnov test procedures with applications to arbitrary right-censored data. *Biometrics*, **36**, 607–625.

Gail, M. and Gart, J.J. (1973). The determination of sample sizes for use with the exact conditional test in 2 × 2 comparative trials. *Biometrics*, **29**, 441–448.

Gail, M.H. and Green, S.B. (1976). Critical values for the one-sided two-sample Kolmogorov-Smirnov statistic. *J. Amer. Statist. Assoc.*, **71**, 757–760.

Gart, J.J. (1963). A median test with sequential applications. *Biometrika*, **50**, 55–62.

Gastwirth, J. (1968). The first median test: a two-sided version of the control median test. *J. Amer. Statist. Assoc.*, **63**, 692–706.

Gastwirth, J.L. and Wang, J. (1998). Control percentile test procedures for censored data. *J. Statist. Plan. Infer.*, **18**, 267–276.

Gehan, E.A. (1965). A generalized Wilcoxon test for comparing arbitrarily singly censored samples. *Biometrika*, **52**, 203–223,

Gehan, E.A. and Schneiderman, M.A. (1973). Experimental design of clinical trials, Chapter VIII in *Cancer Medicine*, Edited by J.F. Holland and E. Frei III, Lea & Febiger, Philadelphia.

Gilbert, J.P. (1962). Random Censorship, Ph.D. Dissertation, University of Chicago, Chicago.

Haseman, J.K. (1978). Exact sample sizes for use with the Fisher-Irwin test for 2 × 2 tables. *Biometrics*, **34**, 106–109.

Hettmansperger, T.P. (1984). *Statistical Inference Based on Ranks*, John Wiley & Sons, New York.

Hoeffding, W. (1951). Optimum nonparametric tests. *Proc. 2nd Berkeley Symp.*, 83–92.

Hoel, D.G. (1972). A representation of mortality data by competing risks. *Biometrics*, **28**, 475–488.

Kalbfleisch, J.D. and Prentice, R.C. (1980). *The Statistical Analysis of Failure Time Data*, John Wiley & Sons, New York.

Katz, D., Baptista, J. Azen, S.P., and Pike, M.C. (1978). Obtaining confidence intervals for the risk ratio in cohort studies. *Biometrics*, **34**, 469–474.

Kim, P.J. and Jennrich, R.I. (1973). Tables of the exact sampling distribution of the Kolomogorov-Smirnov criterion, $D_{mn}, m \leq n$. In *Selected Tables in Mathematical Statistics*, Volume 1, Edited by the Institute of Mathematical Statistics, American Mathematical Society, Providence, Rhode Island.

Kimball, A.W., Burnett, W.T., Jr., and Doherty, D.G. (1957). Chemical protection against ionizing radiation I. Sampling methods for screening compounds in radiation protection studies with mice. *Radiation Research*, **7**, 1–12.

Klotz, J. (1963). Small sample power and efficiency for one-sample Wilcoxon and normal scores tests. *Ann. Math. Statist.*, **34**, 624–632.

Koopman, P.A.R. (1984). Confidence intervals for the ratio of two binomial proportions. *Biometrics*, **40**, 513–517.

Lee, E.T. and Desu, M.M. (1972). A computer program for comparing K samples with right-censored data. *Computer Programs in Biomedicine*, **2**, 315–321.

Lehmann, E.L. (1953). The power of rank tests. *Ann. Math. Statist.*, **24**, 23–43.

Lehmann, E.L. (1998). *Nonparametrics: Statistical Methods Based on Ranks, Revised First Edition*, Prentice-Hall, Upper Saddle River, New Jersey.

Levin, B. and Chen, X. (1999). Is the one-half continuity correction used once or twice to derive a well known approximate sample size formula to compare two independent binomial distributions? *Amer. Statist.*, **53**, 62–66.

Mann, H.B. and Whitney, D.R. (1947). On a test whether one of two random samples is stochastically larger than the other. *Ann. Math. Statist.*, **18**, 50–60.

Mantel, N. (1966). Evaluation of survival data and two new rank order statistics arising in its consideration. *Cancer Chemotherapy Rep.*, **50**, 163–170.

Mantel, N. and Haenszel, W. (1959). Statistical aspects of the analysis of data from retrospective studies of disease. *J. Nat. Cancer Inst.*, **22**, 719–748.

Mathisen, H.C. (1943). A method of testing the hypothesis that two samples are from the same population. *Ann. Math. Statist.*, **14**, 188–194.

Milton, R.C. (1964). An extended table of critical values for the Mann-Whitney (Wilcoxon) two-sample statistic. *J. Amer. Stat. Assoc.*, **59**, 925–934.

Mood, A.M. (1950). *Introduction to the Theory of Statistics*, McGraw-Hill Book Co., New York.

Nelson, L.S. (1963). Tables for a precedence life test. *Technometrics*, **5**, 491–499.

Nelson, W. (1972). Theory and applications of hazard plotting for censored data. *Technometrics*, **14**, 945–966.

Newcombe, R.G. (1998). Interval estimation for the difference between independent proportions: comparison of eleven methods. *Statist. Med.*, **17**, 873–890.

Newcombe, R.G. (2001). Estimating the difference between differences: measurement of additive scale interaction for proportions. *Statist. Med.*, **20**, 2885–2893.

Noether, G.E. (1967). *Elements of Nonparametric Statistics*, John Wiley & Sons, New York.

Noether, G.E. (1987). Sample size determination for some common nonparametric tests. *J. Amer. Statist. Assoc.*, **82**, 645–647.

Peto, R. and Peto, J. (1972). Asymptotically efficient rank invariant test procedures. *J. Royal Statist. Soc., Series A*, **135**, 185–207.

Pike, M. (1966). A suggested method of analysis of a certain class of experiments in carcinogenesis. *Biometrics*, **22**, 142–161.

Pratt, J.W. and Gibbons, J.D. (1981). *Concepts of Nonparametric Theory*, Springer-Verlag, New York.

Prentice, R.L. (1978). Linear rank statistics with right censored data. *Biometrika*, **65**, 167–179.

Radhakrishna, S. (1965). Combination of results from several 2×2 contingency tables. *Biometrics*, **21**, 86–99.

Rao, C.R. (1973). *Linear Statistical Inference and Its Applications*, 2nd edition, John Wiley & Sons, New York.

Rodary, C., Com-Nougue, C., and Tournade, M. (1989). How to establish equivalence between treatments: a one-sided clinical trial in pediatric oncology. *Statist. Med.*, **8**, 595–598.

Rosenbaum, S. (1953). Tables for a nonparametric test of dispersion. *Ann. Math. Statist.*, **24**, 663–668.

Rosenbaum, S. (1954). Tables for a nonparametric test for location. *Ann. Math. Statist.*, **25**, 146–150.

Sahai, H. and Kurshid, A. (1996). Formulae and tables for the determination of sample sizes and power in clinical trials for testing differences in proportions for the two-sample design: a review. *Statist. Med.*, **15**, 1–21.

Santner, T.J. and Snell, M. K. (1980). Small sample confidence limits for $p_1 - p_2$ and p_1/p_2 in 2×2 contingency tables. *J. Amer. Statist. Assoc.*, **75**, 386–394.

Savage, I.R. (1956). Contributions to the theory of rank order statistics – the two-sample case. *Ann. Math. Statist.*, **27**, 1397–1409.

Schenker, N. and Gentleman, J.F. (2001). On judging the significance of differences by examining the overlap confidence intervals. *Amer. Statist.*, **55**, 182–186.

Serfling, R.J. (1980). *Approximation Theorems of Mathematical Statistics*, John Wiley & Sons, New York.

Shorack, R.A. (1967). On the power of precedence life tests. *Technometrics*, **9**, 154–158.

StatXact, Version 3 (1996). Cytel Software Corporation, Cambridge, Massachusetts.

Tarone, R. and Ware, J. (1977). On distribution-free tests for equality of survival distributions. *Biometrika*, **64**, 156–160.

Thomas, D.G. and Gart, J.J. (1977). A table of exact confidence limits for difference in ratios of two proportions and their odds ratios. *J. Amer. Statist. Assoc.*, **72**, 73–76 (see also Corrigenda (1978): **73**, 233).

van der Waerden, B.L. (1952, 1953). Order tests for the two sample problem and their power, I, II, III. *Proc. Koninklijke Nederlandse Akademie van Wetenschappen (A)*, **55** *(Indag. Math.,* **14***)*, 453–458 (1952); *Indag. Math.*, **15**, 303–310, 311–316 (1953); correction, *Indag. Math.*, **15**, 80 (1953).

Verdooren, L.R. (1963). Extended tables of critical values for Wilcoxon's test statistic. *Biometrika*, **50**, 177–186.

Wald, A. (1943). Tests of statistical hypotheses concerning several parameters when the number of observations is large. *Trans. Amer. Math. Soc.*, **54**, 426–482.

Wilcoxon, F. (1945). Individual comparisons by ranking methods. *Biometrics*, **1**, 80–83.

Wilcoxon, F., Katti, S.K., and Wilcox, R.A. (1973). Critical values and probability levels for the Wilcoxon rank sum test and Wilcoxon signed rank test. In *Selected Tables in Mathematical Statistics*, Volume 1, Edited by the Institute of Mathematical Statistics, American Mathematical Society, Providence, Rhode Island.

Wilk, M.B. and Gnanadesikan, R. (1968). Probability plotting methods for the analysis of data. *Biometrika*, **55**, 1–17.

Wilson, E.B. (1927). Probable inference, the law of succession, and statistical inference. *J. Amer. Statist. Assoc.*, **22**, 209–212.

CHAPTER 3

Procedures for paired samples

3.1 Introduction

Let X and Y be the random variables denoting the responses under treatments A and B, respectively. Let $F(.)$ and $G(.)$ be the cdf's of the X and Y variables. We want to test the null hypothesis

$$H_0 : F(.) = G(.).$$

In Chapter 2 we considered this testing problem in relation to independent samples on X and Y. However, in some situations the data are obtained in blocks. Often, it is somewhat difficult to get many homogeneous experimental units for obtaining independent samples. In those cases, it may be convenient to obtain a small number of homogeneous units and use a set of such groups of homogeneous units, which are usually called *blocks*. In this chapter we consider studies where such blocks are used to investigate the effects of two treatments. We only consider the case of blocks consisting of two units. The two treatments are randomly assigned to the units in each block.

An important scenario, where two treatments are tested in blocks of size two, is a study where the treatments are given to the same subjects. In other words, each subject receives both treatments in a random order. First we describe the methods for binary response studies, then the methods for continuous response cases. These methods cover both complete and censored data situations.

3.2 Analysis of paired binary responses

Here we consider the case where the two response variables X and Y are binary variables. The marginal distributions are

$$P(X = 1) = \theta_1, \quad P(X = 0) = 1 - \theta_1; \tag{3.1}$$

and

$$P(Y = 1) = \theta_2, \quad P(Y = 0) = 1 - \theta_2. \tag{3.2}$$

As these variables are not independent, we need to consider the joint distribution, which is given in Table 3.1.

<u>Remark 3.1.</u> The θ_{ij} represents the $P[X = i, Y = j]$, where θ_{00}, θ_{01} and θ_{10} are positive and $\theta_{00} + \theta_{01} + \theta_{10} < 1$. The marginal distributions are defined

Table 3.1 *Joint distribution of X and Y*

	Y		Row Total
	0	1	
$X = 0$	θ_{00}	θ_{01}	$1 - \theta_1$
$X = 1$	θ_{10}	θ_{11}	θ_1
Col. Total	$1 - \theta_2$	θ_2	1

Table 3.2 *Data on (X, Y)*

	Y		Row Total
	0	1	
$X = 0$	m_{00}	m_{01}	$m_{0.}$
$X = 1$	m_{10}	m_{11}	$m_{1.}$
Col. Total	$m_{.0}$	$m_{.1}$	n

by the probabilities $\theta_1 = P(X = 1)$ and $\theta_2 = P(Y = 1)$; thus $\theta_1 = \theta_{10} + \theta_{11}$ and $\theta_2 = \theta_{01} + \theta_{11}$.

The statistical problem is that of comparing the marginal distributions of the variables X and Y. As X and Y are responses under two different treatments, we are interested in testing the null hypothesis of no difference between the two treatments. Another problem of interest is to establish the *clinical equivalence* of the two treatments. These testing problems will be considered now.

3.2.1 McNemar's large sample test for the equality of marginal distributions

In this subsection a test for the equality of marginal distribution hypothesis will be described. The null hypothesis is $H_0 : F(.) = G(.)$, which is equivalent to $H_0 : \theta_1 = \theta_2$. This hypothesis is usually expressed in terms of the difference

$$\Delta = \theta_2 - \theta_1 = \theta_{01} - \theta_{10}, \tag{3.3}$$

so that the null hypothesis can be restated as

$$H_0 : \Delta = 0. \tag{3.4}$$

The n observations on (X, Y) are summarized in Table 3.2. In this table, m_{ij} denotes the frequency of the outcome $(X = i, Y = j)$. We estimate Δ and construct tests using this estimator. Using the fact that the joint distribution of the frequencies m_{ij} is a multinomial distribution, we obtain the maximum likelihood estimators of θ_{01} and θ_{10} as

$$\hat{\theta}_{01} = \frac{m_{01}}{n}, \quad and \quad \hat{\theta}_{10} = \frac{m_{10}}{n}.$$

Table 3.3 *McNemar large sample tests*

Alternative Hypothesis		
θ-Version	Δ-Version	Critical Region
$H_{A1} : \theta_1 < \theta_2$	$\Delta > 0$	$Z_\Delta > z_{(1-\alpha)}$
$H_{A2} : \theta_1 > \theta_2$	$\Delta < 0$	$Z_\Delta < z_\alpha$
$H_A : \theta_1 \neq \theta_2$	$\Delta \neq 0$	$\mid Z_\Delta \mid > z_{\alpha/2}$

Using these estimators we get an estimator of Δ, namely

$$\hat{\Delta} = \hat{\theta}_{01} - \hat{\theta}_{10} = \frac{(m_{01} - m_{10})}{n}. \tag{3.5}$$

Since the estimators of θ's are unbiased, the estimator $\hat{\Delta}$ is also an unbiased estimator of Δ. The variance of this estimator is

$$var(\hat{\Delta}) = \frac{[\theta_{01} + \theta_{10} - (\theta_{01} - \theta_{10})^2]}{n} = \frac{[2\theta_{10} + \Delta - \Delta^2]}{n}. \tag{3.6}$$

The test proposed by McNemar (1947) is based on the large sample distribution of the standardized version of the estimator $\hat{\Delta}$. To define the test statistic, we need to estimate the null variance of the estimator $\hat{\Delta}$. From (3.6), the null variance turns out to be

$$var(\hat{\Delta} \mid H_0) = \frac{[\theta_{01} + \theta_{10}]}{n}.$$

An estimator of this null variance is v^2, where

$$v^2 = \frac{\hat{\theta}_{01} + \hat{\theta}_{10}}{n} = \frac{(m_{01} + m_{10})}{n^2},$$

assuming that $(m_{01} + m_{10}) > 0$. Thus the test statistic of McNemar is

$$\boxed{Z_\Delta = \frac{\hat{\Delta}}{\sqrt{v^2}} = \frac{m_{01} - m_{10}}{\sqrt{m_{01} + m_{10}}}.} \tag{3.7}$$

Under the null hypothesis the distribution of this statistic Z_Δ can be approximated by the standard normal distribution. The critical regions of approximate size α tests are given in Table 3.3.

Stuart (1955) considered the general case where X and Y are categorical variables with $m(> 2)$ values.

Remark 3.2. In the case of the two-sided alternatives, sometimes the critical region is stated in terms of Z_Δ^2, which is approximately a chi-square random variable with one degree of freedom. In general the approximate P-values can be computed using the approximate normal distribution of Z_Δ.

Table 3.4 *Number of patients experiencing pain*

Before Hypnosis (X)	After Hypnosis (Y)		
	No Pain (0)	Pain (1)	Row Total
No Pain (0)	18	4	22
Pain (1)	12	5	17
Col. Total	30	9	39

Example 3.1. May and Johnson (1997) reported the results of a study in which investigators wanted to determine the effect of hypnosis in reducing pain associated with venopuncture experienced by juvenile cancer patients. The data are given in Table 3.4.

For these data $m_{01} = 4$, and $m_{10} = 12$. Hence

$$|Z_\Delta| = |(4 - 12)|/\sqrt{(4 + 12)} = 2.0.$$

We are interested in the two-sided alternatives, so the P-value is approximated by $P[\chi_1^1 > 4.0] = 0.0455$. Hence we reject the null hypothesis that hypnosis does not have any effect. In this context, the following estimates provide useful additional information:

$$\hat{P}(pain\ before\ hypnosis) = (17/39),$$

and

$$\hat{P}(pain\ after\ hypnosis) = (9/39).$$

A computer program for performing McNemar's test is given in Appendix B.

3.2.2 Exact test for equality of marginal distributions

The tests considered in the previous subsection used the asymptotic distribution of the estimator $\hat{\Delta}$. We now explore the testing problem without considering the asymptotics. In general the main task is to assess whether or not $\hat{\Delta}$ is significantly different from zero. One method of doing this assessment is to find out how likely the observed value of $\hat{\Delta}$ is under the null hypothesis. To find this probability we examine the probability function of (M_{01}, M_{10}), the random vector corresponding to the frequencies (m_{01}, m_{10}). This function is

$$f(a, b) = Pr(M_{01} = a, M_{10} = b \mid H_0)$$

$$= \frac{n!}{a!b!(n - a - b)!}\theta^{a+b}(1 - 2\theta)^{n-a-b}$$

$$= \binom{n}{a + b}(2\theta)^{a+b}(1 - 2\theta)^{n-a-b}\binom{a + b}{a}(1/2)^{a+b},$$

Table 3.5 *P-values for exact tests of equality of marginals*

Alternative	P-Value
H_{A1}	$1 - P[Bin(m, 1/2) \le m_{01} - 1](= P_1)$
H_{A2}	$P[Bin(m, 1/2) \le m_{01}](= P_2)$
H_A	$2 \min(P_1, P_2)$

where θ is the common value of θ_{01} and θ_{10}, under H_0. This probability function depends on θ, which is unknown. The first factor is the probability function of $M = M_{01} + M_{10}$ and the second factor can be seen to be the conditional probability function of M_{01}, given $M = a + b$. This conditional distribution is the binomial distribution with parameters $(a + b)$ and $(1/2)$ and it can be used to draw conclusions about H_0. The exact (conditional) P-values are computed as indicated in Table 3.5, where $m = m_{01} + m_{01}$. The cdf of the binomial distribution is a function in the SAS system.

3.2.3 Testing for clinical equivalence

In Subsection 2.1.3 of Chapter 2, we considered the problem of testing the clinical equivalence hypothesis. This testing problem is also of interest here. In this discussion the $X(Y)$-variable is the response under the new (standard) treatment. The clinical equivalence hypothesis states that the new treatment is *not more than* $100\Delta_0\%$ *inferior*. This is taken as the alternative hypothesis and the null hypothesis is that the new treatment is $100\Delta_0\%$ inferior. Noting that $\Delta = \theta_S - \theta_N$, the problem is to test

$$H_0^* : \Delta = \Delta_0 \quad against \quad H_{A2}^* : \Delta < \Delta_0, \qquad (3.8)$$

where $\Delta_0 > 0$. Nam (1997) suggested a test procedure, which is a generalization of McNemar's test. To describe this test, we need to evaluate the mean and the variance of $\hat{\Delta}$, under the null hypothesis $\Delta = \Delta_0$. Clearly

$$E(\hat{\Delta} \mid H_0^*) = \Delta_0,$$

and from (3.6) it follows that

$$var(\hat{\Delta} \mid H_0^*) = \frac{[2\theta_{10} + \Delta_0 - \Delta_0^2]}{n}. \qquad (3.9)$$

To obtain the standardized statistic, we need to estimate the null variance. Thus we need to find the maximum likelihood estimate (MLE) of θ_{10} under the null hypothesis $\Delta = \Delta_0$. This MLE, $\tilde{\theta}_{10}$, is (see Appendix A3 for details)

$$\tilde{\theta}_{10} = [\sqrt{(b^2 - 4ac)} - b]/(2a), \qquad (3.10)$$

where

$$a = 2n, \qquad b = (2n + m_{01} - m_{10})\Delta_0 - (m_{10} + m_{01}),$$

and

$$c = -m_{01}\Delta_0(1 - \Delta_0). \tag{3.11}$$

Using this estimator, the null variance estimator, $v^2(\Delta_0)$, is computed as

$$v^2(\Delta_0) = \frac{[2\tilde{\theta}_{10} + \Delta_0 - \Delta_0^2]}{n}. \tag{3.12}$$

Hence the test statistic is

$$Z_{CE} = \frac{[\hat{\Delta} - \Delta_0]}{v(\Delta_0)}. \tag{3.13}$$

An approximate size α test is to

$$\textit{reject } H_0 \textit{ if } Z_{CE} \leq z_\alpha. \tag{3.14}$$

It should be noted that rejecting the null hypothesis H_0^* of (3.8) amounts to establishing *clinical equivalence*.

Remark 3.3. When $\Delta_0 = 0$, the null hypothesis of (3.8) is the same as the null hypothesis (3.4). For this case the test (3.14) is the same as the test for H_{A2} given in Table 3.3.

Remark 3.4. If $m_{01} = 0$, then $c = 0$, which implies that $\tilde{\theta}_{10} = 0$. This case needs careful consideration. The relevant details are given in Tango (1998).

3.2.4 Confidence interval for the difference Δ

In some instances, a confidence interval for Δ, the difference in response rates, is needed. Various proposals have been put forward. Newcombe (1998) examined various proposals for this confidence interval and made certain recommendations. Tango (1998) also considered a test for equivalence and suggested a method for calculating a confidence interval for the difference Δ. We discuss the method of Tango (1999)(see also Newcombe (1999)).

Tango's method can be related to the test we discussed in Subsection 3.2.3 and so we describe it now. Since we want to construct a confidence interval for Δ, let us consider the problem of testing

$$H_0 : \Delta = \Delta_0 \quad \textit{against} \quad H_A : \Delta \neq \Delta_0, \tag{3.15}$$

for some arbitrary Δ_0. Now we can construct a test using the same statistic of Subsection 3.2.3. As in Subsection 3.2.1, an approximate α-level test is to

$$\textit{reject } H_0 \textit{ if } |Z_{CE}| > z_{(1-\alpha/2)}, \tag{3.16}$$

where Z_{CE} is defined in (3.13). Now it is easy to see that this test does not reject Δ_0 values that belong to the set

$$\frac{|\hat{\Delta} - \Delta_0|}{v(\Delta_0)} \leq z_{(1-\alpha/2)},$$

where $v^2(\Delta_0)$ is given by (3.12), and $\tilde{\theta}_{10}$ is the mle of θ_{10}, under the hypothesis $H_0 : \Delta = \Delta_0$. In other words, the above set is a confidence region for Δ and this region will turn out to be an interval. To find the end points of this interval we need to solve the following equation,

$$\frac{[\hat{\Delta} - \Delta]}{v(\Delta)} = \pm z_{(1-\alpha/2)}, \tag{3.17}$$

for Δ. We cannot solve this equation explicitly, since $\tilde{\theta}_{10}$ is a function of Δ, however, we can use an iterative procedure. A computer program for calculating this confidence interval is given in Appendix B3.

Example 3.2. In the Example 3.1 study, the ability of hypnosis to reduce pain is the main interest. So we are concerned with the difference

$$\Delta^* = P(pain|before) - P(pain|after).$$

Since we denoted the before response by X and the after response by Y, $\Delta^* = -Delta$ of (3.3). A point estimate of $\Delta^* = (17/39) - (9/39) = (8/39) = 0.2051$. Further, from the output of the computer program, a 95% confidence interval for Δ is $(-0.3926, -0.0044)$. Thus a 95% confidence interval for Δ^* is $(0.0044, 0.3926)$.

Remark 3.5. The above analysis is valid when $m_{01} > 0$, and $m_{10} > 0$. For other cases, see Tango (1999).

3.2.5 Sample size for equivalence trials

To design a study for testing the clinical equivalence hypothesis, we will use a test of significance level α to test

$$H_0^* : \Delta = \Delta_0 \quad versus \quad H_{A2}^* : \Delta < \Delta_0. \tag{3.18}$$

This test should have a power $(1 - \beta)$ for $\Delta = \Delta_1 (< \Delta_0)$. The problem is to determine the sample size so that we can achieve the stated objective. We use the test of Subsection 3.2.3. The approximate power function of this test is

$$\pi(\Delta) = Pr[N(0, 1) < u(\Delta)] = \Phi(u(\Delta)),$$

where

$$u(\Delta) = \frac{z_\alpha v(\Delta_0) + \Delta_0 - \Delta}{v(\Delta)}, \tag{3.19}$$

$v^2(\Delta)$ being the variance of $\hat{\Delta}$. For this power to be equal to $(1-\beta)$ at $\Delta = \Delta_1$, we have to equate $u(\Delta_1)$ to $z_{1-\beta}$. By doing so and solving for n, we get

$$n = [\sqrt{(2\check{\theta}_{10} + \Delta_0 - \Delta_0^2)}z_\alpha + \sqrt{(2\theta_{10} + \Delta_1 - \Delta_1^2)}z_\beta]^2/(\Delta_0 - \Delta_1)^2, \tag{3.20}$$

where $\check{\theta}_{10}$ is the large sample approximation to the MLE $\tilde{\theta}_{10}$ of (3.10). It is obtained as the solution

$$\check{\theta}_{10} = [-b^* + (b^{*2} - 4a^*c^*)^{1/2}]/(2a^*)$$

of the quadratic equation

$$a^*\check{\theta}_{10}^2 + b^*\check{\theta}_{10} + c^* = 0,$$

where

$$a^* = 2, \quad b^* = (2+\Delta_0)\Delta_0 - (2\theta_{01} - \Delta_0), \quad c^* = -\theta_{01}\Delta_0(1 - \Delta_0). \tag{3.21}$$

In (3.21) the θ_{01} value is based on prior information and $\theta_{10} = \theta_{01} - \Delta_1$. Further details about this derivation are available in Nam (1997).

Remark 3.6. When $\Delta_0 = 0$, the test considered here is the same as McNemar's test. So by considering the appropriate specialization of (3.20), we obtain an expression for the sample size for a one-sided problem discussed in Subsection 3.2.1. This result is given as Problem 3.

3.2.6 Estimation of the ratio of marginal probabilities

The equality of the marginal distributions can also be formulated using the ratio $\psi = (\theta_1/\theta_2)$, as in Chapter 2. Here we study the problem of the interval estimation of ψ. This estimation is useful in evaluating the effect of an event on the success probability. For example, we may be interested in knowing the amount of reduction in the smoking rate among mothers after the birth of a child. We may be interested in evaluating the effect of an exercise program on the extent of obesity. A possible experiment is to study a group of subjects for the presence of an attribute before a program and to study the same group for the presence of the same attribute after they have participated in a program that has some effect on the extent of the presence of the attribute. Here X represents the presence or absence of the attribute before a program and Y represents the presence or absence of the attribute after the program. The data can be arranged as in Table 3.2 and the objective is to compare θ_1 with θ_2, using ψ. Peritz (1971) studied the problem of estimation of ψ. We will now give the details.

Using the estimators of θ_1 and θ_2, we estimate ψ by

$$\hat{\psi} = \frac{\hat{\theta}_1}{\hat{\theta}_2} = \frac{m_{1.}}{m_{.1}}. \tag{3.22}$$

For the calculation of a confidence interval for ψ we need an estimate of the variance of $\hat{\psi}$. Using this variance estimator and using an approximate normal distribution we can compute a confidence interval. Instead, we start with the parameter ϕ, which is the logarithm of ψ, and obtain a confidence interval for ϕ. This interval is used to obtain an interval for ψ.

First we note that $\phi = \log\psi = log\theta_1 - log\theta_2$. Thus a natural estimator of ϕ is

$$\hat{\phi} = log\hat{\theta}_1 - log\hat{\theta}_2 = logm_{1.} - logm_{.1}. \tag{3.23}$$

Further, an estimator of the asymptotic variance of $\hat{\phi}$ is

$$s^2 = (m_{10} + m_{01})/(m_{1.}m_{.1}). \tag{3.24}$$

Some details regarding the estimation of the asymptotic variance are given in Appendix A3. From the large sample theory of maximum likelihood estimators, it follows that the distribution of

$$\boxed{Z = (\hat{\phi} - \phi)/s}$$

can be approximated by the standard normal distribution. This result leads to a confidence interval (ϕ_L, ϕ_U) for ϕ, where

$$\phi_L = \hat{\phi} - z_{(1-\alpha/2)}s; \quad \phi_U = \hat{\phi} + z_{(1-\alpha/2)}s. \tag{3.25}$$

This interval gives a confidence interval for ψ as (e^{ϕ_L}, e^{ϕ_U}).

Remark 3.7. The confidence interval for ϕ can be used for testing the null hypothesis $H_0 : \theta_1 = \theta_2$, which is the same as $H_0 : \phi = 0$. The ratio ψ has also been used to formulate the clinical equivalence problem. A detailed discussion of this formulation is available in Tang et al. (2002).

We will now illustrate this method using an artificial data set.

Example 3.3. A study is undertaken to evaluate the effectiveness of an exercise program in reducing the obesity rate in a community. A person is classified as obese (1) or non-obese (0) based on the body mass index. A group of 500 randomly selected individuals participated in the study and each was classified as obese or non-obese before and after the exercise program. The results are given in Table 3.6. An estimate of the obesity rate before the program is $\hat{\theta}_1 = (420/500) = 0.84$, whereas after the program it is $\hat{\theta}_2 = (350/500) = 0.70$. Thus an estimate of the ratio is $\hat{\psi} = (0.84/0.70) = 1.2$. This means that 20%

Table 3.6 *Extent of obesity*

Before	After 0	After 1	Row Total
0	60	20	80
1	90	330	420
Col. Total	150	350	500

more are obese before the program than after the program. An estimate of ϕ is $\hat{\phi} = \log(1.2) = 0.18$. Further, the variance estimate of $\hat{\phi}$ is

$$s^2 = (20 + 90)/[(420)(350)] = 0.00075,$$

and hence $s = 0.0274$. Thus a 95% confidence interval for ϕ turns out to be $(0.1287, 0.2359)$, and hence a 95% confidence interval for ψ is $(1.1374, 1.2661)$. The null hypothesis of the equality of the obese rates will be rejected at the 5% level.

A computer program for calculating these confidence limits is given in Appendix B3.

3.3 Complete data on continuous responses

When X and Y are independent random variables with equal distributions, the joint distribution function, $H(x, y)$, of X and Y is a symmetric function of the arguments, that is, $H(x, y) = H(y, x)$ for all (x, y). This result is not necessarily true when X and Y are dependent. However, if the joint distribution function, $H(x, y)$, is symmetric then the marginal distributions of X and Y are equal. This can be seen by taking the limit of $H(x, y)$ as $y \to \infty$. Assuming symmetry for the joint distribution function implies that the joint density function, $h(x, y)$, is also a symmetric function. Furthermore, in this case the distribution of $X - Y$ is symmetric at about 0 (see Appendix A3), thereby implying that the mean and the median of $X - Y$ are zero.

We now turn our attention to the present setting where the variables X and Y are dependent and the distribution of X is the same as the distribution of $Y + \Delta$. We are interested in testing the null hypothesis $H_0 : \Delta = 0$. The following model summarizes the discussion thus far.

Model
   Let

$$X = \mu_1 + U, \quad and \quad Y = \mu_2 + V,$$

where the joint distribution of (U, V) is defined by a density function $h_0(x, y)$, which is a symmetric function, that is,

$$h_0(x, y) = h_0(y, x), \quad for\ all\ (x, y). \tag{3.26}$$

Further, we have

$$D = X - Y = \mu_1 - \mu_2 + (U - V) \equiv \Delta + W.$$

Since X and Y have a joint distribution for developing tests of hypothesis about Δ, the difference $D = X - Y$ is a natural variable to consider. Let $L(.)$ be the distribution function of W. It is shown in Appendix A3 that the distribution function L is symmetric about zero, that is,

$$L(x) = 1 - L(-x), \quad \text{for all } x.$$

Thus the distribution of D is symmetric about Δ and hence Δ is the median of the distribution of D. Thus testing the null hypothesis $H_0 : F(.) = G(.)$ is the same as testing $H_0^* : \Delta = 0$, where Δ is the median of the distribution of D. Since the hypothesis of interest has been formulated in terms of a hypothesis about the median, we can use the methods described in Section 1.3 of Chapter 1. We first consider the sign test, which only takes into account the signs of D. Next we consider the test proposed by Wilcoxon, which takes into account not only the sign but also the magnitude of D. This test is known as the *Wilcoxon signed rank test*.

The basic problem now is making inference about the parameter Δ, which is the median of the difference variable D. The null hypothesis of interest is

$$H_0 : \Delta = 0. \tag{3.27}$$

Since $E(D) = \Delta + E(W) = \Delta$, we have that $\Delta = E(X) - E(Y)$. We assume that we have a random sample (X_i, Y_i), for $(i = 1, 2, \ldots, n)$. Let $D_i = X_i - Y_i$. The D's will constitute a random sample from the distribution defined by the cdf $L(x - \Delta)$.

3.3.1 Sign test for complete paired data

The test statistic is

$$S_n = number\ of\ positive\ D_i. \tag{3.28}$$

The null distribution of the statistic S_n is the binomial distribution with parameters n and $(1/2)$ and hence it is a symmetric distribution. However, the distribution under the alternative is the binomial distribution with parameters n and $\pi_1(\Delta) = P(D > 0)$.

The alternatives and the critical regions of the sign tests are given in Table 3.7, where $S_n(obs)$ is the observed value of S_n. The critical values are chosen as explained in Chapter 1.

Table 3.7 *Sign tests for paired data studies*

Alternative	Critical Region	P-Value
H_{A1}	$S_n \geq c_1$	$P(Bin(n, 1/2) \geq S_n(obs))(= P_1)$
H_{A2}	$S_n \leq c_2$	$P(Bin(n, 1/2) \leq S_n(obs))(= P_2)$
H_A	$S_n \leq n - c_3\ or\ S_n \geq c_3$	$2 \cdot min(P_1, P_2)$

3.3.2 Wilcoxon signed rank test

In the sign test we only used the signs of D's. Now we will consider a different procedure that uses the magnitudes as well as the signs of D's. Let us assume that all the differences are *nonzero*.

Data without ties among $|D|$ values

The absolute values $|D_i|$ are ranked and to each of these ranks a sign is attached, the sign being the sign of the difference D_i. The resulting quantities are called *signed ranks* of the differences. The test statistic is V_+, which is the sum of positive signed ranks. In order to decide upon the appropriate critical regions, we calculate the mean of V_+. This calculation is relatively easy once we express the statistic in terms of the averages $(D_i + D_j)/2$, for $1 \leq i \leq j \leq n$. These averages are called *Walsh averages*. The relevant expression is

$$V_+ = number\ of\ \left(\frac{D_i + D_j}{2} > 0 \right). \tag{3.29}$$

In other words, V_+ is the number of positive Walsh averages. A proof of this result is given in Appendix A3. This representation is also useful for obtaining a confidence interval for Δ, which is discussed in Subsection 3.3.4.

To derive the mean and the variance we introduce the variables

$$U(i,j) = \begin{cases} 1, & \text{if } \frac{D_i + D_j}{2} > 0, \\ 0, & \text{otherwise.} \end{cases}$$

Now, we have

$$V_+ = \sum_{j=1}^{n} \sum_{i=1}^{j} U(i,j) = \sum_{i=1}^{n} U(i,i) + \sum_{j=2}^{n} \sum_{i=1}^{j-1} U(i,j),$$

where there are n terms in the first sum and $\binom{n}{2}$ terms in the second sum. It is easy see that

$$E[U(i,i)] = P(D_i > 0) \equiv \pi_1(\Delta),$$

and

$$E[U(i,j)] = P[D_i + D_j > 0] \equiv \pi_2(\Delta).$$

It can be shown that these two expectations do not depend on (i,j) (see Appendix A3); thus, we have

$$E[V_+] = n\pi_1(\Delta) + \binom{n}{2}\pi_2(\Delta). \tag{3.30}$$

Using the fact that $\pi_1(0) = (1/2)$ and $\pi_2(0) = (1/2)$, we get

$$E[V_+ \mid H_0] = \frac{n + \binom{n}{2}}{2} = \frac{n(n+1)}{4}. \tag{3.31}$$

Table 3.8 *Wilcoxon signed rank tests*

Alternative	Critical Region
$H_{A1} : \Delta > 0$	$V_+ \geq c_1$
$H_{A2} : \Delta < 0$	$V_+ \leq c_2$
$H_A : \Delta \neq 0$	$V_+ \leq c_3$ or $V_+ \geq [n(n+1)/2] - c_3$

It is possible to establish an ordering between the null expectation and the expectation under the alternative $H_{A1} : \Delta > 0$ (see Appendix A3 for details). This ordering for $\Delta > 0$ is

$$E[V_+ \mid H_{A1}] > E[V_+ \mid H_0].$$

So we reject H_0 in favor of H_{A1} for large values of V_+. We can also establish an appropriate inequality for the case $H_{A2} : \Delta < 0$. The derivation of the variance from this representation is involved and so we omit the same. However, the variance under the null hypothesis is derived in Appendix A3.

For future reference the alternatives and the critical regions are summarized in Table 3.8.

The constants c's are the appropriate percentiles of the null distribution of the test statistic V_+. Since the null distribution of V_+ is symmetric about its mean an equi-tailed test is used for the two-sided alternatives. McCornack (1965) gives tables of critical values to implement these tests. Wilcoxon et al. (1973) produced an extensive tabulation of the critical values.

Exact null distribution of V_+

Under the null hypothesis, the distribution of D is the same as the distribution of W and hence it is symmetric about zero. This means that the null distribution of D is the same as the null distribution of $-D$. The distribution of the ranks of $|D|$ is uniform over the set of all permutations of integers $(1, 2, \ldots, n)$. The number of positive D's can vary from 0 to n. So for obtaining the distribution of V_+, we need to consider various size subsets of $(1, 2, \ldots, n)$, which are 2^n in number. These subsets represent the possible values for the positive signed ranks and they are associated with the same probability. By enumeration we can develop the distribution of V_+.

We will illustrate these calculations for $n = 3$. The various subsets and the values of V_+ are given below.

Subset	V_+	Subset	V_+
Empty Set	0	$\{1, 2\}$	3
$\{1\}$	1	$\{1, 3\}$	4
$\{2\}$	2	$\{2, 3\}$	5
$\{3\}$	3	$\{1, 2, 3\}$	6

Thus the null distribution of V_+ is

v	0	1	2	3	4	5	6
$P_0(V_+ = v)$	(1/8)	(1/8)	(1/8)	(2/8)	(1/8)	(1/8)	(1/8)

It may be noted that this distribution is symmetric about 3, which is the mean of the distribution. This symmetry can be established in general.

In large samples, approximations to the percentiles can be obtained by considering the approximate normal distribution of the standardized statistic

$$Z_V = \frac{[V_+ - E(V_+ \mid H_0)]}{\sqrt{var(V_+ \mid H_0)}}. \tag{3.32}$$

The mean required is given in (3.31) and the variance is

$$var(V_+ \mid H_0) = [n(n+1)(2n+1)/24]. \tag{3.33}$$

This normal approximation can also be used to obtain approximations to P-values that are adequate whenever $n > 20$.

Remark 3.8. If there are few zero differences, they are discarded and the analysis proceeds with the remaining nonzero differences. For other methods of handling zero differences see Pratt (1959).

Data with ties among $|D|$ values

In the case of tied $|D_i|$ values, we assign the average ranks R_i^*. The test statistic is denoted by V_+^*, which is the sum of the ranks associated with positive D values. The critical regions are similar to the ones defined earlier, where the test statistic V_+ is replaced by V_+^*. The null distribution of this statistic is not tabulated. One can generate the null distribution by enumeration as indicated earlier. We suggest the use of the asymptotic normal distribution of the standarized statistic

$$Z_V^* = \frac{[V_+^* - E(V_+^* \mid H_0\{n(n+1)/4\}]}{\sqrt{var(V_+^* \mid H_0)}} \tag{3.34}$$

to approximate the critical values. It can be shown that

$$E(V_+^* \mid H_0) = n(n+1)/4,$$

and

$$var(V_+^* \mid H_0) = [n(n+1)(2n+1)/24] - (1/48) \sum_{i=1}^{k} t_i(t_i^2 - 1), \tag{3.35}$$

COMPLETE DATA ON CONTINUOUS RESPONSES

Table 3.9 *Sign test and Wilcoxon signed rank test on FEV(1) data*

Subject	Treatment A	B	Sign of D_i	Rank of D_i	Signed $|D_i|$	Ranks
1	1.28	1.33	−0.05	−	2	−2
2	1.60	2.21	−0.61	−	7	−7
3	2.46	2.43	0.03	+	1	1
4	1.41	1.81	−0.40	−	5	−5
5	1.40	0.85	0.55	+	6	6
6	1.12	1.20	−0.08	−	3	−3
7	0.90	0.90	0			
8	2.41	2.79	−0.38	−	4	−4

Table 3.10 *Outcomes for which $V_+ \leq 7$*

Subset	Empty Set	{1}	{2}	{3}	{1,2}	{4}
Value of V_+	0	1	2	3	3	4
Subset	{1,3}	{5}	{1,4}	{2,3}	{6}	{1,5}
Value of V_+	4	5	5	5	6	6
Subset	{2,4}	{1,2,3}	{7}	{1,6}	{2,5}	{3,4}
Value of V_+6	6	7	7	7	7	7
Subset	{1,2,4}					
Value of V_+	7					

where t_i's are multiplicities of the k distinct values among $|D_i|$ (for a proof see Lehmann (1998, Appendix, p. 331)). It can also be shown that

$$var(V_+^* \mid H_0) = \frac{1}{4}\sum_i (R_i^*)^2.$$

Example 3.4. Patel (1983) considered a study with two active treatments A and B for treating acute bronchial asthma. The response is the forced expired volume in one second (FEV_1). A part of the data is reproduced in Table 3.9 and is used for illustrating this procedure. The full details of the analysis of the entire data will be discussed in a later section.

In the analysis of D_i, we delete the zero value. So $n = 7$. For the sign test on D_i, the test statistic $S_n = 2$. The two-sided P-value is $2P[Bin(7, 0.5) \leq 2] = 0.4532$. So we do not reject the null hypothesis.

The Wilcoxon signed rank statistic is $V_+ = 7$. The two-sided P-value is $2P[V_+ \leq 7 \mid H_0] = 0.2969$. The exact P-value can be calculated by enumerating the outcomes that correspond to values of V_+, which are less than 7. This enumeration is given in Table 3.10.

From the table, we note that the number of subsets for which the value of V_+ is not greater than 7 is 19. Thus $P[V_+ \leq 7|H_0] = (19/2^7) = 0.1484$. Hence the two-sided P-value is $2 \times 0.1484 = 0.2968$. This tallies with the result given by the computer program (see Appendix B3).

3.3.3 Rank transformed t-test

Let $(X_i, Y_i), i = 1, 2, \ldots, n$ be the sample. We form a string of $2n$ values, $(X_1, \ldots, X_n, Y_1, \ldots, Y_n)$ and rank them by giving average ranks, if necessary. Let $R(X_i)$ and $R(Y_i)$ be the ranks of X_i and Y_i, respectively. Further, let

$$D_i^* = R(X_i) - R(Y_i). \tag{3.36}$$

We compute the one-sample t-statistic for the data $D_i^*, i = 1, \ldots, n$, which is given by

$$t = \frac{\sum D_i^*}{\{[n \sum D_i^{*2} - (\sum D_i^*)^2]/(n-1)\}^{1/2}}, \tag{3.37}$$

which is approximately distributed as a t-variable with $(n-1)$ degrees of freedom. Using this approximate distribution, one can test the equality of the marginal distributions (see Conover and Iman (1981)).

Example 3.4 (cont'd.). The data of Patel used earlier will be used to illustrate this rank transformed t-test. The required quantities appear in Table 3.11. The t-statistic of (3.37), based on the data D_i^*, is

$$t = \frac{-1}{[(8/7)(69) - (1/7)]^{1/2}} = -0.1127.$$

Using the t-distribution approximation, the two-sided P-value is 0.9134.

All these tests indicate that there is no difference in the treatment effects of A and B.

3.3.4 Confidence interval for Δ corresponding to Wilcoxon signed rank test

Let $M = n(n+1)/2$. From the Walsh averages $A_{ij} = (D_i + D_j)/2$, for $1 \leq i \leq j \leq n$. Arrange these averages in increasing order of magnitude, breaking ties

Table 3.11 *Data on FEV(1)*

Subject	Treatment A	B	D_i	$R(X_i)$	$R(Y_i)$	D_i^*
1	1.28	1.33	−0.05	6	7	−1
2	1.60	2.21	−0.61	10	12	−2
3	2.46	2.43	0.03	15	14	1
4	1.41	1.81	−0.40	9	11	−2
5	1.40	0.85	0.55	8	1	7
6	1.12	1.20	−0.08	4	5	−1
7	0.90	0.90	0	2.5	2.5	0
8	2.41	2.79	−0.38	13	16	−3

at random, if necessary. The ordered A values are $A_{(1)} < A_{(2)} < \cdots < A_{(M)}$. Let c be the largest integer such that

$$P(V_+ \le c \mid \Delta = 0) \le (\alpha/2), \tag{3.38}$$

and let $l = c + 1$. Then a $100(1 - \alpha)\%$ confidence interval for Δ is the open interval

$$[A_{(l)},\ A_{(M-l+1)}). \tag{3.39}$$

In large samples the required quantity l can be approximated as

$$l = c + 1 \approx \lfloor n(n+1)/4 + z_{(\alpha/2)} \sqrt{n(n+1)(2n+1)/24} \rfloor + 1. \tag{3.40}$$

The details about the derivation are given in Appendix A3. A computer program for doing these calculations is given in Appendix B3.

Example 3.4 (cont'd.). Let us compute a confidence interval for Δ using the data of Example 3.4. From the program output, we have $l = 3$, and a 95% confidence interval derived in Subsection 3.3.4 is $(-0.59, 0.25)$. Since this interval includes zero, we do not reject the null hypothesis $\Delta = 0$ at the 5% level.

3.3.5 Analysis of cross-over designs

In studies involving two treatments, sometimes cross-over designs are used, especially when the experimental units are humans. With these designs, each subject is given both the treatments. Suppose we are going to use n subjects in the study. These subjects are divided into two groups of n_1 and n_2 subjects. Subjects in group 1 will receive treatment A in the first period and treatment B in the second period. In contrast, subjects in group 2 will receive treatment B first and then treatment A. In this type of experiment the responses in the first period are only influenced by the direct effects of the treatments received, whereas the responses in the second period are not only influenced by the direct effects of the treatments, but also by the carry-over effects of the treatments received in the first period. We are mainly concerned with the comparison of the direct effects. In many cases a large wash-out period is used between the administration of the two treatments, so that the carry-over effects are practically eliminated. We will consider the analysis, assuming that the carry-over effects are not present.

If we had only one group, we would have considered the differences in response A—response B, since the two responses are correlated, and we would have used one of the procedures considered earlier. Now we have two independent groups. The hypothesis that the two treatment effects are equal will imply that the two sets of differences are samples from the same distribution. Thus we can apply the WMW test to the two samples of differences. The details will be explained in full, using necessary notation.

Table 3.12 *Data from a cross-over experiment*

Group	Period 1	Period 2	D_{ik}	Rank
1	1.28	1.33	−0.05	7
1	1.60	2.21	−0.61	1
1	2.46	2.43	0.03	9
1	1.41	1.81	−0.40	2
1	1.40	0.85	0.55	14
1	1.12	1.20	−0.08	6
1	0.90	0.90	0	8
1	2.41	2.79	−0.38	3
2	3.06	1.38	1.68	17
2	2.68	2.10	0.58	15
2	2.60	2.32	0.28	13
2	1.48	1.30	0.18	11
2	2.08	2.34	−0.26	4
2	2.72	2.48	0.24	12
2	1.94	1.11	0.83	16
2	3.35	3.23	0.12	10
2	1.16	1.25	−0.09	5

Suppose that Y_{ijk} is the response of the kth subject in group i and period j, $k = 1, 2, \ldots, n_i$; $i = 1, 2$; and $j = 1, 2$. Now the differences are defined as

$$D_{1k} = Y_{11k} - Y_{12k}, \quad for\ k = 1, 2, \ldots, n_1, \tag{3.41}$$

and

$$D_{2k} = Y_{21k} - Y_{22k}, \quad for\ k = 1, 2, \ldots, n_2. \tag{3.42}$$

We apply the WMW test for testing the hypothesis that the two independent samples $\{D_{11}, D_{12}, \ldots, D_{1n_1}\}$ and $\{D_{21}, D_{22}, \ldots, D_{2n_2}\}$ come from the same distribution. The following example will illustrate the procedure.

Example 3.5. Patel (1983) considered a study with two active drugs A and B for treating acute bronchial asthma. The response is the *forced expired volume in one second* (FEV_1). The data appear in Table 3.12.

Now W_1, the rank sum for group 1, is 50. From a computer program output, the two-sided exact P-value is found to be 0.036. So we declare that the treatments are not equally effective.

3.4 Asymptotic relative efficiency

To test the null hypothesis (3.27) against the alternative $H_{A1} : \Delta > 0$, we use the sample $D_i = X_i - Y_i$, $(i = 1, 2, \ldots, n)$. In the parametric analysis, we usually assume that the joint distribution of X and Y is a bivariate normal, so that the distribution of $D = X - Y$ is a univariate normal and thus we use the paired t-test. In this chapter we developed two other tests, namely, the sign test and the Wilcoxon signed rank test. Now we compare the two nonparametric tests with the paired t-test.

Following an analysis similar to the one used in Section 2.5, we get the following results assuming that D has a normal distribution with variance σ^2.

Efficacies

Let T, S, and V_+ be the test statistics of the paired t-test, sign test, and Wilcoxon signed rank test, respectively. Then

1. $e(T) = \frac{1}{\sigma}$,

2. $e(S) = \sqrt{\frac{2}{\pi} \frac{1}{\sigma}}$,

3. $e(V_+) = \sqrt{\frac{3}{\pi} \frac{1}{\sigma}}$.

Now we will summarize the asymptotic relative effficiences.

1. $\text{ARE}(S, T) = \frac{2}{\pi}$,

2. $\text{ARE}(V_+, T) = \frac{3}{\pi}$.

As indicated in Section 2.5, the sample sizes for the sign test and the Wilcoxon signed rank test can be determined from the sample size needed for the paired t-test in the case of normal distribution. This sample size problem has also been considered by Noether (1987).

3.5 Analysis of censored data

Two tests, namely, the sign test and the Wilcoxon signed rank test, have been discussed for the analysis of continuous responses assuming the responses are complete. As indicated in Chapter 2, we will come across situations where the responses may be censored. Here we describe extensions of the sign and Wilcoxon signed rank tests, proposed for the censored data situations.

3.5.1 A sign test for censored data

When the observations on X and Y were complete, we considered the differences $X - Y$ and used the signs and magnitude of these differences to construct the tests. In this section also we consider the differences $D = X - Y$. The median of the D is Δ and the null hypothesis is that $\Delta = 0$. The various alternatives are defined in Table 3.8.

Finding the signs of the D's is equivalent to establishing an order between X and Y. This task of establishing order can be done through the use of the sign function, defined as follows.

$$sgn(x, y) = \begin{cases} 1 & \text{if } x \text{ is decidedly greater than } y; \\ -1 & \text{if } x \text{ is decidedly less than } y; \\ 0 & \text{otherwise.} \end{cases}$$

Using this function we compare each X_i with the corresponding Y_i. Now we compute the statistic

$$S = \sum_{i=1}^{n} sgn(X_i, Y_i),\qquad (3.43)$$

and use it as a test criterion. The mean of this statistic under the null hypothesis is zero. Further, this statistic is equal to

$$S = U_+ - U_-,\qquad (3.44)$$

where $U_+ (U_-)$, is the number of pairs for which $sgn(x_i, y_i)$ is 1 (the number of pairs for which $sgn(x_i, y_i)$ is -1). The test statistic is

$$Z = \frac{(U_+ - U_-)}{(U_+ + U_-)^{1/2}}.\qquad (3.45)$$

Under certain assumptions about the censoring distributions, the null distribution of this statistic can be approximated by the standard normal distribution. So critical regions of approximate size α can be constructed. These are listed in Table 3.12.

This test was proposed by Wei (1980) and also by Woolson and Lachenbruch (1980). For further details see these references.

Remark 3.9. When all observations are complete and when there are no zeros among D_i, the statistics U_+ and U_- are equal to S_n and $n - S_n$, respectively, of Subsection 3.2.1. Thus the statistic Z of (3.45) reduces to $\frac{(S_n - (n/2))}{\sqrt{(n/4)}}$, which is the standardized version of S_n. Thus the above test is a generalization of the sign test considered earlier.

3.5.2 A generalized signed rank test

Woolson and Lachenbruch (1980) proposed various rank tests for censored matched pairs. One of them is a generalization of the Wilcoxon signed rank test. It will be described now. We consider the right-censored data case. In these cases we have four types of observations: (1) X and Y are uncensored; (2) X is uncensored and Y is right censored; (3) X is right censored and Y is uncensored; and (4) X and Y are both right censored. Here a common censoring time is assumed and additional followup is ignored. Consequently, we will not consider pairs of observations where $min(x, y)$ is censored. As before, the variable of interest is $D = X - Y$. A shift model is assumed. Under this model the variable D is distributed like $W + \Delta$, where W is a continuous random variable with a distribution symmetric about zero. The null hypothesis is

$$H_0 : \Delta = 0.\qquad (3.46)$$

For observations of type (1), the true value of D is known, and for observations of type (2) and type (3) the value of D is censored. However, for observations of type (4), the value of D is unknown. Further, the sign of D is unknown for the type (4) pair, whereas the sign is known for the other types of pairs. Let k denote the number of pairs of type (1) and let the values of D for these pairs be $D_1, D_2, \ldots, D_k$. We denote their ordered absolute values by $Z_{(1)}, < Z_{(2)} < \ldots < Z_{(k)}$. Let $Z_{(0)} = 0$ and we denote $+\infty$ by $Z_{(k+1)}$. Let n_j denote the number of the absolute values of the censored type (2) or type (3) differences in the interval $[Z_{(j)}, Z_{(j+1)})$, for $j = 0, 1, \ldots, k$. Let p_j be the number of n_j censored differences of type (1) and type (2), which are positive. Further, let

$$m_j = \sum_{l=j}^{k}(n_j + 1), \quad P_j = \prod_{l=1}^{j} \frac{m_l}{(m_l + 1)}. \tag{3.47}$$

Let

$$\delta_j = \begin{cases} 1 & \text{if } D_j > 0; \\ -1 & \text{if } D_j < 0. \end{cases}$$

The test statistic is the sum of the scores attached to the absolute values of the differences. It turns out to be (see Woolson and O'Gorman (1992), p. 196)

$$T = T_u + T_c, \tag{3.48}$$

where

$$T_u = \sum_{j=1}^{k}(2\delta_j - 1)(1 - P_j),$$

and

$$T_c = \frac{1}{2}(2p_0 - n_0) + \sum_{j=1}^{k}(2p_j - n_j)(1 - \frac{1}{2}P_j). \tag{3.49}$$

The critical regions are defined by the standardized variable

$$Z_T = \frac{T}{\sqrt{\sigma_T^2}},$$

where

$$\sigma_T^2 = \sum_{j=1}^{k}(1 - P_j)^2 + \sum_{j=1}^{k} n_j[1 - (P_j/2)]^2 + \frac{1}{4}n_0. \tag{3.50}$$

The distribution of Z_T is approximated by the standard normal distribution. The critical regions and the P-values are listed in Table 3.13, where $Z_T(obs)$ is the observed value of Z_T.

Table 3.13 *Sign tests for censored data*

Alternative	Critical Region
H_{A1}	$Z > z_{1-\alpha}$
H_{A2}	$Z < z_{\alpha}$
H_{A3}	$|Z| > z_{\alpha/2}$

Table 3.14 *Artificial censored paired data*

Pair #	1	2	3	4	5	6
X	30	25	42+	35	41	28
Y	40+	20	36	38	34	26
D	$-4+$	5	6+	-3	7	2
$sgn(X,Y)$	-1	1	1	-1	1	1

Table 3.15 *Generalized signed rank tests*

Alternative	Critical Region	P-Value
$H_{A1} : \Delta > 0$	$Z_T > z_{(1-\alpha)}$	$P(N(0,1) > Z_T(obs))$
$H_{A2} : \Delta < 0$	$Z_T < z_{\alpha}$	$P(N(0,1) < Z_T(obs))$
$H_A : \Delta \neq 0$	$|Z_T| > z_{(1-\alpha/2)}$	$2P(N(0,1) > |Z_T(obs)|)$

Remark 3.10. When there is no censoring, $T_c = 0$ and T_u reduces to the usual Wilcoxon signed-rank statistic. So T_c can be viewed as the sum of scores for the censored differences, and T_u can be viewed as the sum of the scores for the uncensored differences.

Example 3.6. We illustrate the two methods of testing $\Delta = 0$ using an artificial censored data set given in Table 3.14.

Sign test
  From the values of sgn, we have $U_+ = 4$ and $U_- = 2$. Hence the test statistic (3.45) for the sign test is

$$Z = \frac{4 - 2}{\sqrt{(4 + 2)}} = 0.8165.$$

The approximate P-value for a two-sided alternative is 0.4142. Thus we do not reject the hypothesis $\Delta = 0$.

Generalized signed rank test
  Here $k = 4$ and the remaining details are given in Table 3.15.

Table 3.16 *Generalized signed rank test calculations*

j	0	1	2	3	4	5
$Z_{(j)}$	0	2	3	5	7	∞
n_j	0	0	1	1	0	-
p_j	0	0	0	1	0	-
$n_j + 1$	-	1	2	2	1	-
m_j	-	6	5	3	1	-
$\frac{m_j}{m_j+1}$	-	(6/7)	(5/6)	(3/4)	(1/2)	-
P_j	-	(6/7)	(5/7)	(15/28)	(15/56)	-
δ_j	-	1	-1	1	1	-

Now

$$T_u = (1)\left(\frac{1}{7}\right) + (-3)\left(\frac{2}{7}\right) + (1)\left(\frac{13}{28}\right) + (1)\left(\frac{41}{56}\right) = 0.4821.$$

$$T_c = \frac{1}{2}(0) + 0 + (-1)\left(1 - \frac{5}{14}\right) + (1)\left(1 - \frac{15}{56}\right) = 0.0893.$$

Hence

$$T = 0.4821 + 0.0893 = 0.5714.$$

Now we calculate the variance estimate:

$$\sigma_T^2 = \left[(1/7)^2 + (2/7)^2 + (13/28)^2 + (41/56)^2\right]$$
$$\left[0 + (1)(9/14)^2 + (1)(41/56)^2 + 0\right] + 0 = 1.8029.$$

Hence the test statistic is

$$Z_T = \frac{0.5714}{\sqrt{1.8029}} = 0.4255.$$

The approximate *P*-value for two-sided alternatives is 0.6704 and we do not reject the null hypothesis $\Delta = 0$.

3.5.3 Paired Prentice-Wilcoxon test

Several other tests for paired censored data are available, and Woolson and O'Gorman (1992) made a comparative study. They recommended the use of the paired Prentice-Wilcoxon test proposed by O'Brien and Fleming (1987). This is a generalization of the procedure discussed in Subsection 3.3.3 for the censored data, where instead of ranks, Prentice-Wilcoxon scores are used. For further details, see O'Brien and Fleming (1987).

3.6 Appendix A3: Mathematical supplement

A3.1 Maximum likelihood estimation of θ_{10}

In the clinical equivalence problem, we need an MLE of θ_{10} under the null hypothesis $\Delta = \Delta_0$. We combine the two frequencies m_{00} and m_{11} and then the likelihood is

$$L = c(m_{ij})\theta_{10}^{m_{10}}\theta_{01}^{m_{01}}(1 - \theta_{10} - \theta_{01})^{(n - m_{10} - m_{01})}, \tag{A1}$$

so that the log of the likelihood, under H_0, is

$$logL = logc(m_{ij}) + m_{10}log\theta_{10} + m_{01}log(\theta_{10} + \Delta_0)$$
$$+(n - m_{10} - m_{01})log(1 - 2\theta_{10} - \Delta_0).$$

Here we are assuming that both m_{10} and m_{01} are positive. Taking the derivative of $logL$ with respect to θ_{10} and equating it to zero, we get the equation

$$\frac{m_{10}}{\theta_{10}} + \frac{m_{01}}{\theta_{10} + \Delta_0} - \frac{2(n - m_{10} - m_{01})}{1 - 2\theta_{10} - \Delta_0} = 0. \tag{A2}$$

This equation is the same as

$$\frac{m_{10}(\theta_{10} + \Delta_0) + m_{01}\theta_{10}}{\theta_{10}(\theta_{10} + \Delta_0)} - \frac{2(n - m_{10} - m_{01})}{1 - 2\theta_{10} - \Delta_0} = 0.$$

We can rewrite this equation as

$$(1 - 2\theta_{10} - \Delta_0)[m_{10}\Delta_0 + (m_{10} + m_{01})\theta_{10}] - 2(n - m_{10} - m_{01})\theta_{10}(\theta_{10} + \Delta_0) = 0.$$

It is easy to see that this equation is a quadratic equation, and to calculate the appropriate root, we state it as

$$a\theta_{10}^2 + b\theta_{10} + c = 0,$$

where

$$a = 2n, \ b = (2n + m_{01} - m_{10})\Delta_0 - (m_{01} + m_{10}),$$

and

$$c = -m_{10}\Delta_0(1 - \Delta_0). \tag{A3}$$

We need to take the positive solution, namely,

$$\check{\theta}_{10} = [\sqrt{(b^2 - 4ac)} - b]/(2a). \tag{A4}$$

A3.2 Approximate variance of $\hat{\phi}$

First we note that

$$var(\hat{\phi}) = var(log\,m_{1.}) + var(log\,m_{.1}) - 2cov(log\,m_{1.}, log\,m_{.1}). \tag{A5}$$

Now we compute approximations to the three terms on the right side of (A5). Following the delta method discussed in Appendix A1 in Chapter 1, we have

$$var(log\, m_{1.}) \approx var(m_{1.})/[E(m_{1.})]^2 = (1 - \theta_1)/n\theta_1. \qquad (A6)$$

We also have

$$var(log\, m_{.1}) \approx (1 - \theta_2)/n\theta_2. \qquad (A7)$$

Now

$$cov(log\, m_{1.}, log\, m_{.1}) \approx cov(m_{1.}/n\theta_1, m_{.1}/n\theta_2) = cov(m_{1.}, m_{.1})/(n^2\theta_1\theta_2). \qquad (A8)$$

It can be shown that

$$cov(m_{1.}, m_{.1}) = n(\theta_{11}\theta_{00} - \theta_{10}\theta_{01}). \qquad (A9)$$

The proof of this result is left as a problem.

Using the results (A6), (A7), (A8) and (A9) in (A5) and after some algebraic simplification we get

$$var(\hat{\phi}) \approx (\theta_{10} + \theta_{01})/(n\theta_1\,\theta_2). \qquad (A10)$$

Using the estimates of θ's, we get an estimate of the approximate variance as

$$s^2 = (m_{10} + m_{01})/m_{1.}m_{.1}. \qquad (A11)$$

A3.3 Symmetric property of the distribution of W

Recall that $W = U - V$, where the joint pdf of U and V is $h_0(u, v)$, which satisfies the relation $h_0(u, v) = h_0(v, u)$, for all (u, v). Now

$$P(W \le w) = \iint_{(x-y\le w)} h_0(x, y)dxdy$$

$$= \iint_{(y'-x'\le w)} h_0(y', x')dy'dx'$$

$$= \iint_{(y'-x'\le w)} h_0(x', y')dx'dy'$$

$$= P(-W \le w).$$

Thus the variables W and $-W$ follow the same distribution. In other words, the distribution of W is symmetric about zero, when $h_0(u, v)$ is a symmetric function of its arguments.

A3.4 Mean and variance of V_+, under the null hypothesis

To get a different expression for the statistic V_+, let us consider the variables

$$\psi_i = \begin{cases} 1, & \text{if } D_i > 0, \\ 0, & \text{if } D_i < 0. \end{cases}$$

Let R_i be the rank of $|D_i|$ among $|D_j|$'s. Then the test statistic is

$$V_+ = \sum_{i=1}^{n} \psi_i R_i, \tag{A12}$$

and $P(\psi_i = 1|H_0) = (1/2)$. Recall that $L(.)$ is the cdf of D_i, under H_0. Now we establish the *independence* of ψ_i and $|D_i|$. To show this we consider the joint distribution under H_0. For positive d, we have

$$
\begin{aligned}
P(\psi_i = 1, |D_i| \le d|H_0) &= P(0 < D_i \le d|H_0)\\
&= L(d) - L(0) = L(d) - (1/2)\\
&= (1/2)[2L(d) - 1] = (1/2)[L(d) - (1 - L(d))]\\
&= (1/2)[L(d) - L(-d)]\\
&= P(\psi_i = 1|H_0)P(|D_i| \le d|H_0).
\end{aligned}
$$

Using this independence property, we calculate the conditional mean and the conditional variance of V_+ given the rank vector $\mathbf{R}$. It is easy to see that

$$
\begin{aligned}
E(V_+|\mathbf{R}, H_0) &= \sum_{i=1}^{n} R_i E(\psi_i|H_0) = (1/2)\sum_{i=1}^{n} R_i\\
&= (1/2)\left(\sum_{i=1}^{n} i\right) = (1/2)[n(n+1)/2] = n(n+1)/4.
\end{aligned}
$$

Further,

$$var(V_+|\mathbf{R}, H_0) = \sum_{i=1}^{n} R_i^2 Var(\psi_i|H_0) = (1/4)\sum_{i=1}^{n} R_i^2.$$

From the conditional mean, we get

$$E(V_+|H_0) = n(n+1)/4. \tag{A13}$$

To obtain the variance, we need to use the formula connecting the conditional variance and the unconditional variance. Now

$$
\begin{aligned}
var(V_+|H_0) &= E[var(V_+|R_i, H_0)] + var[E(V_+|R_i, H_0)]\\
&= (1/4)E\left[\sum_{i=1}^{n} R_i^2|H_0\right] + 0\\
&= (1/4)\sum_{i=1}^{n} i^2 = [n(n+1)(2n+1)/24].
\end{aligned}
$$

A3.5 Statistic V_+ expressed in terms of Walsh averages

Let us assume that there are no zeros and no ties among the D values. Let the ordered D values be $D_{(1)} < D_{(2)} < \cdots < D_{(m)} < 0 < D_{(m+1)} < \cdots < D_{(n)}$. Note that we have m negative D values. Let $r_1 < r_2 < \cdots < r_{n-m}$ be the ranks of the absolute values of the positive D values, which are $D_{(m+1)}, D_{(m+2)}, \ldots$, $D_{(n)}$. It is easy to see that, for $j = 1, \ldots, n - m$,

$$r_j = j + [number\ of\ D_{(1)} < D_{(2)} < \cdots < D_{(m)}\ with\ |D_{(k)}| < D_{(m+j)}]$$

$$= j + \sum_{k=1}^{m} \psi(D_{m+j} + D_{(k)}),$$

where $\psi(t) = 1$, *for $t > 0$*, and $\psi(t) = 0$, *otherwise*. The Wilcoxon signed rank statistic is $V_+ = \sum_{j=1}^{n-m} r_j$ and this sum is

$$\sum_{j=1}^{n-m} r_j = \sum_{j=1}^{n-m} j + \sum_{k=1}^{m} \sum_{j=1}^{n-m} \psi(D_{(m+j)} + D_{(k)}) \equiv T_1 + T_2.$$

However,

$$T_2 = \sum_{k=1}^{m} \left[\sum_{j=k}^{n} \psi(D_{(k)} + D_{(j)}) - \sum_{j=k}^{m} \psi(D_{(k)} + D_{(j)}) \right].$$

It can be seen that the second sum on the right is zero since each $D_{(k)} + D_{(j)}$ is negative for $1 \leq k < j \leq m$. Further, the sum T_1 can be identified as $\sum_{k=m+1}^{n} \sum_{j=k}^{n} \psi(D_{(k)} + D_{(j)})$. Finally, we have

$$V_+ = \sum_{k=1}^{n} \sum_{j=k}^{n} \psi(D_{(k)} + D_{(j)})$$

$$= \sum_{k=1}^{n} \sum_{j=k}^{n} \psi(D_k + D_j)$$

$$= \sum_{k=1}^{n} \sum_{j=k}^{n} \psi[(D_k + D_j)/2].$$

The right-hand side sum is the number of positive Walsh averages.

A3.6 Some general results about $E(V_+)$

Let $L(.)$ be the distribution function of $W = D - \Delta$, which is symmetric about zero. So the distribution function of D is

$$P[D \leq d|\Delta] = L(d - \Delta).$$

Hence

$$\pi_1(\Delta) = P(D > 0|\Delta) = 1 - L(-\Delta) = L(\Delta).$$

This implies that for $\Delta > 0$,

$$\pi_1(\Delta) > L(0) = \pi_1(0).$$

Further,

$$\begin{aligned}
\pi_2(\Delta) &= P[D_i + D_j > 0|\Delta] \\
&= P[D_i - \Delta > -2\Delta - (D_j - \Delta)|\Delta] \\
&= 1 - P[D_i - \Delta \le -2\Delta - (D_j - \Delta)|\Delta] \\
&= 1 - \int L(-2\Delta - x)L'(x)dx = 1 - \int [1 - L(x + 2\Delta)]L'(x)dx \\
&= \int L(x + 2\Delta)L'(x)dx,
\end{aligned}$$

where $L'(.)$ is the density function of the distribution function $L(.)$. From this result we have, for $\Delta > 0$,

$$\pi_2(\Delta) > \pi_2(0) = (1/2).$$

Using these orderings, it follows that, for $\Delta > 0$,

$$E[V_+|H_{A1}] > E[V_+|H_0].$$

A similar analysis will give the result, that for $\Delta < 0$,

$$E[V_+|H_{A2}] < E[V_+|H_0].$$

A3.7 Confidence interval for Δ, using Wilcoxon signed rank test

A confidence interval for Δ can be obtained from the acceptance region of the Wilcoxon test of $H_0 : \Delta = \Delta_0$ against the two-sided alternative $H_A : \Delta \ne \Delta_0$. This testing problem is equivalent to testing $H_0 : \Delta^* = 0$ against $H_{A3} : \Delta^* \ne 0$, in relation to $D_i^* = D_i - \Delta_0$. Let the statistic V_+ computed from the transformed data, D_i^*, be V_+.

It is convenient to restate the test statistic in terms of Walsh averages. So we recall the representation of V_+ in terms of the Walsh averages $A_{ij} = (D_i + D_j)/2$, which is

$$V_+ = A_+; \tag{A14}$$

where A_+ is the number of positive Walsh averages A_{ij}. The Walsh averages, A_{ij}^*, for D_i^* are related to the averages A_{ij} and this relationship is

$$A_{ij}^* = A_{ij} - \Delta_0. \tag{A15}$$

The quantity $A_\dagger^*$, which is the number of positive averages A_{ij}^*, is equal to the number of averages A_{ij} that are greater than Δ_0. Since $V_\dagger = A_\dagger^*$, the acceptance region of the Wilcoxon test can be stated as

$$c_3 + 1 \leq V_\dagger \leq [n(n+1)/2] - c_3 - 1. \qquad (A16)$$

This region can be restated as

$$l \leq A_\dagger^* \leq [n(n+1)/2] - l, \qquad (A17)$$

where $l = c_3 + 1$ and c_3 is the largest integer such that

$$P(V_+ \leq c_3 | H_0) = \alpha/2. \qquad (A18)$$

Let M denote the number of Walsh averages, so that $M = [n(n+1)/2]$. Thus all values Δ_0 that satisfy (A17) will not be rejected by the test. It is possible to reinterpret the acceptance set as an interval.

Let $A_{(k)}$ denote the ordered values of the averages A_{ij}. When the left-side equality holds, there are exactly l of the averages $(D_i + D_j)/2$ that are greater than Δ_0. This configuration means that the l largest averages are above Δ_0, which is the same as $A_{(M-l+1)} > \Delta_0$. When the right side holds, we have $A_{(l)} < \Delta_0$. This observation leads to the conclusion that all Δ_0 values in the interval $[A_{(l)}, A_{(M-l+1)})$ will be accepted. In other words, this interval is a confidence interval for Δ, with confidence coefficient $1 - \alpha$.

3.7 Appendix B3: Computer programs

B3.1 McNemar test

We calculate the McNemar test. We use the data of Example 3.1. In the output the statistic is denoted by S. It is the square of Z of (3.7). The P-value is for the two-sided alternatives.

```
data one;
input x y freq;
lines;
0 0 18
0 1 4
1 0 12
1 1 5
;
proc freq data=one;
weight freq;
tables x*y/agree norow nocol nopercent;
run;
```

Output
_____

The FREQ Procedure

Table of x by y

```
   x            y

   Frequency|        0|        1|  Total
   ---------+---------+---------+
         0 |    18 |     4 |     22
   ---------+---------+---------+
         1 |    12 |     5 |     17
   ---------+---------+---------+
   Total          30        9       39
```

Statistics for Table of x by y

McNemar's Test

```
Statistic (S)    4.0000
DF                    1
Pr > S           0.0455
```

B3.2 Confidence interval for the ratio ψ

Here we calculate the confidence interval for the ratio of marginal probabilities. We use the data of Example 3.3.

```
data one;
cc=0.95;*confidence coefficient*
alpha=1-cc;
p2=(350/500);
p1=(420/500);
r=p1/p2;
lr=log(r);
var=(90+20)/(350*420);
s=sqrt(var);
z=probit(1-alpha/2);
phiL1=lr-z*s;
phiU1=lr+z*s;
phiL=round(phil1,.0001);
phiU=round(phiU1,.0001);
psiL1=exp(phiL1);
psiU1=exp(phiU1);
```

```
psiL=round(psiL1,.0001);
psiU=round(psiU1,.0001);
phi=round(r,.0001);
psi=round(lr,.0001);
proc print;
title 'Data of Example 3.3';
var p1 p2 cc alpha ;
proc print ;
title 'Computed statistics';
var phi psi var s ;
proc print ;
title 'confidence interval for phi';
var phiL phiU;
proc print;
title 'confidence interval for psi';
var psiL psiU;
run;
```

Output

Data of Example 3.3

p1	p2	cc	alpha
0.84	0.7	0.95	0.05

Computed statistics

phi	psi	var	s
1.2	0.1823	.000748299	0.027355

confidence interval for phi

phiL	phiU
0.1287	0.2359

confidence interval for psi

psiL	psiU
1.1374	1.2661

B3.3 Confidence interval for risk difference Δ

Here we are computing a confidence interval for the difference between two
correlated proportions. The method has been proposed by Tango (1999). This
is an iterative method. The secant method is used to solve a nonlinear equation
in delta, which is the parameter of interest. The calculations are done using
the notation in Tango (1999). First, we enter data as per the notation in the
text and relate them to b and c. We use the data of Example 3.1.

```
proc iml;
alpha=0.05;
cc=1-alpha;
m01=4;
b=m01;
m10=12;
c=m10;
m00=18;
m11=5;
n=m00+m11+b+c;
Deltahat=(m01-m10)/n;
print 'Results for Data of Example 3.1';
print 'Data and specifications';
print m00 m01 m10 m11 n Deltahat cc;
start tango(z) global(n,k,b,c,p);
k=z*(sqrt(n));
p=2*n;
x1=0.5;
x2=(b-c-0.5)/n;
f1=fun(x1);
do inc=1 to 10
  while(abs(f1) > 0.00005);
  f2=fun(x2);
  h=-f2*(x2-x1)/(f2-f1);
  x1=x2;
  x2=x2+h;
  f1=f2;
end;
return (x2);
finish tango;

start fun(x) global(n,k,b,c,p);
q=-b-c+(p-b+c)*x;
 r=-x*(1-x);
d=(sqrt(q*q-4*p*r*c)-q)/(2*p);
if 2*d-r < 0 then t=0;
```

```
else t=k*sqrt(2*d-r);
f=(b-c-n*x)-t;
return(f);
finish fun;
print 'confidence limits for Delta';
do;
z=probit(0.025);
 xs= tango(z);
 UL=round(xs,0.0001);
 print "Upper limit" UL;
end;
do;
z=probit(0.975);
xs= tango(z);
LL=round(xs,0.0001);
print "Lower limit" LL;
end;
quit;
```

Output

```
                    Results for Data of Example 3.1
                        Data and specifications
        MOO   MO1    M10   M11   N    DELTAHAT    CC

        18    4      12    5     39   -0.205128   0.95


                    confidence limits for Delta
                                            UL
                        Upper limit     -0.0044

                                            LL

                        Lower limit     -0.3926
```

B3.4 Sign and signed rank procedures

Using the data from Table 3.9, we illustrate the calculation of the sign test and the signed rank test. We create the D values and omit zero differences. If there are ties among $|D|$, average ranks are used. We calculate U to be 1 or -1 depending on whether D is positive or negative. SRANK are signed ranks. The calculations follow the steps outlined in the text. For the Wilcoxon signed rank, the normal approximation for the P-value is given. We can obtain some of this output from the univariate procedure, but we have to transform the statistics of the output to get the statistics given in the text.

```
data one;
input X Y @@;
d=X-Y;
lines;
1.28 1.33 1.60 2.21 2.46 2.43 1.41 1.81 1.40 0.85
1.12 1.20 0.90 0.90 2.41 2.79
;
data two;
set one;
if d=0 then delete;
ABSD=ABS(d);
U=SIGN(d);
sd=(u+1)/2;
proc rank data=two out=C;
     var ABSD;
     ranks RABSD;
data three;
     set C;
     SRANK=RABSD*U;
     ws=rabsd*sd;
proc print data=three;
var x y d ;
title 'data';
proc print data=three;
var u sd srank ws;
title ' signs and signed ranks ';
proc summary data=three ;
 var ws ;
output out=c1 sum=wsign n=n1;
proc summary data=three;
var sd;
output out=c2 sum=sign n=n2;
data four;
set c1;
mu=n1*(n1+1)/4;
var=n1*(n1+1)*(2*n1+1)/24;
sd=sqrt(var);
z=(wsign-mu)/sd;
p2=2*(1-probnorm(abs(z)));
data five;
set c2;
```

```
p31=probbnml(1/2,n2,sign);
p32=1-probbnml(1/2,n,sign-1);
p3=2?*min(p31,p32);
proc print data=five;
var n2 sign p3;
title 'results of sign test';
 proc print data=four;
var n1 wsign p2;
title 'results of Wilcoxon test';
proc univariate data=two;
var d;
 run;
```

Output

data

X	Y	d
1.28	1.33	-0.05
1.60	2.21	-0.61
2.46	2.43	0.03
1.41	1.81	-0.40
1.40	0.85	0.55
1.12	1.20	-0.08
2.41	2.79	-0.38

signs and signed ranks

U	sd	SRANK	ws
-1	0	-2	0
-1	0	-7	0
1	1	1	1
-1	0	-5	0
1	1	6	6
-1	0	-3	0
-1	0	-4	0

results of sign test

Obs	n2	sign	p3
1	7	2	0.45313

results of Wilcoxon test

Obs	n1	wsign	p2
1	7	7	0.23672

The UNIVARIATE Procedure
Variable: d

Tests for Location: Mu0=0

Test	-Statistic-		-----p Value------	
Student's t	t	-0.93704	Pr > |t|	0.3849
Sign	M	-1.5	Pr >= |M|	0.4531
Signed Rank	S	-7	Pr >= |S|	0.2969

B3.5 Rank transformed t-test

Here we calculate the rank transformed *t*-test. We use the data of Example 3.4.

```
data one;
input x y;
lines;
1.28 1.33
1.60 2.21
2.46 2.43
1.41 1.81
1.40 0.85
1.12 1.20
0.90 0.90
2.41 2.79
;
proc iml;
use one;
read all variables{x} into M1;
read all variables{y} into M2;
z=j(16,1,0);
z[1:8,1]=M1[1:8,1];
z[9:16,1]=M2[1:8,1];
create two from z[colname={'Z'}];;
  append from Z;
quit;
```

```
proc rank data=two out=three;
var Z;
ranks rz;

proc print data=three;
proc iml;
use three;
read all variables{rz} into M4;
R=j(8,2,0);
R[1:8,1]=M4[1:8,1];
R[1:8,2]=M4[9:16,1];
print R;
D=j(8,1,0);
print R;
D=j(8,1,0);
do i=1 to 8;
d[i,1]=R[i,1]-R[i,2];
end;

create four from D [colname={'D'}];
    append from D;
quit;
proc print data=four;
proc means mean std stderr t prt;
var D;
run;
```

Output
──────
The set Z and their ranks are as follows.

Z	rz
1.28	6.0
1.60	10.0
2.46	15.0
1.41	9.0
1.40	8.0
1.12	4.0
0.90	2.5
2.41	13.0
1.33	7.0
2.21	12.0
2.43	14.0
1.81	11.0
0.85	1.0

```
                              1.20      5.0
                              0.90      2.5
                              2.79     16.0
```

The ranks are paired.

```
                               R
                               6        7
                              10       12
                              15       14
                               9       11
                               8        1
                               4        5
                              2.5      2.5
                              13       16
```

The rank differences D are calculated.

```
                          Obs      D
                           1      -1
                           2      -2
                           3       1
                           4      -2
                           5       7
                           6      -1
                           7       0
                           8      -3
```

The results of the t-test on D are calculated.

```
                        The MEANS Procedure
                       Analysis Variable : D
```

Mean	Std Dev	Std Error	t Value	Pr > |t|
-0.1250000	3.1367636	1.1090134	-0.11	0.9134

B3.6 Confidence interval for Δ difference in means

We are observing pairs (X, Y) and X has the same distribution as $Y - \Delta$. We want a confidence interval for Δ. This procedure is related to the Wilcoxon signed rank procedure. We use the paired data given in Table 3.9 of Desu and Raghavarao (1990). The approximation for c given by (3.40) is used. cc is the confidence coefficient. Below we calculate integers l and u. The confidence interval is the open interval from a(l) to a(u). We enter the differences as x and y in order to calculate the Walsh averages a=(x+y)/2.

```
data aver;
n=7;
input x1-x7 y1-y7;
array X(I) x1-x7;
array Y(J) y1-y7;
do I = 1 to n;
do J= i to n;
a=(x+y)/2;output;
end;
end;
cards;
-0.05 -0.61 0.03 -0.40 0.55 -0.08 -0.78
-0.05 -0.61 0.03 -0.40 0.55 -0.08 -0.78
  ;
proc sort data=aver(keep=a) out=two;
by a;
run;
```

We calculate the order statistics we need to construct the end points of the confidence interval.

```
data one;
n=7;
cc=0.95;
 mean=n*(n+1)/4;
 M=2*mean;
 var=n*(n+1)*(2*n+1)/24;
 sd=sqrt(var);
 alpha=(1-cc);
 c1= mean +sd*probit(alpha/2);
 l=floor(c1)+1;
 c=l-1;
 u1=M-l+1;
 u=floor(u1);
 call symput('ll',l);
 call symput('uu',u);
 proc print;
 title 'Sample size cc No.of Averages critical value';
 var n cc M c ;
 proc print;
 title 'Required OS of Walsh Averages';
 var l u ;
 run;
  data four;
 set two;
  if( _n_=symget('ll') or _n_=symget('uu'));
 run;
```

```
proc transpose data=four out=output(rename=(col1=lower
                            col2=upper)drop=_name_);
run;
proc print data=output;
title 'end points of confidence interval';
run;
```

Output

Sample size cc No.of Averages
critical value

n	cc	M	c
7	0.95	28	2

Required OS of Walsh Averages

l	u
3	26

end points of confidence interval

lower	upper
-0.61	0.25

3.8 Problems

1. A study to evaluate the performance of a screening test relative to a standard method was undertaken. Each individual is given test 1(QS test) and the standard culture test. The tests indicate the presence (1) or absence (0) of group B streptococcus (GBS) bacteria in the genital tract of women. We are observing a bivariate binary response. The resulting frequencies are listed in Table 3.17. Is there reason to believe that test 1 is as good as the standard test for detecting the presence of the bacteria GBS? (These data have been derived from the data reported by Viana and Pereira (2000).)

2. Prove the result (A9).

Table 3.17 *Results of two diagnostic tests*

Test 1	Culture Test	
	0	1
0	416	19
1	48	41

3. *Sample size for one-sided testing problem using McNemar's test:* Suppose we are testing the null hypothesis of equality, that is, $\Delta = 0$ against the alternative $\Delta > 0$. We want to determine the sample size so that an appropriate size α test has a power not less than $1 - \beta$ for $\Delta = \Delta_1 > 0$. Under H_0,

$$\theta_{10} = \theta_{01} = \theta,$$

where θ is an unknown. However, using the maximum likelihood method we get that an estimator $\theta_{ML} = (m_{01} + m_{10})/2n$. The null variance of $\hat{\Delta}$ is estimated by $v_0^2 = 2\theta_{ML}/n = v^2$. The test does depend on Z_Δ of (3.7). Using the approximate normal distribution, we equate the power function at Δ_1 to $1 - \beta$. This analysis is similar to the analysis of Subsection 3.1.5. We also use the fact that the large sample limit of v^2 is $\pi_d/n = (\theta_{01} + \theta_{10})/n$. The required sample size is given by

$$n = [z_\alpha \sqrt{\pi_d} + z_\beta \sqrt{\pi_d - \Delta_1^2}]^2 / \Delta_1^2,$$

(A useful reference in this connection is Royston (1993). For a slightly different formulation see Desu and Raghavarao (1990, p. 84).)

4. A study for comparing two laboratories with respect to the accuracy of the measurement of blood sugar levels was conducted. Blood samples for 10 normal individuals were collected. Each sample was split into two equal parts. Lab 1 analyzed one part and lab 2 analyzed the other part. The resulting data appear in Table 3.18. Is there evidence that the distributions of the blood sugar levels are different? (Use $\alpha = 0.05$.)

Table 3.18 *Lab determination of blood sugar levels*

Lab	Individuals									
	1	2	3	4	5	6	7	8	9	10
1	118	105	124	110	115	119	121	105	110	100
2	121	103	118	108	117	117	119	109	108	102

5. A company arranged a training program to increase the productivity of the workers, so that the workers have to spend less time on different projects. To evaluate the benefits of the program 6 workers were recruited for a study. The times (in minutes) to complete a task before the training and to complete a similar task after the training were collected. The data appear in Table 3.19. Do the data support that the training program is effective? (Use $\alpha = 0.05$.)

6. To test the effectiveness of a gas additive to improve the gas mileage (m/gl), 10 similar cars were chosen. The gas mileage for these 10 cars, without and with the additive, were determined. The same driver used the car under the two different conditions. The data appear in Table 3.20. Is there enough evidence that the gas additive increases the gas mileage? (Use $\alpha = 0.05$.)

Table 3.19 *Times to complete a task*

	Employees					
	1	2	3	4	5	6
Before	50	45	60	55	45	50
After	40	48	55	40	50	45

Table 3.20 *Gas mileage results*

Additive	Cars									
	1	2	3	4	5	6	7	8	9	10
Without	18	20	19	20	22	18	19	21	20	22
With	20	21	22	19	24	18	21	20	23	25

Table 3.21 *Gas prices on two holiday weekends*

	Gas Stations					
	1	2	3	4	5	6
Labor Day	1.48	1.50	1.45	1.55	1.55	1.40
Memorial Day	1.55	1.60	1.50	1.50	1.60	1.55

7. Gas prices of regular unleaded gas at six gas stations during Memorial Day weekend and Labor Day weekend in a year were gathered. The data appear in Table 3.21. Do the data indicate the price distributions are the same over the two weekends? (Use $\alpha = 0.05$.)

8. Develop the null distribution of V_+ for $n = 4$.

9. *Calculation of the standardized Wilcoxon statistic* Z_V^*: It is convenient to consider the signed ranks. These are $(2\psi_i - 1)R_i^*$, where ψ_i is 1 or 0 depending on whether D_i is positive or negative. Let T be the sum of these signed ranks. Observing that

$$T = 2\sum_i \psi_i R_i^* - \sum_i R_i^*,$$

verify that

$$T = 2[V_+^* - E(V_+^*|H_0)].$$

Recalling that $4var(V_+^*|H_0) = \sum_i (R_i^*)^2$, verify that

$$Z_V^* = \frac{T}{\sqrt{\sum_i (R_i^*)^2}}.$$

Table 3.22 *Age at onset of diabetes*

	Twin #									
	1	2	3	4	5	6	7	8	9	10
First	40	35	50	32	48	50+	45+	48	37	42
Second	42	45	48	45+	46	45	42	35	39	45

Instead of using Z_V^* for testing purposes, we can use

$$t_R = Z_V^* \sqrt{n-1}/[n - (Z_V^*)^2]^{1/2}.$$

The null distribution can be approximated by t-distribution with $(n-1)$ degrees of freedom. The univariate procedure of SAS uses this approximation for $n > 20$.

10. *Sample size determination*: Consider an α-level paired t-test for testing $H_0 : \Delta = 0$ against the alternative $H_{A1} : \Delta > 0$. The sample size needed for this test to have a power of at least $1 - \beta$ at the alternative $\Delta_1 (> 0)$ is

$$n_T = \frac{\sigma^2 (z_{1-\alpha} + z_\beta)^2}{\Delta_1^2},$$

where σ^2 is the variance of $D = X - Y$. Hence derive the sample sizes for the sign test and Wilcoxon signed rank test with the same specifications. (Hint: Use the normal approximation for the distribution of the t-statistic.)

11. Ten sets of twins born to parents of whom at least one is a diabetic were observed for the age of onset of diabetes. The data collected are given in Table 3.22. During the course of the investigation three twins died of other causes and hence the times for them are censored. Does this constitute evidence that the distributions of onset of diabetes are the same for each twin of a set?

3.9 References

Conover, W.J. and Iman, R.L. (1981). Rank transformations as a bridge between parametric and nonparametric statistics. *Amer. Statist.*, **35**, 124–129.

Desu, M.M. and Raghavarao, D. (1990). *Sample Size Methodology*, Academic Press, New York.

May, W.L. and Johnson, W.D. (1997). The validity and power of tests for equality of two correlated proportions. *Statist. Med.*, **16**, 1081–1096.

McCornack, R.L. (1965). Extended tables of the Wilcoxon matched pair signed rank statistic. *J. Amer. Statist. Assoc.*, **60**, 864–871.

McNemar, I. (1947). Note on the sampling error of the difference between correlated proportions or percentages. *Psychometrika*, **12**, 153–157.

Nam, J. (1994). Sample size requirements for stratified prospective studies with null hypothesis of non-unity relative risk using score test. *Statist. Med.*, **13**, 79–86.

Nam, J. (1997). Establishing equivalence of two treatments and sample size requirements in matched-pairs design. *Biometrics*, **53**, 1422–1430.

Newcombe, R.G. (1998). Improved confidence intervals for the difference between bionomial proportions based on paired data. *Statist. Med.*, **17**, 2635–2650.

Newcombe, R.G. (1999). Author's reply. *Statist. Med.*, **18**, 3513.

Noether, G.E. (1987). Sample size determination for some common nonparametric tests. *J. Amer. Statist. Assoc.*, **82**, 645-647.

O'Brien, P.C. and Fleming, T.R. (1987). A paired Prentice-Wilcoxon test for censored paired data. *Biometrics*, **43**, 169–180.

Patel, H.I. (1983). Use of baseline measurements in the two-period crossover design. *Comm. Statist.*, **A 12**, 2693–2712.

Peritz, E. (1971). Estimating the ratio of two marginal probabilities in a contingency table. *Biometrics*, **27**, 223–225; Correction (1971): *Biometrics*, **27**, 1104.

Pratt, J.W. (1959). Remarks on zeros and ties in the Wilcoxon signed-rank procedures. *J. Amer. Statist. Assoc.*, **54**, 655–667.

Royston, P. (1993). Exact conditional and unconditional sample size for pair-matched studies with binary outcome: A practical guide. *Statist. Med.*, **12**, 699–712.

Stuart, A. (1955). A test for homogeneity of the marginal distributions in a two-way classification. *Biometrika*, **42**, 412–416.

Tang, M.L., Tang, N.S., Chan, I., and Chan, B.P. (2002). Sample size determination for establishing equivalence/noninferiority via ratio of two proportions in matched-pair design. *Biometrics*, **58**, 957–963.

Tango, T. (1998). Equivalence test and confidence interval for the difference in proportions for the paired-sample design. *Statist. Med.*, **17**, 891–908.

Tango, T. (1999). Letter to the editor. *Statist. Med.*, **18**, 3511–3513.

Viana, M.A.G. and Pereira, C.A.De B. (2000). Statistical assessment of jointly observed screening tests. *Bio. J.*, **42**, 855–864.

Wei, L.J. (1980). A generalized Gehan and Gilbert test for paired observations that are subject to arbitrary right censorship. *J. Amer. Statist. Assoc.*, **75**, 634–637.

Woolson, R.F. and Lachenbruch, P.A. (1980). Rank tests for censored matched pairs. *Biometrika*, **67**, 597–606.

Woolson, R.F. and O'Gorman, T.W. (1992). A comparison of several tests for censored paired data. *Statist. Med.*, **11**, 193–208.

Wilcoxon, F., Katti, S.K., and Wilcox, R.A. (1973). Critical values and probability levels for the Wilcoxon rank sum test and the Wilcoxon signed rank test. *Selected Tables in Mathematical Statistics*, Vol. 1, American Mathematical Society, Providence, Rhode Island.

CHAPTER 4

Procedures for several independent samples

4.1 Introduction

In Chapter 2 we considered procedures for comparing two distributions, using independent samples. Here we consider a similar problem in relation to $k(> 2)$ distributions. We denote these distributions as $F_1, F_2, \ldots, F_k$. The general k-sample problem is that of testing the null hypothesis of homogeneity, namely,

$$H_0 : F_1 = F_2 = \ldots = F_k. \tag{4.1}$$

First we entertain all alternatives, which are called *global alternatives*. These are stated as

$$H_{GL} : not\ H_0,\ i.e.,\ there\ exists\ at\ least\ one\ pair\ (F_i, F_j)\ such\ that\ F_i \neq F_j. \tag{4.2}$$

This discussion will be followed by procedures for certain types of restricted alternatives, which will be defined now. Unlike the case of two distributions, here two types of one-sided alternatives, called *completely ordered alternatives* and *partially ordered alternatives*, are of interest. The completely ordered alternatives are

$$H_{CO1} : F_1\ st < F_2\ st < \ldots st < F_k, \tag{4.3}$$

or

$$H_{CO2} : F_1\ st > F_2\ st > \ldots st > F_k. \tag{4.4}$$

The partially ordered alternatives is a very large class. We consider only one type of partially ordered alternatives, which are called *simple tree alternatives*. These are appropriate ones to entertain when one of the treatments is a control treatment and we want to compare each of the other $k - 1$ treatments with the control. In this context, without loss of generality, F_1 is taken as the distribution of the response variable under the control treatment. The alternatives of interest are

$$H_{TR1} : F_1\ st < F_i,\ for\ at\ least\ one\ i(i = 2, \ldots, k), \tag{4.5}$$

or

$$H_{TR2} : F_1\ st > F_i,\ for\ at\ least\ one\ i(i = 2, \ldots, k). \tag{4.6}$$

It may be noted that for $k = 2$, H_{CO1} is the same as H_{TR1} and H_{CO2} is the same as H_{TR2}.

As in Chapter 2, we will consider both complete samples and samples with right-censored observations. In the case of complete samples, the two situations of interest are (1) binary response and (2) continuous response. Of course, in the discussion about samples that may contain some censored observations, the response variable is a continuous variable. Procedures for the binary studies and categorical data situations will be considered first. This discussion is limited to tests for global alternatives; however, the case of continuous response restricted alternatives also will be discussed. We will also consider complete and censored data settings with a continuous response variable.

4.2 Discrete responses

In this section we consider studies where the response is a discrete variable. The case of binary data will be discussed in Subsection 4.2.1. Subsection 4.2.2 deals with the analysis of categorical data with $c(> 2)$ categories.

4.2.1 Binary response studies

When the response variable is a binary variable, the data can be summarized as in Table 4.1.

Let

$$\theta_i = P(response\ is\ 1\ in\ pop.i), \quad for\ i = 1, 2, \ldots, k.$$

The null hypothesis can now be stated as

$$H_0 : \theta_1 = \theta_2 = \ldots = \theta_k. \tag{4.7}$$

We consider only the global alternatives (4.2), which can be restated as

$$H_{GL} : \theta_i \neq \theta_j, \quad for\ at\ least\ one\ pair\ (i, j). \tag{4.8}$$

As a first step, we estimate each θ_i and also Θ, the common value of θ's under H_0. The test criterion is a function of the differences between the individual

Table 4.1 *Sample frequencies*

Sample ID	No. of 0's	No. of 1's	Total
1	n_{10}	n_{11}	n_1
2	n_{20}	n_{21}	n_2
.	.	.	.
i	n_{i0}	n_{i1}	n_i
.	.	.	.
k	n_{k0}	n_{k1}	n_k
Total	$n_{.0}$	$n_{.1}$	N

$\hat{\theta}$'s and $\hat{\Theta}$. The ML estimates are

$$\hat{\theta}_i = n_{i1}/n_i; \quad \hat{\Theta} = n_{.1}/N. \tag{4.9}$$

The test criterion, proposed by Cochran (1952), is

$$T_C = \sum_{i=1}^{k} n_i(\hat{\theta}_i - \hat{\Theta})^2/[\hat{\Theta}(1 - \hat{\Theta})]. \tag{4.10}$$

The null distribution of this statistic can be approximated by a chi-square distribution with $(k-1)$ degrees of freedom. Thus a test of approximate size α for the two-sided alternatives (4.8) is to

$$reject\ H_0\ if\ T_C > \chi^2_{(k-1)}(1 - \alpha). \tag{4.11}$$

The P-value can be approximated as

$$P\text{-value} \approx P\big[\chi^2_{(k-1)} > T_C(obs)\big], \tag{4.12}$$

where $T_C(obs)$ is the observed value of the statistic T_C.

Remark 4.1. A simplified formula for T_C is useful. It is

$$T_C = \left[\sum_{i=1}^{k} n_{i1}\hat{\theta}_i - n_{.1}\hat{\Theta}\right]\bigg/[\hat{\Theta}(1 - \hat{\Theta})].$$

Example 4.1. Three pain killers A, B, and C were tested for their effectiveness in relieving pain in 20 minutes. There were 150 volunteers recruited and the pain killers were randomly assigned to the subjects so that each treatment was assigned to 50 people. The number of subjects reporting pain relief in 20 minutes were $n_A(= n_{11}) = 32, n_B(= n_{21}) = 24$, and $n_C(= n_{31}) = 13$. Do the data indicate that the three treatments are not equally effective?

We compute the (ML) estimates

$$\hat{\theta}_A = (32/50) = 0.64, \hat{\theta}_B = (24/50) = 0.48, \hat{\theta}_C = (13/50) = 0.26.$$

Further, here, $\hat{\Theta} = \frac{32+24+13}{150} = 0.46$. Thus the test statistic is

$$T_C = \frac{[\{32(0.64) + 24(0.48) + 13(0.26)\} - 69(0.46)]}{(0.46)(1 - 0.46)}$$

$$= \frac{3.64}{0.2484} = 14.65.$$

Thus an approximate P-value is 0.0007 and hence we conclude that the treatments have different effects in relieving pain in 20 minutes.

A FORTRAN program for analyzing binomial data using randomization tests for various restricted alternatives is given by Soms and Torbeck (1982). A recent work on testing for ordered alternatives is the paper by

Table 4.2 *Frequencies in various categories*

Sample ID	Category						Total
	1	2	$\ldots$	j	$\ldots$	c	
1	n_{11}	n_{12}	$\ldots$	n_{1j}	$\ldots$	n_{1c}	n_1
2	n_{21}	n_{22}	$\ldots$	n_{2j}	$\ldots$	n_{2c}	n_2
$\cdot$	$\cdot$	$\cdot$	$\cdot$	$\cdot$	$\cdot$	$\cdot$	$\cdot$
i	n_{i1}	n_{i2}	$\ldots$	n_{ij}	$\ldots$	n_{ic}	n_i
$\cdot$	$\cdot$	$\cdot$	$\cdot$	$\cdot$	$\cdot$	$\cdot$	$\cdot$
k	n_{k1}	n_{k2}	$\ldots$	n_{kj}	$\ldots$	n_{kc}	n_k
Total	$n_{.1}$	$n_{.2}$	$\ldots$	$n_{.j}$	$\ldots$	$n_{.c}$	N

Peddada et al. (2001). A reference in connection with the determination of sample sizes is Desu and Raghavarao (1990, p. 49).

4.2.2 Categorical data with c categories

It is obvious that the procedure for the binary case can also be used when the response can belong to two categories. In some cases we have to deal with response variables with more than two categories. Thus we now consider the general case where the response can belong to c categories. Here the data can be arranged in a $k \times c$ table, as in Table 4.2.

Let $\theta_{ij} = P(response\ belongs\ to\ cat.j\ in\ pop.i)$ for $j = 1, 2, \ldots, c$ and $i = 1, 2, \ldots, k$. The null hypothesis of homogeneity is

$$H_0 : \theta_{1j} = \theta_{2j} = \ldots = \theta_{kj} = \Theta_j, \quad for\ j = 1, 2, \ldots, c, \qquad (4.13)$$

where the Θ_j's are unknown constants. Here again we need to estimate the common values, Θ_j, assuming the null hypothesis is true. Using these estimates we compute the expected frequencies for the various cells and then compute a discrepancy measure, which is the test statistic. The ML estimates are

$$\hat{\Theta}_j = n_{.j}/N. \qquad (4.14)$$

The expected values of n_{ij}, under H_0, are $n_i \Theta_j$. The estimates of these quantities are called *expected frequencies* and they are

$$e_{ij} = n_i \hat{\Theta}_j = n_i n_{.j}/N. \qquad (4.15)$$

Now the test statistic is

$$T = \sum_{i=1}^{k} \sum_{j=1}^{c} \frac{(n_{ij} - e_{ij})^2}{e_{ij}}. \qquad (4.16)$$

This statistic is asymptotically distributed as a chi-square variable with $(k-1)(c - 1)$ degrees of freedom (see Appendix A4). A test of approximate size α

Table 4.3 *Duration of TV viewing among three age groups*

Age Group	Viewing Categories (hours)			Total
	1	2	3	
1	35(52.5)	95(67.5)	20(30)	150
2	60(35)	30(45)	10(20)	100
3	10(17.5)	10(22.5)	30(10)	50
Total	105	135	60	300

is to

$$\text{reject } H_0 \text{ if } T > \chi^2_{(k-1)(c-1)}(1-\alpha). \tag{4.17}$$

Here again an approximate P-value is

$$P\text{-value} \approx P[\chi^2_{(k-1)(c-1)} > T(obs)], \tag{4.18}$$

where $T(obs)$ is the observed value of the statistic T.

Example 4.2. We are interested in comparing the TV watching habits in different age groups, namely (less than 25), [25 to 50], and (above 50). These groups are denoted by 1, 2, and 3, respectively. Three samples of 150, 100, and 50 subjects were chosen from the three age groups. Each person was asked to report the average number of hours of TV watched per day. The responses were grouped into three categories: (1) less than 1, (2) 1 to 3, and (3) above 3. The data collected appear in Table 4.3.

Are there significant differences in the viewing habits among the age groups?

We will use the test (4.17). To calculate the statistic T, we need to calculate the expected frequencies, e_{ij}, and these are shown in parentheses in Table 4.3. Now the test statistic is

$$T = \frac{(35-52.5)^2}{52.5} + \frac{(95-67.5)^2}{67.5} + \cdots + \frac{(30-10)^2}{10} = 98.38.$$

The associated approximate P-value is less than 0.0001. So we conclude that the TV viewing habits are significantly different among the age groups.

Remark 4.2. When the expected frequency in any cell is less than 5, the chi-square approximation for the P-value should not be used. A method similar to Fisher's exact test as given in Appendix A2 of Chapter 2 is recommended. A computer program for doing these calculations is given in Appendix B4. This exact method can be implemented using SAS or StatXact packages, among others.

Table 4.4 *Data on continuous response*

Sample		
ID	Size	Observations
1	n_1	$X_{11}, X_{12}, \ldots, X_{1n_1}$
2	n_2	$X_{21}, X_{22}, \ldots, X_{2n_2}$
.	.	$\ldots$
i	n_i	$X_{i1}, X_{i2}, \ldots, X_{in_i}$
.	.	$\ldots$
k	n_k	$X_{k1}, X_{k2}, \ldots, X_{kn_k}$

4.3 Continuous responses with complete data

Here we consider studies in which the response variable is a continuous variable and there are no censored observations. It is convenient to display the data in a tabular form as in Table 4.4.

First we describe various tests in relation to *global alternatives*. These tests are called the *omnibus tests*. This discussion will be followed by *multiple comparison (test) procedures*.

4.3.1 Kruskal-Wallis test

An extension of the Wilcoxon rank sum statistic has been proposed by Kruskal and Wallis (1952). To compute this statistic, we rank all the observations in the combined sample. Tied observations are assigned average ranks. Let R_{ij} be the rank of the observation X_{ij}. Then we obtain the rank averages $\bar{R}_{i.}$ for the k samples. Further, we calculate the variance of all the ranks as

$$S_R^2 = \frac{1}{(N-1)} \sum_{i=1}^{k} \sum_{j=1}^{n_i} [R_{ij} - \bar{R}_{..}]^2, \qquad (4.19)$$

where N is the total sample size and $\bar{R}_{..}$ is the average of all ranks. We also calculate the statistic

$$B_R = \sum_{i=1}^{k} n_i [\bar{R}_{i.} - \bar{R}_{..}]^2, \qquad (4.20)$$

which is a measure of variability of the rank averages $\bar{R}_{i.}$. The proposed test criterion is

$$T_{KW} = B_R / S_R^2, \qquad (4.21)$$

and the procedure to test H_0 against the global alternatives is to

reject H_0 if $T_{KW} > c$,

where c is the appropriate percentile of the null distribution of T_{KW}. These percentiles have been tabulated by Kruskal and Wallis (1952). An extensive tabulation of some percentiles has been made by Iman et al. (1975).

Using a chi-square distribution approximation to the null distribution of the statistic T_{KW}, an approximation to the P-value is

$$P\text{-value}(T_{KW}) \approx P\left[\chi^2_{(k-1)} > T_{KW}(obs)\right], \qquad (4.22)$$

where $T_{KW}(obs)$ is the observed value of T_{KW}. Iman and Davenport (1976) suggested the use of the statistic

$$F_R = \frac{B_R/(k-1)}{\left[(N-1)S_R^2 - B_R\right]/(N-k)}, \qquad (4.23)$$

and the proposed test rejects the null hypothesis H_0 for large values of F_R. Further, the null distribution of F_R can be approximated by the $F(k-1, N-k)$ distribution. The test based on F_R may be called the *rank F-test*. An approximation to the P-value is

$$P\text{-value}(F_R) \approx P[F(k-1, N-k) > F_R(obs)], \qquad (4.24)$$

where $F_R(obs)$ is the observed value of the statistic F_R. It has been mentioned that the approximation (4.24) to the P-value is closer to the exact P-value compared to the approximation (4.22). For further details about this approximation readers should refer to Iman and Davenport (1976).

<u>Remark 4.3.</u> When there are no ties, S_R^2 simplifies to $[N(N+1)/12]$. Using this value, we can rewrite the expression for T_{KW}, in terms of rank sums $R_{i.} = n_i \bar{R}_{i.}$, as

$$T^*_{KW} = \frac{12}{N(N+1)} \sum_{i=1}^{k} \frac{R_{i.}^2}{n_i} - 3(N+1).$$

In general, we have

$$S_R^2 = \frac{N(N+1)}{12} \left[1 - \frac{\sum_j d_j \left(d_j^2 - 1\right)}{N(N^2 - 1)}\right].$$

where d_j is the number of values in the jth tied group. Thus the test statistic can be expressed as

$$T_{KW} = \frac{T^*_{KW}}{1 - \frac{\sum_j d_j (d_j^2 - 1)}{(N^3 - N)}}.$$

<u>Remark 4.4.</u> It may be noted that the statistic F_R is the F statistic in the context of the one-way analysis of variance (ANOVA) model computed from the ranks. This observation is useful in computing the statistic from the one-way ANOVA program, when the relevant nonparametric rank program is not available.

Remark 4.5. From equations (4.21) and (4.23), we get a relationship between the statistics T_{KW} and F_R. This relationship is

$$F_R = \frac{(N-k)}{(k-1)} \frac{T_{KW}}{(N-1-T_{KW})}.$$

Remark 4.6. When $k = 2$, the statistic T_{KW} will become $[Z_{W*}]^2$, and the test is the same as the Wilcoxon-Mann-Whitney test, discussed in Chapter 2.

4.3.2 Savage test

Here we consider the extension of the Savage test discussed in Chapter 2. For the two-sample case, the procedure can be viewed as attaching scores to the observations and then considering the sum of the scores for one sample as the test criterion. A natural extension is to combine all the samples and rank the observations. Using these ranks, Savage scores given by (2.114) are calculated. Let S_{ij} be the score for X_{ij}. It is

$$S_{ij} = E(V_{R_{ij}}) - 1 = \hat{\Lambda}(X_{ij}) - 1,$$

where V_i is the ith order statistic in a sample of size N, from the exponential distribution with mean 1. These scores will be used in the same manner as the ranks were used in the Kruskal-Wallis procedure. In other words, we calculate the average scores $\bar{S}_{i\cdot}$ for various samples and a variability measure of the scores

$$S_E^2 = \frac{1}{(N-1)} \sum_{i=1}^{k} \sum_{j=1}^{n_i} S_{ij}^2, \qquad (4.25)$$

since the $\bar{S}_{\cdot\cdot}$, the average of all scores, is zero. We also calculate a measure of variability among the average scores, which is

$$B_E = \sum_{i=1}^{k} n_i \bar{S}_{i\cdot}^2. \qquad (4.26)$$

Now the test statistic is

$$T_E = B_E / S_E^2, \qquad (4.27)$$

and for large values of this statistic, the null hypothesis is rejected. The P-value can be approximated by a chi-square probability. In particular,

$$P\text{-value}(T_E) \approx P\left[\chi_{(k-1)}^2 > T_E(obs)\right], \qquad (4.28)$$

where $T_E(obs)$ is the observed value of the statistic T_E. An example to illustrate the implementation of the Kruskal-Wallis procedure and the Savage scores procedure is given at the end of Subsection 4.3.3.

Table 4.5 $2 \times k$ table for Mood's median test

Sample ID	1	2	...	i	...	k	Total
No. $> M$	v_1	v_2	...	v_i	...	v_k	V
No. $\leq M$	$n_1 - v_1$	$n_2 - v_2$	...	$n_i - v_i$	...	$n_k - v_k$	$N - V$
Total	n_1	n_2	...	n_i	...	n_k	N

Next we explore the two median procedures, namely the extensions of Mood's median test and Mathisen's median test.

4.3.3 Mood's median test

We first find M, the median of the combined sample. Then we compare the individual observations in each sample with M and determine how many exceed M. Let us denote these counts by v_i. Now the information provided by these counts will be used to find a test criterion. It is convenient to set up a summary display of these counts as a $2 \times k$ table (see Table 4.5).

The null hypothesis can be interpreted as the homogeneity of the distributions from which the above two-category samples are taken. So we can use the procedure suggested in Subsection 4.2.1.

An equivalent way (see D'Agostino, 1972) is to score the observations as 1 or 0, depending on whether the observation is *above M* or *not above M*. These scores are called *Mood's median scores*. We can process these scores in a manner similiar to the one used in Subsection 4.3.2. Then we compute the average scores for each sample, which are

$$MS_i = (v_i/n_i), \tag{4.29}$$

and the average of all scores, which is (V/N). Further we compute a variability measure of the sample averages, namely,

$$B_M = \sum_{i=1}^{k} n_i [MS_i - (V/N)]^2, \tag{4.30}$$

and a variability measure of all scores, namely,

$$S_M^2 = \frac{V(N-V)}{N(N-1)}. \tag{4.31}$$

Now the test statistic is

$$T_M = B_M/S_M^2, \tag{4.32}$$

and for large values of T_M the null hypothesis is rejected. An approximation to the relevant P-value is

$$P\text{-value}(T_M) \approx P\left[\chi^2_{(k-1)} > T_M(obs)\right], \tag{4.33}$$

Table 4.6 *Times (in months) to expiration date*

Process						
A	30	29	34	31	32	30
B	35	25	28	35	30	31
C	32	23	24	29	29	28
D	27	29	28	31	27	26

Table 4.7 *Summary of tests for data of Table 4.6*

Test	Test Statistic	Approx. P-value
Kruskal-Wallis	6.4417	0.0920
Savage	5.7819	0.1227
Mood's Median	8.3056	0.0401

where $T_M(obs)$ is the observed value of the statistic T_M. An example will clarify all the details. We use one data set to illustrate all three procedures discussed thus far.

Example 4.3. Determination of the expiration date for a drug consists of three steps, namely, (1) measuring the potency at 0, 3, 6, 9, 12, 18, 24, 36 months after manufacture, (2) fitting a regression line, and (3) finding the time when the 95% lower confidence band reaches 90% potency. A company is interested in comparing four manufacturing processes of making a particular drug with respect to the expiration date. Six batches of the product made under each of the processes A, B, C, and D were used and times to the expiration dates for them have been determined. The resulting data appear in Table 4.6.

The NPAR1WAY procedure of SAS (see Appendix B4 for the program code) will provide the needed results. The summary results are given in Table 4.7.

Mood's test indicates that the processes are not all equal with respect to the expiration date. However, the other two tests do not reject the null hypothesis that the processes are similar.

Remark 4.7. Exact P-values can be obtained from the SAS system, but these calculations take a long time. We can also obtain the exact P-values using a package such as StatXact.

4.3.4 Extension of Mathisen's test

We will now give an extension of Mathisen's test of Chapter 2. Here we use sample 1 as the control sample and determine the median M_1 of sample 1. This is taken as the ath order statistic, where $a = \lfloor n_1/2 \rfloor + 1$. Then we determine in each of the other samples how many observations are above the median.

Table 4.8 $2 \times (k-1)$ *table for Mathisen's test*

Sample ID	2	3	...	i	...	k	Total
No. $> M_1$	m_{21}	m_{31}	...	m_{i1}	...	m_{k1}	$m_{.1}$
No. $\leq M_1$	m_{20}	m_{30}	...	m_{i0}	...	m_{k0}	$m_{.0}$
Total	n_2	n_3	...	n_i	...	n_k	$N - n_1$

As before, the current summarization of the data is done through these counts. It is convenient to display this summarization as a $2 \times (k-1)$ table (see Table 4.8).

Now the null hypothesis implies that the $(k-1)$ two-category samples are from a common distribution, with probabilities $\lambda(= \frac{a}{n_1+1})$ and $1 - \lambda$. Now the expected frequencies are

$$e_{i0} = E(m_{i0}) = n_i\,\lambda, \quad and \quad e_{i1} = E(m_{i1}) = n_i(1 - \lambda).$$

So we can use the procedure of Subsection 4.2.2 to analyze this table. The test statistic (4.16) will be simplified for use in this context:

$$T_{MA} = \sum_{i=2}^{k} \left\{ \frac{(m_{i0} - n_i\,\lambda)^2}{n_i\,\lambda} + \frac{(m_{i1} - n_i(1 - \lambda))^2}{n_i(1 - \lambda)} \right\}$$

$$= \frac{1}{\lambda(1 - \lambda)} \sum_{i=2}^{k} \frac{(m_{i0} - n_i\,\lambda)^2}{n_i}.$$

The final expression is the same as

$$T_{MA} = \frac{(n_1 + 1)^2}{a(n_1 + 1 - a)} \sum_{i=2}^{k} \frac{(m_{i0} - n_i\,\lambda)^2}{n_i}. \tag{4.34}$$

Large values of T_{MA} lead to the rejection of H_0. The approximate P-value is

$$P\text{-value}(T_{MA}) \approx P\left[\chi^2_{(k-1)} > T_{MA}(obs)\right], \tag{4.35}$$

where $T_{MA}(obs)$ is the observed value of the statistic T_{MA}. This generalization was proposed by Sen (1962).

This test can be used in any context; it is especially appropriate when sample 1 is a gold standard (or reference) sample.

When the null hypothesis of homogeneity is rejected by any one of the tests discussed above, we usually want to compare two distributions at a time so as to determine which pairs are unequal. Procedures that can be used for this purpose are called *multiple comparison procedures*. If the objective of the study is to compare all pairs of distributions, we may skip performing the tests discussed in this section and implement a multiple comparison procedure.

It may happen that the global tests give significant results and the multiple comparison procedures may not detect a significantly different pair of distributions. The global tests can be performed for the unequal sample size situations; however, the multiple comparison procedures to be discussed assume *equal sample sizes.*

4.4 Multiple comparison procedures

We start with redefining the null hypothesis as an intersection of several null hypotheses, where each subhypothesis involves only two distributions. Then we consider the relevant two-sample test statistics and combine them into a single test criterion. This is the idea behind the procedures to be described below. Two books that give a full account of these procedures are Miller (1981) and Hochberg and Tamhane (1987). In an earlier section we assumed stochastic alternatives. In this section we restrict the discussion to the *shift alternatives.* In other words, we assume that

$$F_i(x) = F(x - \Delta_i) \tag{4.36}$$

Thus the null hypothesis of homogeneity (4.1) is the same as

$$H_0 : \Delta_1 = \Delta_2 = \cdots = \Delta_k. \tag{4.37}$$

The one-sided alternative (4.3) reads as

$$H_{CO1} : \Delta_1 < \Delta_2 < \cdots < \Delta_k, \tag{4.38}$$

whereas the alternative (4.4) is the same as

$$H_{CO2} : \Delta_1 > \Delta_2 > \cdots > \Delta_k. \tag{4.39}$$

The global alternative (4.2) reads as

$$H_{GL} : \text{At least for one pair } (\Delta_i, \Delta_j), \ \Delta_i \neq \Delta_j. \tag{4.40}$$

The one-sided alternatives are called *completely ordered alternatives.* In addition to these, the class of *simple tree alternatives* will be considered. We take F_1 as the distribution of the responses under the control. Thus the alternatives of interest can be restated as

$$H_{TR1} : \Delta_1 < \Delta_j \quad \text{for at least one } j \ (j = 2, 3, \ldots, k), \tag{4.41}$$

or

$$H_{TR2} : \Delta_1 > \Delta_j \quad \text{for at least one } j \ (j = 2, 3, \ldots, k), \tag{4.42}$$

A test for pairwise comparisons

Here we derive a test using the union-intersection principle. First we express the null hypothesis (4.1) as the intersection of H_{ij0}. In other words,

$$H_0 : \bigcap H_{ij0},$$

where

$$H_{ij0} : \Delta_i = \Delta_j, \text{ for the pair } (i, j), \qquad (4.43)$$

and the global alternatives (4.2) can be viewed as the union

$$H_{GL} : \bigcup H_{ij1} = \bigcup (\Delta_i \neq \Delta_j). \qquad (4.44)$$

Now to obtain a test, we argue as follows. Consider various two-sample test statistics, T_{ij}, based on samples (i, j) relevant for testing

$$H_{ij0} : \Delta_i = \Delta_j, \quad against \quad H_{ij1} : \Delta_i \neq \Delta_j,$$

and suppose the relevant test has the critical region

$$C_{ij} : T_{ij} > c. \qquad (4.45)$$

Since the sample sizes are equal and we want the sizes of all tests to be the same, the same critical values are used for all the tests. Thus from the union-intersection principle, the overall test is defined by the critical region

$$C = \cup(C_{ij}) = max(T_{ij}) > c.$$

In other words, the overall test is to

$$reject \ H_0 \ if \ max(T_{ij}) > c, \qquad (4.46)$$

where c is the $100(1 - \alpha)$ percentile of the null distribution of the test statistic, $max(T_{ij})$. One can analyze the other cases of pairwise comparison of all treatments in a similar manner. This method of test construction can also be used in relation to the problem of comparing several treatments with a control.

In the next subsection we describe in detail one special case of this general method, where the statistics T_{ij} are Wilcoxon rank sums. The completely ordered alternatives case is discussed in Section 4.5. Section 4.6 deals with procedures for tree alternatives, three of which can be derived from the union-intersection principle.

4.4.1 Steel-Dwass procedure based on pairwise rankings

Distribution pairs (F_i, F_j) are compared by constructing the Wilcoxon statistic using the samples i and j. We calculate the sum of ranks of sample i observations. This sum is called the rank sum for sample i relative sample j and is denoted by W_{ij}. A multiple of the standardized version of the Wilcoxon rank sum statistic, namely,

$$W_{ij}^* = \sqrt{2} \frac{[W_{ij} - E_0(W_{ij})]}{\sqrt{var_0(W_{ij})}}, \qquad (4.47)$$

Table 4.9 *A short table of $q_{k,\infty}^{0.95}$ values**

k	2	3	4	5	6	7	8	9	10
q	2.77	3.31	3.63	3.86	4.03	4.17	4.29	4.39	4.47

will be used to define the test. The critical region of an appropriate two-sample test for the two-sided alternatives is

$$|W_{ij}^*| > c. \tag{4.48}$$

The overall test statistic is

$$W^{**} = max_{i,j}|W_{ij}^*| \tag{4.49}$$

and the test is to

reject H_0 if $W^{**} \geq c^*$.

Under this procedure we can infer that $F_i \neq F_j$ for which $|W_{ij}^*|$ is greater than c^*. Here we can state the reasons for the rejection of the null hypothesis. Such an explanation cannot be made without further analysis when we use the global (Kruskal-Wallis) test.

Some tabulations of these critical values are available in Steel (1960, 1961) and Critchlow and Fligner (1991). However, a large sample approximation to the critical value is useful and is given here. To obtain this approximation we argue as follows. For large n, the joint null distribution of W_{ij}^* is a multivariate normal with zero means and variances equal to 2. The null correlation structure of these variables W_{ij}^* is the same as the correlation structure of $\binom{k}{2}$ pairwise differences $Z_p - Z_q$, where $(Z_1, Z_2, \ldots, Z_k)$ are independent $N(0, 1)$ variables. So the distribution of W^{**} can be approximated by the distribution of $Q_{k,\infty}$ the range of Z's. This statistic is a limiting case of Tukey's studentized range statistic, which is denoted by $Q_{k,\nu}$ (see Appendix A4 for more details). In other words,

$$c^*(1 - \alpha) \approx q_{k,\infty}^{1-\alpha} \tag{4.50}$$

where q is the $100(1 - \alpha)$ percentile of Tukey's studentized range statistic (see Appendix A4).

It may be noted that Critchlow and Fligner have extended the Steel-Dwass procedure for unequal sample sizes. They also derived simultaneous confidence intervals for all the pairwise location differences.

A table of exact percentiles of Q is available in the *CRC Handbook of Tables*, and an extract from this handbook is given in Table 4.9.*

---

* Reproduced from *Handbook of Tables for Probability and Statistics*, 2nd edition, CRC Press, Boca Raton, Florida, p. 364, by permission of the publisher.

Table 4.10 *Various two-sample statistics*

i	j	W_{ij}	$W_{ij}^*/\sqrt{2}$	$|W_{ij}^*|$
1	2	213	1.9231	2.7196
1	3	225	2.5385	3.5899
2	3	189	0.6923	0.9791

Table 4.11 *Critical regions of Jonckheere tests*

Alternatives	Critical Region
H_{CO1}	$W_J \geq c_1$
H_{CO2}	$W_J \leq c_2$

Example 4.4. The data are part of the set reported by Fine and Bosch (2000). The response is change in weight (grams) for rats in the Aconiazide study. Various groups correspond to different doses of the drug Aconiazide. We used only three groups. Also, we modified some observations and added three observations to each group. The data, used for illustration purposes, are as follows.
Group 1: 5.7, 10.2, 13.9, 10.3, 1.3, 12.0, 14.0, 15.1, 8.8, 12.7, 15.0, 17.0, 15.2.
Group 2: 8.3, 12.3, 6.1, 10.1, 6.3, 12.1, 13.0, 13.4, 11.9, 9.9, 5.0, 6.0, 4.0.
Group 3: 9.5, 8.1, 7.0, 7.8, 9.3, 12.2, 6.7, 10.6, 6.6, 7.0, 8.0, 9.0, 3.0.

The Kruskal-Wallis statistic is found to be 7.1680 with an approximate two-sided P-value of 0.0278. So let us apply the Steel-Dwass procedure.

The standardized two-sample Wilcoxon statistics W_{ij} and W_{ij}^* are given in Table 4.10. From Table 4.9, the critical value is $c^* = 3.31$. The Steel-Dwass statistic is 3.5899, which exceeds the critical value. Only the statistic $W_{1,3}^*$ exceeds the critical value. So we declare that groups 1 and 3 are different.

4.5 Jonckheere's test for completely ordered alternatives

Now we consider the problem of testing the null hypothesis of equality (4.37) against the alternatives H_{CO1} or H_{CO2}, which are restated in (4.38) and (4.39). Both Terpstra (1952) and Jonckheere (1954) considered this problem. Their proposal will be described now. First, consider pairs of samples $\{X_{i1}, \ldots, X_{in_i}\}$ and $\{X_{j1}, \ldots, X_{jn_j}\}$ for $i < j$. Let U_{ij} be the Mann-Whitney statistic (2.62) for these two samples. The proposed test statistic is

$$W_J = \sum_{i<j} U_{ij}. \tag{4.51}$$

The critical regions are given in Table 4.11. The null distribution of W_J can be approximated by a normal distribution. Using this normal distribution, approximate P-values can be computed. For this purpose we need the expectation and variance of W_J under H_0. Let $N_i = \sum_{j=1}^{i} n_j$, for $i = 1, 2, \ldots, k$. It can be shown that

$$E(W_J|H_0) = \frac{1}{4}\left(N_k^2 - \sum n_i^2\right), \tag{4.52}$$

Further, when there are no ties, the variance is

$$var(W_J|H_0) = \frac{1}{12}\sum_{i=2}^{k} n_i N_{i-1}(N_i + 1). \tag{4.53}$$

When there are ties, a modified expression for the variance is needed, and this expression is given in Lehmann (1998, 5.64 on p. 235). Some details about the derivation of the expression (4.53) for the variance are given in Appendix A4.

Tables and normal approximations

Odeh (1971) prepared a table of critical values needed for the implementation of this test. One table deals with equal sample size case. For this case, the normal approximation for the critical values seems adequate for $n \geq 4$. For example, when $n = 4, k = 4$, and $\alpha = 0.05$, we have

$$c_{1-\alpha} \approx \lfloor E(W_J|H_0) + z_{1-\alpha}\sqrt{var(W_J|H_0)}\rfloor + 1 = 67.$$

From Odeh's table we have

$$P(W_J \geq 67) = 0.04198 \quad and \quad P(W_J \geq 66) = 0.05142.$$

Hence the exact value of $C_{0.95}$ is 67.

An alternative test was proposed by Chacko (1963) for equal sample sizes and extended by Shorack (1967) to the general case. The discussion of these procedures is beyond the scope of this book. The interested reader may refer to these papers for details.

Example 4.5. In a feeding experiment 18 chicks were used. They were divided into three groups at random and the different groups were fed different diets that differed in protein content. Diet 1 had the lowest level of protein, diet 2 had a medium level of protein, and diet 3 had the highest level of protein. After 6 weeks of feeding, weights (in grams) were recorded. The data are shown in Table 4.12. We want to examine the hypothesis that weight gain increases with the increased amount of protein. Thus we are interested in testing the

Table 4.12 *Weight (in grams) of chicks*

Diet 1	120	170	130	150	126	148
Diet 2	200	160	210	155	220	154
Diet 3	230	145	180	175	165	250

null hypothesis of the equality of weight distributions, namely,

$$H_0 : F_1 = F_2 = F_3,$$

against the completely ordered alternative

$$H_{CO1} : \Delta_1 < \Delta_2 < \Delta_3.$$

From computer program output (see Appendix B4), we have $U_{12} + U_{13} = W_{123} = 65, U_{13} = 32$, and hence $W_J = 86$. Further,

$$E(W_J|H_0) = 54, \quad and \quad var(W_J|H_0) = 153.$$

Thus the test statistic is $(86 - 54)/\sqrt{153} = 2.587$, and the approximate P-value is $P(N(0, 1) > 2,587) = 0.0048$. Hence we reject H_0 and conclude that the weights increase with increase in protein intake.

4.6 Comparison of several treatments with a control

In some experiments, one treatment is a control treatment and the researcher is interested in comparing the control with each of the other treatments, so as to ascertain which treatments will produce the desired results. We will formulate this objective as a multiple testing problem. We treat F_1 as the control treatment. Thus the objective is to test the null hypothesis (4.37), which can be viewed as

$$H_0 : \Delta_1 = \Delta_j, \quad for\ j = 2, 3, \ldots, k. \tag{4.54}$$

In this context alternatives H_{TR1} or H_{TR2} are relevant. If large values for the response are desirable, we use the alternative H_{TR1}, which is

$$H_{TR1} : \Delta_1 \leq \Delta_j, \quad with\ strict\ inequality\ for\ at\ least\ one\ j.$$

In some applications smaller values for the response are desirable; in these cases we use the alternative H_{TR2}, which is

$$H_{TR2} : \Delta_1 \geq \Delta_j, \quad with\ strict\ inequality\ for\ at\ least\ one\ j.$$

A natural way to proceed is to compute $(k - 1)$ two-sample statistics T_{1j}, based on sample j and the control sample for $(j = 2, 3, \ldots, k)$ and combine the information provided by them to construct an overall test. Two different ways of combining the information provided by these two-sample statistics are discussed in the literature. They are described below. One method uses the union-intersection principle to come up with the test statistic, and the other method uses a linear combination of T_{1j}'s as the test statistic. One of the earliest procedures from Steel (1959) is based on $max_j T_{1j}$, which can be obtained from the union-intersection principle. A procedure similar to Steel's was studied by Slivka (1970) and by Spurrier (1992). For this problem, another procedure of interest is the one proposed by Fligner and Wolfe (1982), which is based on $\sum_j T_{1j}$. The Chakraborti and Desu (1988) proposal uses the sum of different statistics T_{1j}. In the next subsections we describe these procedures.

4.6.1 Steel's multiple comparison test

It is assumed that all treatment samples are of the same size n and the control sample size is m. The jth treatment sample is combined with the control sample and the observations are ranked. Then the rank sum of the treatment j observations W_j is computed. The two-sample statistics are these rank sum statistics. The union-intersection principle results in the test that

$$\text{rejects } H_0 \text{ in favor of } H_{TR1} \text{ if } Max(W_2, \ldots, W_k) > c_1, \qquad (4.55)$$

where c_1 is an appropriate percentile of the null distribution of the test criterion. Obviously the critical value c_1 depends on k, n, and α. Any treatment j for which $W_j > c_1$ is declared as the one that produces larger response values compared to the control treatment.

If the alternatives H_{TR2} are relevant, we use the statistics

$$W_j^* = n(n + m + 1) - W_j, \qquad (4.56)$$

and the test is to

$$\text{reject } H_0 \text{ in favor of } H_{TR2} \text{ if } max_j(W_j^*) > c_2, \qquad (4.57)$$

where c_2 is an appropriate percentile. The statistic W_j^* is the sum of ranks for treatment j, when the ranks are assigned in the reverse order (1 for the largest, $\ldots, m + n$ for the smallest). Like before, any treatment j for which $W_j^* > c_2$, is declared as the one that produces smaller response values compared to the control response values.

Steel gave a table of critical values, using an approximation to the null distribution of the test statistic. Miller (1980) reproduced these approximate critical values in Table VIII of his Appendix B.

4.6.2 Spurrier's procedure

Spurrier (1992) proposed a different procedure for the alternatives H_{TR1} and considered various generalizations of Steel's procedure. One test uses the rank sums of the control observations, which are denoted as W_j^c, and the test is to

$$\text{reject } H_0 \text{ in favor of } H_{TR2} \text{ if } min_j(W_j^c) < c_3, \qquad (4.58)$$

where c_3 is an appropriate percentile. Further, Spurrier developed algorithms for computing the exact distribution of the test statistics he proposed, and gave tables of critical values. Baker and Spurrier (1998) extended this procedure for normal scores and Savage scores. The recent work of van de Wiel (2002) deals with new efficient algorithms for computing the exact distribution for rank statistics to make multiple comparisons.

4.6.3 Slivka's control quantile test

Slivka (1970) used control quantile statistics and developed a method similar to that of Steel. He assumed that the sample sizes for the treatments are all equal to n, that is, $n_2 = \ldots = n_k = n$ and the control sample size, m, may be different from n. Let Z_s be the sth order statistic of the control sample. Now for each treatment sample $j(j = 2, 3, \ldots, k)$ compute the statistic, T_j, the number of observations that do not exceed Z_s. Slivka considered only the alternatives H_{TR2}. The test statistic is $T_S = min_j(T_j)$ and the test is to

$$\text{reject } H_0 \text{ in favor of } H_{TR2} \text{ if } T_S \leq c, \tag{4.59}$$

where c is a percentile of the null distribution of the test criterion. It is easy to see that this test also can be derived using the union-intersection principle. An approximation to the P-value is

$$P\text{-value} \approx 1 - G[-T_S(obs); k - 1, \rho], \tag{4.60}$$

where $\rho = \frac{n}{m+n+1}$ and $T_S(obs)$ is the observed value of the statistic T_S. The function G has been tabulated by Gupta (1963). This approximation is based on the fact that the asymptotic null distribution of the vector, $(T_2, T_3, \ldots, T_k)$, is a multivariate normal distribution.

Hogg (1965) gives a method of constructing a simple procedure when the alternatives are restricted. The two procedures to be described may have been motivated by Hogg's work.

4.6.4 Fligner and Wolfe test

Fligner and Wolfe (1982) proposed the use of MW statistics, U_{1j}^*, and the test statistic is

$$T_{FW} = \sum_{j=2}^{k} U_{1j}^*. \tag{4.61}$$

The critical regions are given in Table 4.13. The implementation of this test is easy once we observe that the statistic counts how many times the treatment

Table 4.13 *Critical regions for Fligner and Wolfe tests*

Alternatives	Critical Region
H_{TR1}	$T_{FW} > c_1$
H_{TR2}	$T_{FW} < c_2$

observation exceeds the control observation. We can calculate the required quantities by applying the two-sample WMW test to the control sample and the combined sample of all treatment samples. If, in this application, W is the sum of the ranks for the treatment observations then

$$T_{FW} = W - (1/2)(N - n_1)(N - n_1 + 1), \tag{4.62}$$

where $N = \sum_{i=1}^{k} n_i$, which is the total sample size. Thus this testing procedure is the same as the testing problem using the Wilcoxon statistic. The critical values or the associated P-values can be approximated by using a normal approximation for the null distribution of the test statistic. To use this approximation one needs to find the mean and the variance of the statistic under the null hypothesis. From the results for the two-sample statistics, we have

$$E(T_{FW}|H_0) = \sum_{j=2}^{k} n_1 n_j/2 = n_1(N - n_1)/2,$$

and

$$var(T_{FW}|H_0) = var(W|H_0). \tag{4.63}$$

Thus the Fligner and Wolfe procedure is an adaptation of the two-sample Wilcoxon procedure. Chakraborti and Desu (1988b) proposed a similar adaptation of Mathisen's median procedure for the current problem. This is described in the next subsection.

4.6.5 Chakraborti and Desu test

Chakraborti and Desu (1988b) use the same statistics T_j as Slivka and they assume that $s = \lfloor (m/2) \rfloor + 1$. Their test criterion is

$$T_{CD} = \sum_{j=2}^{k} T_j. \tag{4.64}$$

The test for the alternatives H_{TR1} is to

reject H_0 in favor of H_{TR1} if $T_{CD} \leq c_-$,

and the test for the alternatives H_{TR2} is to

reject H_0 in favor of H_{TR1} if $T_{CD} \leq c_+$, \tag{4.65}

where the c's are appropriate percentiles of the null distribution of the test statistic, T_{CD}.

The advantage of this procedure is that the critical values of the test statistic can be obtained from the critical values of the two-sample control median test statistic. Unlike Slivka's procedure, no new tables are needed. The exact P-value can be calculated using the hypergeometric distribution. Using a simulation study, Chakraborti and Desu compared the power functions of their test with that of Slivka. This study showed for certain alternatives the powers are close to each other, so we can use this simpler procedure without sacrificing power.

We can use a normal approximation to the studentized statistic

$$Z_{T_{CD}} = [T_{CD} - E(T_{CD}|H_0)]/\sqrt{var(T_{CD}|H_0)},$$

for computing an approximate P-value. It may be noted that

$$E(T_{CD}|H_0) = s(N - m)/(m + 1),$$

and

$$var(T_{CD}|H_0) = [s(m - s + 1)(N - m)(N + 1)]/[(m + 1)^2(m + 2)],$$

where $N = n(k - 1) + m$.

Example 4.6. A social scientist undertook an investigation for comparing the effects of two training programs A and B on the efficiency of workers in a company. Twenty-one workers were chosen and were divided into three equal size groups. Two groups were randomly assigned to the programs A and B, while the third group was the control group. Times (in minutes) to finish identical tasks were observed for the 21 workers after the intervention with the programs. The resulting data appear in Table 4.14. We want to test the null hypothesis that the programs are not effective against the TR2 alternative. In other words,

$$H_0 : F_0 = F_A = F_B; \quad and \quad H_{TR2} : \Delta_0 > \Delta_A, \quad or \quad \Delta_0 > \Delta_B.$$

Here from the computer output we have $W = 124.50$, so that the statistic $T_{FW} = 125.5 - (14)(15)/2 = 19.5$. Further,

$$E(T_{FW}|H_0) = 7(14)/2 = 49,$$

and from the output

$$std(T_{FW}|H_0) = 13.3516.$$

The P-value for T_{FW} is approximately 0.0084, so this test rejects the null hypothesis.

Table 4.14 *Time to finish a job*

Control (0)	20	18	25	17	22	21	15
Program A	15	16	12	19	21	13	17
Program B	12	16	14	19	21	11	15

We also apply Chakraborti and Desu's test. The median of the control group is $M = 20$. Thus the counts $T_A = 6$, and $T_B = 6$, so that $T_{CD} = 6 + 6 = 12$. The null mean and the null standard deviation are

$$E(T_{CD}|H_0) = 4(14)/(7 + 1) = 7, \ std(T_{CD}|H_0) = 2.925.$$

The standardized statistic is

$$Z = (12 - 7)/(2.925) = 1.709,$$

and hence the approximate P-value is 0.0436. Thus we do reject the null hypothesis.

4.7 Censored data

The three procedures for testing the null hypothesis (4.1) against global alternatives can be viewed as scoring the observations and analyzing the scores as in a one-way classification model. The test statistic has a chi-square distribution and permutation distribution is used to calculate the approximate P-values. In the case of censored data methods, in Chapter 2 we noted that the statistics obtained earlier are particular members of a class considered by Tarone and Ware. A similar setting prevails in the k-sample case and we now present the discussion in the framework of Tarone and Ware.

Consider the censored data setting as in Chapter 2 for k-samples. Let $t_{(1)} < t_{(2)} < \cdots < t_{(I)}$ be the distinct ordered uncensored times in the combined sample. Censored observations that occur before $t_{(1)}$ are noninformative and hence are omitted from the discussion.

At each time $t_{(i)}$, we set up a $2 \times k$ table as in Table 4.15. It may be noted that r_i is the number at risk at $t_{(i)}$ in the combined sample of which r_{ij} are from the jth sample. Let a_i be the score for an uncensored observation at $t_{(i)}$ and A_i be the score assigned to an observation censored in the interval $[t_{(i)}, t_{(i+1)})$, for $i = 1, 2, \ldots, I$. For $i = I$ the interval is taken as $[t_{(I)}, \infty)$. We compute the sum of scores received by each sample. The total score for sample l is

$$U_l = \sum_{i=1}^{I} [m_{il}a_i + e_{il}A_i], \tag{4.66}$$

Table 4.15 *Number of events at uncensored time $t_{(i)}$*

	Sample ID				
	1	2	$\cdots$	k	Total
Events	m_{i1}	m_{i2}	$\cdots$	m_{ik}	m_i
Survivals	$r_{i1} - m_{i1}$	$r_{i2} - m_{i2}$	$\cdots$	$r_{ik} - m_{ik}$	$r_i - m_i$
At Risk	r_{i1}	r_{i2}	$\cdots$	r_{ik}	r_i

where e_{il} are the number of observations in the lth sample that were censored in the interval $[t_{(i)}, t_{(i+1)})$. It easy to see that

$$e_{il} = r_{il} - m_{il} - r_{i+1l}.$$

For the purpose of finding variances and covariances, it is convenient to reexpress the statistic U_l as

$$U_l = \sum_{i=1}^{I} w_i[m_{il} - r_{il}(m_i/r_i)]. \tag{4.67}$$

This result is proved in Appendix A4, assuming that the scores satisfy the condition

$$A_i - A_{i-1} = -(a_i - A_i)(m_i/r_i). \tag{4.68}$$

Then the weights are given by

$$w_i = (a_i - A_i).$$

Thus the statistic U_l is seen to be a weighted sum of differences between observed number of deaths and conditional expected number of deaths. We use this observation to calculate the null covariance matrix of $\mathbf{U}' = (U_1, U_2, \dots, U_k)$ given the marginal totals. It can be shown that this covariance matrix is

$$\mathbf{\Sigma} = (\sigma_{ll'}),$$

where

$$\sigma_{ll'} = \sum_{i=1}^{I} w_i^2 \frac{m_i(r_i - m_i)}{r_i - 1} \frac{r_{il}}{r_i} \left(\delta_{ll'} - \frac{r_{il'}}{r_i} \right), \tag{4.69}$$

where $\delta_{ll'}$ being the Kronecker delta taking the value 1 if $l = l'$ and 0, if $l \neq l'$ and in the sum we exclude the terms for which $r_i = 1$.

Since the sum of U_l is zero the covariance matrix $\mathbf{\Sigma}$ is singular. So we take the vector statistic

$$\mathbf{U}^* = (U_1, U_2, \dots, U_{k-1}),$$

and construct a test criterion. Let $\mathbf{\Sigma_{11}}$ be the covariance matrix of $\mathbf{U}^*$. The test criterion is the quadratic form

$$T = \mathbf{U}^{*'}\mathbf{\Sigma_{11}^{-1}}\mathbf{U}^*, \tag{4.70}$$

which is approximately a chi-square variable with $(k-1)$ degrees of freedom. An approximate α-level test for global alternatives is to

$$reject \ H_0 \ if \ T > \chi_{k-1}^2(1-\alpha), \tag{4.71}$$

and the associated P-value is

$$P\text{-value}(T) = P(\chi_{k-1}^2 > T(obs)),$$

where $T(obs)$ is the observed value of T.

Permutation tests

If it is appropriate to assume that the *censoring distributions are equal*, one uses the permutation covariance matrix $\mathbf{\Sigma_p}$ to construct the test criterion. The test statistic is

$$T_p = \mathbf{U}'\mathbf{\Sigma_p^-}\mathbf{U} = \frac{(r-1)\sum_l(U_l^2/r_{1l})}{\sum_i w_i^2 m_i(r_i - m_i)/r_i},$$

and the test is defined in the same way as above where T is replaced by T_p. Here $r = \sum_i r_i$.

Remark 4.7. The sets of weights for the popular tests are as follows.

$$w_i = 1, \text{ for logrank test}$$
$$= r_i, \text{ for Gehan Wilcoxon test}$$
$$= \prod_{j=1}^{i} \frac{r_j}{r_j + d_j}, \text{ for Prentice-Peto Wilcoxon test}$$
$$= \sqrt{r_i}, \text{ for Tarone and Ware test} \qquad (4.72)$$

Remark 4.8. The SAS life test procedure gives logrank and Gehan's Wilcoxon tests as defined by (4.72).

Remark 4.9. The statistic T of (4.70) can be expressed as the quadratic form $\mathbf{U}'\mathbf{\Sigma^-}\mathbf{U}$, where $\mathbf{U}' = (\mathbf{U_1}, \mathbf{U_2}, \ldots, \mathbf{U_k})$ and $\mathbf{\Sigma^-}$ is a symmetric g-inverse of $\mathbf{\Sigma}$. We will discuss this type of results in Chapter 5.

A recent work that proposed some other weights is the paper by Leton and Zuluaga (2002).

Breslow (1970) considered the permutation test with Gehan Wilcoxon scores, whereas Schemper (1983) discussed the permutation test with Savage scores.

Generalizations of the median tests have appeared in literature. The references include Brookmeyer and Crowley (1982), Chakraborti (1984), and Gastwirth and Wang (1988).

We now give an example. We need a computer program to do the calculations indicated earlier. The relevant SAS program, given in Appendix B4, calculates the T-statistic of (4.70) for Wilcoxon scores and logrank scores.

Example 4.7. Lee and Desu (1972) presented a data set consisting of three treatments. The response is the survival time of leukemia patients. The censored values are indicated by a "+" sign. The data are as follows.
Group 1: 4.0, 5.0, 9.0, 10.0, 10.0, 12.0, 13.0, 23.0, 28.0, 28.0, 29.0, 31.0, 32.0, 37.0, 41.0, 41.0, 57.0, 62.0, 74.0, 100.0, 139.0, 20.0+, 258.0+, 269.0+.

Group 2: 8.0, 10.0, 10.0, 12.0, 14.0, 20.0, 48.0, 70.0, 75.0, 99.0, 103.0, 162.0, 169.0, 195.0, 220.0, 161.0+, 199.0+, 217.0+, 245.0+.
Group 3: 8.0, 10.0, 11.0, 23.0, 25.0, 25.0, 28.0, 28.0, 31.0, 31.0, 40.0, 48.0, 89.0, 124.0, 143.0, 12.0+, 159.0+, 190.0+, 196.0+, 197.0+, 205.0+, 219.0+.
From the ouput we get the following results.

Test	Statistic	P-Value
Logrank	4.3484	0.1137
Wilcoxon	3.7815	0.1510

From the two-sided P-values, we do not reject the null hypothesis that the treatments are equally effective.

Using Gehan's Wilcoxon scores, Patel and Hoel (1973) extended Jonckheere's test and gave expressions for variances and covariances that are needed for the implementation of the test. Crowley (1979) considered a different test statistic that is a sum of uncorrelated statistics, similar to the discussion of Appendix A4.2, and the test is based on the asymptotic normal distribution of the test statistic. Chakraborti and Desu (1990, 1991) treated the problem of comparison of several treatments with a control. For the treatments versus control problem, Chen (1994) extended Steel's procedure.

4.8 Appendix A4: Mathematical supplement

A4.1 Pearson's χ^2 statistic

In the analysis of categorical data, Pearson's test statistic is frequently used. Here we define this statistic and indicate its asymptotic distribution. Two important applications will be indicated.

It is of interest to test a hypothesis about the cell probabilities, using a categorical data set. For fixing the ideas we display the data and the cell probabilities in the following table.

	1	2	$\cdots$	i	$\cdots$	k
Frequency	m_1	m_2	$\cdots$	m_i	$\cdots$	m_k
Probability	π_1	π_2	$\cdots$	π_i	$\cdots$	π_k

It should be noted that $\sum_i \pi_i = 1$. Let $N = \sum_i m_i$. Two important sampling situations that conform to this set up are (1) r independent multinomial samples, each with c categories, and (2) a multinomial sample of size N with k categories, where the categories are defined by two attributes. One attribute has r categories, whereas the second attribute has c categories. In this case $k = rc$.

We want to test the hypothesis that all the probabilities π_i are specified functions of p independent parameters $(\theta_1, \ldots, \theta_p)$. In other words,

$$H_0 : \pi_i = g_i(\theta_1, \ldots, \theta_p),$$

where g_i are specified functions. The parameter θ's may be known or unknown.

The expected cell frequencies are $N\pi_i$ and under the hypothesis they are $Ng_i(\theta_1, \ldots, \theta_p)$. We estimate the frequencies as

$$e_i = N\hat{\pi}_i = Ng_i(\hat{\theta}_1, \ldots, \hat{\theta}_p).$$

Now we need to ascertain how close the observed frequencies are to the expected frequencies. To get a measure of the closeness, we can consider the residuals, $(m_i - N\hat{\pi}_i)$, and combine them into a single criterion. From the distribution theory considerations, the standardized residuals are used; these are defined as

$$\delta_i = (m_i - N\hat{\pi}_i)/(N\hat{\pi}_i)^{1/2}$$
$$= N^{1/2}(p_i - \hat{\pi}_i)/(\hat{\pi}_i)^{1/2},$$

where $p_i = m_i/N$. These are combined into a single criterion, namely,

$$X^2 = \sum_i \delta_i^2 = \sum_i \frac{(m_i - N\hat{\pi}_i)^2}{N\hat{\pi}_i}.$$

This statistic has an asymptotic χ^2 distribution with ν degrees of freedom. Large values of this statistic leads to the rejection of the hypothesis H_0. The quantity ν is computed as follows.

$\nu = \#$ *of independent* π *parameters* $- \#$ *of independent* θ *parameters.*

Thus the P-value can be approximated as

$$P\text{-value}(X^2) \approx P(\chi^2(\nu) > X^2(obs)),$$

where $X^2(obs)$ is the observed value of the statistic X^2.

Now we apply this general result to the two special cases indicated earlier.

Homogeneity of r multinomial distributions

Here we arrange the frequencies $(n_{i1}, \ldots, n_{ij}, \ldots, n_{ic})$ of the ith sample as the ith row and hence the data appear as an $r \times c$ table. Let n_i be the ith sample size, and $N = \sum_{i=1}^{r} n_i$. In this case the hypothesis states that

$$H_0 : \theta_{1j} = \ldots = \theta_{ij} = \ldots = \theta_{rj}(= \Theta_j), \forall j.$$

Thus all probabilities are defined in terms of c Θ's. However, these parameters have to add up to 1. Hence the number of independent Θ-parameters is $(c-1)$. The number of independent π parameters is $r(c-1)$. Thus

$$\nu = r(c-1) - (c-1) = (r-1)(c-1).$$

The expected frequencies are

$$e_{ij} = n_i \hat{\Theta}_j = n_i n_{.j}/N,$$

where $n_{.j} = \sum_i n_{ij}$. Now using these expressions we see that the statistic X^2 is the same as the statistic T of (4.16).

Independence of two attributes

Here, as in the case discussed above, the data will be displayed as an $r \times c$ table, where the rows represent the categories of the first attribute and the columns represent the categories of the second attribute. The hypothesis of independence is

$$\pi_{ij} = \rho_i \times \gamma_j,$$

where $\sum_{i=1}^{r} \rho_i = 1$ and $\sum_{j=1}^{c} \gamma_j = 1$. Thus the number of independent θ parameters is $(r-1)+(c-1)$. Originally we have $k-1$ independent parameters. Hence

$$\nu = (k-1) - \{(r-1)+(c-1)\} = rc - 1 - \{(r-1)+(c-1)\} = (r-1)(c-1).$$

This quantity is the same as the corresponding quantity for the homogeneity case. Further, the expected frequencies are

$$e_{ij} = N\hat{\pi}_{ij}$$
$$= N\hat{\rho}_i \times \hat{\gamma}_j$$
$$= N(n_i/N)(n_{.j}/N)$$
$$= n_i n_{.j}/N.$$

These are equal to the expected frequencies of the homogeneity case. Thus the statistic here is the same as the test statistic in the previous case. We can use the computer program for testing independence to test the homogeneity hypothesis.

A4.2 Derivation of the variance of W_J

In this discussion we assume that there no ties. To minimize the complexity, we give the details for the case of three samples. In this case

$$W_J = U_{12} + U_{13} + U_{23}, \tag{A1}$$

so that the variance is

$$\text{var}(W_J) = \text{var}(U_{12}) + \text{var}(U_{13}) + \text{var}(U_{23}) + 2\text{cov}(U_{12}, U_{13})$$
$$+ 2\text{cov}(U_{12}, U_{23}) + 2\text{cov}(U_{13}, U_{23}). \tag{A2}$$

The variance of a statistic such as U_{ij} can be calculated from the results given in Chapter 2. The main task here is to derive a formula for the covariance. Let us consider the covariance between U_{ij} and U_{ik}. This covariance can be obtained by considering the variance of $U_{i(j+k)}$, which is the statistic obtained by comparing the ith sample with the jth and kth samples combined. We see that

$$var(U_{i(j+k)}) = var(U_{ij} + U_{ik})$$
$$= var(U_{ij}) + var(U_{ik}) + 2cov(U_{ij}, U_{ik}).$$

Hence

$$cov(U_{ij}, U_{ik}) = (1/2)\{var(U_{i(j+k)}) - var(U_{ij}) - var(U_{ik})\}. \qquad (A3)$$

Using this expression in (A2), the required variance can be expressed in terms of variances of two-sample Mann-Whitney statistics. From now on, we consider the variances under the null hypothesis. Since there are *no ties* by assumption, the expression for the covariance simplifies considerably:

$$cov_0(U_{ij}, U_{ik}) = (1/12)n_i n_j n_k. \qquad (A4)$$

We also see that

$$cov_0(U_{ij}, U_{jk}) = -cov_0(U_{ji}, U_{jk}) = -(1/12)n_j n_i n_k,$$
$$cov_0(U_{ik}, U_{jk}) = cov_0(U_{ki}, U_{kj}) = (1/12)n_k n_i n_j. \qquad (A5)$$

Using the expressions for the variances and covariances, we get

$$var_0(W_J) = (1/12)[n_1 n_2(n_1 + n_2 + 1) + n_1 n_3(n_1 + n_3 + 1)$$
$$+ n_2 n_3(n_2 + n_3 + 1)$$
$$+ 2n_1 n_2 n_3 - 2n_1 n_2 n_3 + 2n_1 n_2 n_3].$$

This expression simplifies to

$$var_0(W_J) = (1/12)[n_1 n_2(n_1 + n_2 + 1) + n_1 n_3(n_1 + n_2 + n_3 + 1)$$
$$+ n_2 n_3(n_1 + n_2 + n_3 + 1].$$

It is convenient to express this in terms of $N_i = \sum_{j=1}^{i} n_j$. Now it is easy to see that

$$var_0(W_J) = (1/12)\sum_{i=2}^{3} n_i N_{i-1}(N_i + 1).$$

In general we have

$$var_0(W_J) = (1/12)\sum_{i=2}^{k} n_i N_{i-1}(N_i + 1). \qquad (A6)$$

Two representations of W_J as a sum of uncorrelated variables

To derive the variance expression (A6) it is convenient to express W_J as a sum of uncorrelated variables. There are two different representations. The first one leads to (A6). The second one leads to a different expression for the variance. The details will be given only for the simplest case considered earlier.

To get one representation, we proceed as follows. The statistic is seen to be

$$W_J = U_{12} + [U_{13} + U_{23}] \equiv U_1 + U_2. \qquad (A7)$$

It can be shown that U_1 and U_2 are uncorrelated. The proof of this result is a problem. These statistics are two-sample MW statistics, so we can find the variances from the results proved in Chapter 2.

Now this idea can be generalized as follows. Let

$$U_i = \sum_{j=1}^{i-1} U_{ji}, \qquad (A8)$$

then we can write W_J as the sum

$$W_J = U_2 + U_3 + \cdots + U_k. \qquad (A9)$$

Since U_i is the MW statistic for comparing combined data of the first $i - 1$ samples with ith sample, we have

$$var_0(U_i) = (1/12)N_{i-1}n_i(N_{i-1} + n_i + 1) = (1/12)n_i N_{i-1}(N_i + 1), \quad (A10)$$

and hence the variance of W_J is the expression (A6).

A second representation

A different way of viewing the statistic is also convenient in some discussions. Again we consider the case of three samples and view the statistic as

$$W_J = [U_{12} + U_{13}] + U_{23} \equiv U^1 + U^2, \qquad (A11)$$

where the statistics U^1 and U^2 are also uncorrelated, under the null hypothesis. So

$$var_0(W_J) = var_0(U^1) + var_0(U^2).$$

We note that the statistic U^1 is the MW statistic for comparing sample 1 with pooled samples 2 and 3, and U^2 is the MW statistic for comparing sample 2 with sample 3. Thus,

$$var_0(W_J) = (1/12)[n_1(n_2 + n_3)(n_1 + n_2 + n_3 + 1) \\ + n_2 n_3(n_2 + n_3 + 1)].$$

To facilitate a generalization, we rewrite it as

$$var_0(W_J) = (1/12) \sum_{i=1}^{2} n_i(N - N_i)(N - N_{i-1} + 1), \qquad (A12)$$

where $N_0 = 0$. Finally, generalizing (A11), we have

$$W_J = U^1 + U^2 + \cdots + U^{k-1},$$

where

$$U^i = \sum_{i=j+1}^{k} U_{ij}. \qquad (A13)$$

These U^i's are pairwise uncorrelated under the null hypothesis. Using this representation, the null variance is expressed as

$$var_0(W_J) = \sum_{j=1}^{k-1} var_0(U^j).$$

Thus we get another (equivalent) expression for the variance, namely,

$$var_0(W_J) = \sum_{j=1}^{k-1} n_i (N_k - N_i)(N_k - N_{i-1} + 1). \qquad (A14)$$

The computer program given in Appendix B4 uses the representation (A11) for computing the statistic and the expression (A14) for computing the variance.

A4.3 Tukey's studentized range statistic

Let $X_1, \ldots, X_r$ be r independent standard normal variables. Let χ_ν^2 be a chi-square random variable with ν degrees of freedom, independent of the X variables. Define Q as

$$Q = \frac{maxX_i - minX_i}{\sqrt{\frac{\chi_\nu^2}{\nu}}}.$$

This variable Q is called the *Tukey's studentized range statistic*. The quantity $q_{r,\nu}^{1-\alpha}$ is the $100(1-\alpha)$ percentile of the distribution of Q. In other words,

$$P[Q < q_{r,\nu}^{1-\alpha}] = 1 - \alpha,$$

and equivalently,

$$P[Q > q_{r,\nu}^{1-\alpha}] = \alpha.$$

Tables of these percentiles are available in Scheffe (1959). Harter (1960) also prepared such tables, which are reproduced in Miller (1981).

When $\nu \to \infty$, the distribution of Q becomes the distribution of the range of samples of size r from the standard normal distribution. The percentiles of this range distribution will be denoted as $q_{r,\infty}^{1-\alpha}$, which were used in (4.49). In other words $q_{r,\nu}^{1-\alpha}$ tend to $q_{r,\infty}^{1-\alpha}$.

A4.4 Null variance of T_{FW}

This derivation does not use the observation that the statistic can be viewed as a two-sample test statistic. From the definition (4.60) of the statistic we have

$$var(T_{FW}|H_0) = \sum_{j=2}^{k} var(U_{1j}|H_0) + 2 \sum_{j<l=2}^{k} cov(U_{1j}, U_{1l}|H_0).$$

So

$$var(T_{FW}|H_0) = \sum_{j=2}^{k} \frac{n_1 n_j (n_1 + n_j + 1)}{12} + 2 \sum_{j<l=2}^{k} \frac{n_1 n_j n_l}{12}$$

$$= \frac{1}{12} \left[n_1^2 (N - n_1) + n_1 (N - n_1)^2 + n_1 (N - n_1) \right].$$

Here $N = \sum_{j=1}^{k} n_j$. Upon simplification, we get

$$var(T_{FW}|H_0) = \frac{n_1(N - n_1)}{12} [n_1 + N - n_1 + 1] = \frac{n_1(N - n_1)(N + 1)}{12}.$$

A4.5 Reformulation of the sum of censored data scores

In discussing tests for censored data we started with a set of scores and computed U_l, the sum of the scores for sample l. Initially, this statistic is computed as

$$U_l = \sum_{i=1}^{I} [m_{il} a_i + e_{il} A_i],$$

and then expressed in a different way.

Substituting for e_{il}, we get

$$U_l = \sum_{i=1}^{I} [m_{il} a_i + (r_{il} - m_{il} - r_{i+1,l}) A_i]$$

$$= \sum_{i=1}^{I} [m_{il}(a_i - A_i) + (r_{il} - r_{i+1,l}) A_i]$$

$$= \sum_{i=1}^{I} [m_{il}(a_i - A_i) + \sum_{i=1}^{I} (r_{il} - r_{i+1,l}) A_i,$$

where $r_{I+1,l} = 0$. By rearranging the terms in the second sum and noting that $A_0 = 0$ we get

$$U_l = \sum_{i=1}^{I} [m_{il}(a_i - A_i) + r_{il}(A_i - A_{i-1})].$$

Let us assume that the scores satisfy the relation

$$A_i - A_{i-1} = -(a_i - A_i)(m_i/r_i).$$

Using this fact we get

$$U_l = \sum_{i=1}^{I} [m_{il}(a_i - A_i) - r_{il}(a_i - A_i)(m_i/r_i)]$$

$$= \sum_{i=1}^{I} (a_i - A_i)[m_{il} - r_{il}(m_i/r_i)]$$

$$= \sum_{i=1}^{I} w_i[m_{il} - E(m_{il}|H_0)].$$

Thus the weights w_i are a function of the scores and are given by

$$w_i = (a_i - A_i), \quad for \ i = 1, 2, \dots, I.$$

It may be noted that, from a given set of weights, we can determine the scores using the relations. First we observe that $A_0 = 0$, then

$$A_i - A_{i-1} = -w_i(m_i/r_i), \quad and \quad a_i = w_i + A_i,$$

for $i = 1, 2, \dots, k$.

4.9 Appendix B4: Computer programs

B4.1 Homogeneity of three samples

This is an illustration of testing for homogeneity of several multinomial samples with three categories. We use the data from Example 4.2.

```
data tv;
input group resp wt@@;
lines;
1 1 35 1 2 95 1 3 20
2 1 60 2 2 30 2 3 10
3 1 10 3 2 10 3 3 30
;
proc freq data=tv;
weight wt;
table group*resp/norow nocol nopercent expected chisq;
title1 'COMPUTATION FOR HOMOGENEITY OF 3 SAMPLES';
title2 'Results for Example 4.2';
run;
```

Output

```
            COMPUTATION FOR HOMOGENEITY OF 3 SAMPLES
                     Results for Example 4.2
                        The FREQ Procedure
                     Table of GROUP by RESP


         GROUP       RESP

         Frequency|
         Expected |      1|        2|       3|  Total
         ---------+--------+--------+--------+
               1 |     35 |     95 |     20 |    150
                 |   52.5 |   67.5 |     30 |
         ---------+--------+--------+--------+
               2 |     60 |     30 |     10 |    100
                 |     35 |     45 |     20 |
         ---------+--------+--------+--------+
               3 |     10 |     10 |     30 |     50
                 |   17.5 |   22.5 |     10 |
         ---------+--------+--------+--------+
         Total         105      135       60      300
```

Statistics for Table of GROUP by RESP

Statistic	DF	Value	Prob
Chi-Square	4	98.3862	<.0001
Likelihood Ratio Chi-Square	4	85.3258	<.0001
Mantel-Haenszel Chi-Square	1	4.5346	0.0332
Phi Coefficient		0.5727	
Contingency Coefficient		0.4970	
Cramer's V		0.4049	

Sample Size = 300

B4.2 Analysis of several independent samples

This is an illustration of testing homogeneity of several samples using the nonparametric F-test. We use the data from Example 4.3. We do the three linear rank tests using the NPAR1WAY procedure.

```
data time;
group='A';
do I=1 to 6;
input T @@;
```

```
output;
end;
group='B';
do I=1 to 6;
input T@@;
output;
end;
group='C';
do I=1 to 6;
input T@@;
output;
end;
group='D';
do I=1 to 6;
input T@@;
output;
end;
drop I;
lines;
30 29 34 31 32 30
35 25 28 35 30 31
32 23 24 29 29 28
27 29 28 31 27 26
;
proc npar1way wilcoxon savage median;
class group;
var T;
title 'Results for Example 4.3';
run;
```

Output

Results for Example 4.3

The NPAR1WAY Procedure

Wilcoxon Scores (Rank Sums) for Variable T
Classified by Variable GROUP

GROUP	N	Sum of Scores	Expected Under H0	Std Dev Under H0
A	6	102.00	75.0	14.918256
B	6	91.00	75.0	14.918256
C	6	54.50	75.0	14.918256
D	6	52.50	75.0	14.918256

Average scores were used for ties.

Kruskal-Wallis Test

Chi-Square	6.4417
DF	3
Pr > Chi-Square	0.0920

Median Scores (Number of Points Above Median) for
Variable T Classified by Variable GROUP

GROUP	N	Sum of Scores	Expected Under H0	Std Dev Under H0
A	6	5.250	3.0	1.013496
B	6	4.000	3.0	1.013496
C	6	1.500	3.0	1.013496
D	6	1.250	3.0	1.013496

Average scores were used for ties.

Median One-Way Analysis

Chi-Square	8.3056
DF	3
Pr > Chi-Square	0.0401

Savage Scores (Exponential) for Variable T
Classified by Variable GROUP

GROUP	N	Sum of Scores	Expected Under H0	Std Dev Under H0
A	6	1.966127	0.0	1.957456
B	6	3.363559	0.0	1.957456
C	6	-2.384585	0.0	1.957456
D	6	-2.945101	0.0	1.957456

Average scores were used for ties.

Savage One-Way Analysis

Chi-Square	5.7819
DF	3
Pr > Chi-Square	0.1227

B4.3 Computation of Jonckheere's test

Here we are interested in constructing the Jonckheere-Terpstra statistic for ordered alternatives. We use the data for three samples given in Example 4.5. We first enter the data from all three samples in one array, x1-x18. We compare the first sample with the second and third samples. Next we compare the second sample with the third sample. We calculate $U_{12} + U_{13}$ and U_{23} and the sum of these two is our statistic. We use the representation (A11) of Appendix A4. The variance is computed using the expression (A12). The mean and variance are used to compute the approximate P-value.

```
data onet;
input nx ny nz x1-x18;
lines;
6 6 6
120 170 130 150 126 148 200 160 210 155 220 154 230 145 180
175 165 250
run;
data oneth;
set onet;
array X(I)x1-x6;
array W(L)x7-x18;
do I=1 to nx;
do L=1 to ny+nz;
if X lt W then c2=1;
if X eq W then c2=0.5;
if X gt W then c2=0;output;
end;
end;
run;
data twoth;
set onet;
npx1=nx+1;
nxpny=1+nx+ny;
array Y(J) x7-x12;
array Z(K) x13-x18;
do J=1 to ny;
do K=1 to nz;
If Y lt Z then c3=1;
if Y eq Z then c3=0.5;
if Y gt Z then c3=0;output;
end;
end;
run;
proc iml;
```

```
use oneth;
read all variables{c2} into M2;
W2=sum(M2);
W123=W2;
print "MW statistic for X < (Y,Z)" W123;
use twoth;
read all variables{c3} into M3;
W3=sum(M3);
print "MW statistic for Y < Z " W3;
T=W123+W3;
print 'J-T Statistic' T;
nx=6;
ny=6;
nz=6;
n=nx+ny+nz;
snx=(nx)*(nx);
sny=(ny)*(ny);
snz=(nz)*(nz);
sqn=snx+sny+snz;
sn=n*n;
et=(sn - sqn)/4;
print 'Mean' et;
tnxx=nx*(ny+nz)*(n+1);
tnyy=ny*nz*(ny+nz+1);
vt=(tnxx+tnyy)/12;
svt=sqrt(vt);
sv=round(svt,.0001);
print 'var and std.' vt sv;
ztt=(T - et)/svt;
ptt=1-probnorm(ztt);
zt=round(ztt,.0001);
pt=round(ptt,.0001);
print 'stat. and p-value' zt pt;
quit;
```

Output

W123

MW statistic for X < (Y,Z)   65

W3

MW statistic for Y < Z	21	
	T	
J-T Statistic	86	
	ET	
Mean	54	
	VT	SV
var and std.	153	12.3693
	ZT	PT
stat. and p-value	2.587	0.0048

B4.4 Comparison of survival in three groups

Here we are using a data set given by Lee and Desu (1972). There are three groups.

```
data two;
input  trt time censor@@;
lines;
1 4 1 1 5 1 1 9 1 1 10 1 1 10 1 1 12 1 1 13 1
1 23 1 1 28 1 1 28 1 1 28 1 1 29 1 1 31 1 1 32 1 1 37 1
1 41 1 1 41 1 1 57 1 1 62 1 1 74 1 1 100 1 1 139 1
1 20 0 1 258 0 1 269 0
2 8 1 2 10 1 2 10 1 2 12 1 2 14 1 2 20 1 2 48 1
2 70 1 2 75 1 2 99 1 2 103 1 2 162 1 2 169 1 2 195 1
2 220 1
2 161 0  2 199 0 2 217 0 2 245 0
3 8 1 3 10 1 3 11 1 3 23 1 3 25 1 3 25 1 3 28 1 3 28 1
3 31 1 3 31 1 3 40 1 3 48 1 3 89 1 3 124 1 3 143 1
3 12 0 3 159 0 3 190 0 3 196 0 3 197 0 3 205 0 3 219 0
;
proc lifetest notable;
time time*censor(0);
strata trt;
run;
```

Output

The SAS System

The LIFETEST Procedure
Testing Homogeneity of Survival Curves
for time over Strata
Rank Statistics

trt	Log-Rank	Wilcoxon
1	6.6349	273.00
2	-3.6934	-170.00
3	-2.9415	-103.00

Covariance Matrix for the Log-Rank Statistics

trt	1	2	3
1	10.1552	-5.1264	-5.0289
2	-5.1264	11.4176	-6.2913
3	-5.0289	-6.2913	11.3201

Covariance Matrix for the Wilcoxon Statistics

trt	1	2	3
1	20330.0	-9556.8	-10773.2
2	-9556.8	19518.0	-9961.2
3	-10773.2	-9961.2	20734.4

Test of Equality over Strata

Test	Chi-Square	DF	Pr > Chi-Square
Log-Rank	4.3484	2	0.1137
Wilcoxon	3.7815	2	0.1510

Table 4.16 *Pain scores under different treatments*

Pain Score	Saline	10 mg	20 mg	40 mg
None	1	5	4	11
Mild	7	12	12	18
Severe	22	13	14	1

Table 4.17 *CGI scores for various groups*

Group	Score													
PE	4	1	2	2	2	1	3	2	5	2	5	2	2	2	6
CL	1	1	2	2	2	2	3	1	1	2	2	1	1	1	2
PBO	5	5	5	5	2	4	1	4	2	4	5	4	5	5	3

4.10 Problems

1. Patients usually receive a certain dose of etomidate for induction of anaesthesia. Etomidate is associated with side effects that include pain on injection. Mixing lidocaine with etomidate may result in the reduction of pain. A study to demonstrate the efficacy of lidocaine in decreasing the pain was undertaken (Yazigi et al. (2002)). In the study four mixtures of etomidate were used. They are created by adding (1) saline, (2) 10 mg of lidocaine, (3) 20 mg of lidocaine, and (4) 40 mg of lidocaine, respectively, to the usual dose of etomidate (20 mg). These four treatments were randomly assigned, so that each treatment was assigned to 30 patients. Each patient was asked to grade the pain in the arm as *none, mild,* or *severe.* The results appear in Table 4.16. Test whether or not the treatments are equally effective.

2. Munzel and Brunner (2000) reported a data set from a psychiatric clinical study. In this study 45 patients with panic disorder were randomly divided into three treatment groups. One group of 15 patients had to perform physical exercise (PE). Another group of 15 patients were given clomipramine (CL) and the third group is the placebo group (PBO). After 4 weeks of treatment, clinical global impression (CGI) scores were assigned by the investigator. The scoring used a seven-point ordinal scale (0 = not ill, 6 = extremely ill). The scores are listed in Table 4.17.

Test whether or not the treatments are equally effective. (This data set is an ordered categorical data set. Use the Kruskal-Wallis test.)

3. Starting from the expression (4.10) for T_C, derive the simplified expression given in Remark 4.1.

4. Verify the result (4.52).

Table 4.18 *Weekly sales of a product*

Campaign	Sales						
Convenience	50	75	80	82	85	76	74
Quality	80	90	85	87	95	60	83
Price	70	65	60	40	50	45	52

Table 4.19 *Time to complete a statistical problem*

Brand	Times									
A	45	50	46	47	45	50	52	30	50	47
B	50	40	25	45	51	60	50	40	55	51
C	54	47	46	50	47	50	46	60	40	40

5. A market researcher wanted to find out whether an advertisement should focus on (a) convenience, (b) quality, or (c) price of a product to increase sales. She randomly chose three cities and ran the advertisement campaign focusing on convenience in the first city, quality in the second city, and price in the third city. Seven stores were randomly selected in each city, and weekly sales (number of items of the product sold) were recorded and are given in Table 4.18. Do these data indicate that the distributions of items sold in a week differ depending on the advertisement focus?

6. We are interested in comparing three statistical packages for the speed in getting the final outputs for a statistical problem. Thirty volunteers were recruited and were randomly divided into three groups of 10 each. Three packages were randomly assigned to the three groups and the time (in min.) for each volunteer to get the final output from the start was noted for each volunteer to complete the same statistical problem. The data appear in Table 4.19. Let F_A, F_B, and F_C be the distribution functions of the time variable. Do these data suggest the following ordering,

$$H_{CO2} : \Delta_A > \Delta_B > \Delta_C,$$

where Δs are the shift parameters?

7. There are eight sections in a basic statistics course at a large university. Based on a common final examination, the median score for all students was determined. The number of students scoring above the median and the number of students in each section are given in Table 4.20. Using Mood's median procedure, test the null hypothesis that the score distributions are the same. Calculate the P-value. (Hint: Prepare a 2×8 table like Table 4.4 and apply the chi-square procedure of Subsection 4.1.1.)

Table 4.20 *Distribution of scores in a final examination*

	Section							
	1	2	3	4	5	6	7	8
Above Median	25	15	16	24	9	21	10	10
Sect. Size	40	32	30	38	20	40	30	30

Table 4.21 *Galactose binding measurements*

Crohn's Disease	Ulcerative Colitis	Controls	
1343	1264	1809	2850
1393	1314	1926	2964
1420	1399	2283	2973
1641	1605	2384	3171
1897	2385	2447	3257
2160	2511	2479	3271
2169	2514	2495	3288
2279	2767	2525	3358
2890	2827	2541	3643
	2895	2769	3657
	3011		
	3013		
	3355		

8. Altman and Bland (1996) presented a data set on measurements of galactose binding in three groups of patients (data from M. Weldon). The object is to compare the means. The data appear in Table 4.21. Find the P-value for the Kruskal-Wallis test.

9. *Alternative to Jonckheere's test:* Consider the problem of testing the null hypothesis (4.1) against the alternative H_{CO1} of (4.3). Assume that all sample sizes are equal to n. Combine all the samples and rank the observations, as in the Kruskal-Wallis procedure. Let $R_{i.}$ be the rank sum for the observations of the ith sample. The test statistic is $T = \sum_{i=1}^{k} iR_{i.}$, and the test is to

$$reject\ H_0\ if\ T > C.$$

For large samples the null distribution of T can be approximated by a normal distribution. To get this approximate distribution, we need to find the null mean and variance of T. Show that

$$E_0(T) = \frac{N(N+1)(k+1)}{4},$$

Table 4.22 *Service time to promotion*

Dept.				Times				
1	60	45	20+	58	15	36		
2	40	36	12	30+	45	25+	30+28	48
3	50	30+	48	40				

and

$$var_0(T) = \frac{(k^2 - 1)N^2(N + 1)}{144}.$$

Using these values give an approximation for C to make the above test an approximate size α test.

10. In the discussion in Appendix A2, it was asserted that W_J can be expressed as the sum of several uncorrelated statistics. Consider the special case of three samples and the representation (A7). Prove that the statistics U_1 and U_2 of (A7) are uncorrelated.

11. (a) Using the results of Appendix A4.5, and using the fact that the weights for the logrank test are $w_i = 1$, show that the logrank scores are

$$a_i = 1 - \sum_{j=1}^{i}(m_j/r_j); \quad and \quad A_i = -\sum_{j=1}^{i}(m_j/r_j),$$

as indicated in Chapter 2. (b) Using the weights $w_i = r_i$, show that the scores for Gehan's Wilcoxon statistic are

$$a_i = r_i - \sum_{j=1}^{i}m_j; \quad and \quad A_i = -\sum_{j=1}^{i}m_j.$$

12. In a large company, the times (in months) for getting a promotion from the time of joining were recorded in three departments for samples of employees. Some employees left the company for personal reasons before they got a promotion. Their service time in the company is indicated by "+" and these data are considered censored observations. The data appear in Table 4.22. Are there differences in the distributions of promotion times in various departments?

4.11 References

Altman, D.G. and Bland, J.M. (1996). Statistics notes: comparing several groups using analysis of variance. *Brit. Med. J.*, **312**, 1472–1473.

Baker, G.R. and Spurrier, J.D. (1998). Comparisons with a control using normal scores and Savage scores. *J. Nonpar. Statist.*, **9**, 123–139.

Brookmeyer, R. and Crowley, J.J. (1982). A k-sample median test for censored data. *J. Amer. Statist. Assoc.*, **77**, 433–440.

Chacko, V.J. (1963). Testing homogeneity against ordered alternatives. *Ann. Math. Statist.*, **34**, 945–956.

Chakraborti, S. (1984). A generalization of the control median test. Unpublished Ph.D. dissertation, State University of New York at Buffalo.

Chakraborti, S. and Desu, M.M. (1988a). A class of distribution-free tests for testing homogeneity against ordered alternatives. *Statist. Prob. Letters*, **6**, 251–256.

Chakraborti, S. and Desu, M.M. (1988b). Generalizations of Mathisen's median test for comparing several treatments with a control. *Comm. Statist. Simul. Comp.*, **17**, 947–967.

Chakraborti, S. and Desu, M.M. (1990). Quantile tests for comparing several treatments with a control under unequal right censoring. *Biom. J.*, **6**, 697–706.

Chakraborti, S. and Desu, M.M. (1991). Linear rank tests for comparing treatments with a control when data are subject to unequal patterns of censorship. *Statist. Neer.*, **45**, 227–254.

Chen, Y. (1994). A generalized Steel procedure for comparing several treatments with a control under random right-censorship. *Comm. Statist. Simul.*, **23**, 1–16.

Cochran, W.G. (1952). A chi-square test for goodness of fit. *Ann. Math. Statist.*, **23**, 315–345.

Critchlow, D.E. and Fligner, M.A. (1991). On distribution-free multiple comparisons in one-way analysis of variance. *Comm. Statist. Theory and Methods*, **20**, 127–139.

Crowley, J. (1979). Some extensions of the logrank test. In *Clinical Trials in Early Breast Cancer*, Edited by H.R. Scheurlen, G. Weckesser, and J. Armbuster, Vol. 4, Springer-Verlag, Berlin, 213–223.

Desu, M.M. and Raghavarao, D. (1990). *Sample Size Methodology*, Academic Press, Boston.

Desu, M.M., Park, S., and Chakraborti, S. (1996). Linear rank statistics for the simple tree alternatives. *Biom. J.*, **38**, 359–373.

Dwass, M. (1960). Some k-sample rank-order tests. In *Contributions to Probability and Statistics*, Edited by Olkin, I. et al., Stanford University Press, Stanford, California.

Fine, J.P. and Bosch, R.J. (2000). Risk assessment via a robust probit model with applications to toxicology. *J. Amer. Statist. Assoc.*, **95**, 375–382.

Fligner, M. and Wolfe, D. (1982). Distribution-free tests for comparing treatments with a control. *Statist. Neer.*, **36**, 119–127.

Gupta, S.S. (1963). Probability integrals of multivariate normal and multivariate t. *Ann. Math. Statist.*, **34**, 792–828.

Harter, H.L. (1960). Tables of range and studentized range. *Ann. Math. Statist.*, **31**, 1122–1147.

Hochberg, Y. and Tamhane, A.C. (1987). *Multiple Comparison Procedures*, John Wiley & Sons, New York.

Hogg, R.V. (1965). On models and hypotheses with restricted alternatives. *J. Amer. Statist. Assoc.*, **60**, 1153–1162.

Iman, R.L., Quade, D., and Alexander, D.A. (1975). Exact probability levels for the Kruskal-Wallis test. In *Selected Tables in Mathematical Statistics*, Vol. 3, Edited by H.L. Harter and D.B. Owen, American Mathematical Society and Institute of Mathematical Statistics, Providence, Rhode Island, 329–384.

Iman, R.L. and Davenport, J.M. (1976). New approximations to the exact distribution of the Kruskal-Wallis test statistic. *Comm. Statist.*, **A5**, 1335–1348.

Jonckheere, A.R. (1954). A distribution-free k-sample test against ordered alternatives. *Biometrika*, **41**, 133–145.

Kruskal, W.H. and Wallis, W.A. (1952). Use of ranks on one criterion variance analysis. *J. Amer. Statist. Assoc.*, **47**, 583–621. (Corrections appeared in Vol. 48, pp. 907–911.)

Lee, E.T. and Desu, M.M. (1972). A computer program for comparing k-samples with right-censored data. *Comp. Prog. Biomedicine*, **2**, 315–321.

Leton, E. and Zuluaga, P. (2002). Survival tests for r groups. *Biom. J.*, **44**, 15–27.

Miller, R.G., Jr. (1981). *Simultaneous Statistical Inference*, 2nd edition, Springer-Verlag, New York.

Munzel, U. and Brunner, E. (2000). Nonparametric tests in the unbalanced multivariate one-way design. *Biom. J.*, **42**, 837–854.

Nemenyi, P. (1963). Distribution-Free Multiple Comparisons, Unpublished doctoral thesis, Princeton University, Princeton, New Jersey.

Patel, K.M. and Hoel, D.G. (1973). A generalized Jonckheere k-sample test against ordered alternatives when observations are subject to arbitrary right censorship. *Comm. Statist.*, *A*, **2**, 373–380.

Peddada, S.D., Prescott, K.E., and Conway, M. (2001). Tests for ordered restrictions in binary data. *Biometrics*, **57**, 1219–1227.

Rao, C.R. (1973). *Linear Statistical Inference and Its Applications*, 2nd edition, John Wiley & Sons, New York.

Scheffe, H. (1959). *Analysis of Variance*, John Wiley & Sons, New York.

Sen, P.K. (1962). On the role of a class of quantile tests in some multisample nonparametric problems. *Cal. Statist. Assoc. Bull.*, **11**, 125–143.

Shorack, G.R. (1967). Testing against ordered alternatives in model analysis of variance: normal theory and nonparametric. *Ann. Math. Statist.*, **38**, 1740–1752.

Slivka, J. (1970). A one-sided nonparametric multiple comparison control percentile test: treatments versus control. *Biometrika*, **57**, 431–438.

Soms, A.P. and Torbeck, L.D. (1982). Randomization tests for k-sample binomial data. *J. Qual. Tech.*, **14**, 220–224.

Spurrier, J.D. (1992). Generalizations of Steel's many-one rank sum test. *J. Nonpar. Statist.*, **1**, 287–299.

Steel, R.G.D. (1959). A multiple comparison rank sum test: treatments versus control. *Biometrics*, **15**, 560–572.

Steel, R.G.D. (1960). A rank sum test for comparing all pairs of treatments. *Technometrics*, **2**, 197–207.

Steel, R.G.D. (1961). Some rank sum multiple comparisons tests. *Biometrics*, **17**, 539–552.

Terpstra, T.J. (1952). The asymptotic normality and consistency of Kendall's test against trend, when ties are present in one ranking. *Indag. Math.*, **14**, 327–333.

van de Wiel, M.A. (2002). Exact distributions of multiple comparisons rank statistics. *J. Amer. Statist. Assoc.*, **97**, 1081–1089.

Yazigi, A., Madi-Jebara, S., Haddal, F., and Hayek, G. (2002). Dose-response relationship for the addition of lidocaine to etomidate for reducing pain on injection. *Amer. J. Anaes.*, **27**, 331–333.

CHAPTER 5

Analysis of block designs

5.1 Introduction

In Chapter 3 we mentioned that sometimes we take groups of homogeneous experimental units called blocks and assign treatments to units in each block at random. These experiments are called experiments with randomized blocks. We explore the cases where each block contains k units, so that each treatment is represented in each block. When in each block all the v treatments are represented, the design is called a *randomized complete block (RCB)* design. In some cases in each block $d(> 1)$ units are assigned to each treatment. Such a design is called a *generalized randomized complete block* (GRCB) design. We also consider incomplete block designs, where not all treatments appear in each block. First, we consider procedures to analyze studies with a binary response and then we give procedures for a continuous response. In both these cases the main objective is to test the hypothesis that all the treatments are equally effective. We also discuss procedures for censored data situations will also be discussed.

In the discussion we consider settings where there are v treatments under study and b blocks of experimental units are to be used. The ith block has k_i experimental units and the jth treatment is assigned to n_{ij} units in the ith block. When $n_{ij} = 1$, for all $i = 1, 2, \ldots, b$ and $j = 1, 2, \ldots, v$, we have a RCB design. When $n_{ij} = d(> 1)$ for all i and j, we get a GRCB design. When some n_{ij} are zero, we get an incomplete block design. Let Y_{ijl} for $l = 1, 2, \ldots, n_{ij}$ be the responses for the jth treatment in the ith block. We suppress the suffix l, while considering RCB designs.

5.2 RCB designs with binary responses

The data are Y_{ij}, ($i = 1, 2, \ldots, b$ and $j = 1, 2, \ldots, v$), where Y_{ij} is the response under treatment j in block i. Here Y_{ij} is either 0 or 1. The probability model is

$$P[Y_{ij} = 1] = \theta_{ij}. \tag{5.1}$$

Further, the response variables are independent. The null hypothesis states that the treatment effects are all equal. This hypothesis is formulated as

$$H_0 : \theta_{i1} = \theta_{i2} = \cdots = \theta_{iv}, \quad \text{for each } i = 1, 2, \cdots, b. \tag{5.2}$$

The procedure, proposed by Cochran (1950), will be described now. For each treatment, we count the number, B_j, of blocks where the response is 1 and for each block we count the number, L_i, of responses that are equal to 1. The quantities B_j indicate the treatment effects and under the null hypothesis the expectations of these quantities are equal to each other. Let τ denote this common mean, which is given by

$$\tau = E(B_j|H_0) = E\left(\sum_i Y_{ij}|H_0\right) = \sum_i \Theta_i, \tag{5.3}$$

where Θ_i is the common value of θ_{ij} under H_0. So a measure of variability among B_j will give us an indication whether to accept or reject the null hypothesis. One such measure is

$$S = \sum_{j=1}^{v} [B_j - \bar{B}]^2, \tag{5.4}$$

where $\bar{B} = \sum_j B_j/v$. To assess whether the observed value of S is to be considered as an extreme value or not, we compute the expected value of S under the null hypothesis. An estimator of this expectation is obtained and the statistic S is compared to the estimator of its expectation.

To find the expectation of S, let us rewrite S as

$$S = \sum_{j=1}^{v} [(B_j - \tau) - (\bar{B} - \tau)]^2$$

$$= \sum_{j=1}^{v} (B_j - \tau)^2 - v(\bar{B} - \tau)^2.$$

Thus

$$E(S|H_0) = \sum_j var(B_j|H_0) - v\, var(\bar{B}|H_0).$$

Noting that

$$var(\bar{B}|H_0) = (1/v^2) \sum_j var(B_j|H_0),$$

and

$$var(B_j|H_0) = \sum_i \Theta_i(1 - \Theta_i) \equiv \sigma^2, \tag{5.5}$$

we have

$$E(S|H_0) = (v - 1)\sigma^2 (= \mu,\, say).$$

An unbiased estimator of the expectation, μ, is

$$\hat{\mu} = \left[v \sum_{i=1}^{b} L_i - \sum_{i=1}^{b} L_i^2 \right] \Big/ v. \tag{5.6}$$

Now if the S value is very different from $\hat{\mu}$, we would tend to reject the null hypothesis. The Cochran's test criterion is a multiple of the ratio of S to $\hat{\mu}$,

$$Q_C = \frac{v(v-1)\sum_{j=1}^{v}[B_j - \bar{B}]^2}{v\sum_{i=1}^{b}L_i - \sum_{i=1}^{n}L_i^2}, \tag{5.7}$$

and the test is to

$$\text{reject } H_0 \text{ if } Q_C > c, \tag{5.8}$$

where c is a percentile of the null distribution of Q_C. The null distribution of Q_C can be approximated by a chi-square distribution with $(v-1)$ degrees of freedom. Thus the P-value can be approximated as

$$P\text{-value} \approx Pr\left[\chi_{v-1}^2 > Q_C(obs)\right],$$

where $Q_C(obs)$ is the observed value of the statistic Q_C.

Patil (1975) derived the exact null distribution of the statistic Q_C. The derivations of the expectation of S and the estimation of this expectation are included in problems.

Remark 5.1. When $v = 2$, the statistic Q_C reduces to the statistic suggested by McNemar.

Example 5.1. Blood samples of 10 subjects were sent to four different laboratories for a diagnostic test. A positive result is denoted by 1, and a negative result is denoted by 0. The results from the various laboratories are given in Table 5.1. Is there enough evidence that the laboratories are equally efficient in detecting the presence of the disease?

Here $\bar{B} = (20/4) = 5$. Hence

$$\sum_j (B_j - \bar{B})^2 = (4 - 5)^2 + (5 - 5)^2 + (5 - 5)^2 + (6 - 5)^2 = 2,$$

and

$$v\sum_i L_i - \sum_i L_i^2 = (4)(20) - 52 = 28.$$

Table 5.1 *Laboratory test results*

Labs	Subjects										Total (B_j)
	1	2	3	4	5	6	7	8	9	10	
A	0	0	1	1	0	1	1	0	0	0	4
B	1	0	0	0	1	1	1	0	0	1	5
C	0	0	0	1	0	1	1	1	0	1	5
D	1	1	1	0	1	0	1	0	0	1	6
Total (L_i)	2	1	2	2	2	3	4	1	0	3	20

Finally, $Q_C = 4(3)(2)/28 = 0.8571$, with an approximate P-value of 0.8358. So we do not reject the null hypothesis that the laboratories are equally efficient in detecting the disease.

5.3 RCB designs with continuous uncensored data

Let F_{ij} be the distribution of the response variable Y_{ij}. The response variables Y_{ij} are assumed to be independent. We are interested in testing the null hypothesis that the treatment effects are equal and it is formulated as

$$H_0 : F_{i1} = F_{i2} = \cdots = F_{iv}, \quad \text{for each } i = 1, 2, \ldots, b. \tag{5.9}$$

One of the earliest procedures is the test proposed by Friedman (1937), which depends on the ranks of the observations. Brown and Mood (1951) proposed a median procedure. Downton (1976) proposed a procedure based on the exponential scores. First we describe the rank procedure of Friedman and then we give a class of rank tests. We describe a special member of this class, the median procedure. We state the proportional hazard model and we describe Downton's procedure.

5.3.1 Friedman's test

Observations are ranked within each block. Let R_{ij} be the rank of Y_{ij}, the jth treatment observation in the ith block. The sum of the ranks received by the jth treatment is

$$R_{\cdot j} = \sum_{i=1}^{b} R_{ij}. \tag{5.10}$$

Under H_0,

$$E(R_{\cdot j}|H_0) = \sum_{i=1}^{b} E(R_{ij}|H_0) = b(v+1)/2,$$

which does not depend on j. So the differences, $T_j = R_{\cdot j} - b(v+1)/2$, will give us information that can be used to test the null hypothesis. Let $\mathbf{T}' = (T_1, T_2, \ldots, T_v)$. One way to combine the information provided by the differences is to consider the measure

$$D = \sum_{j=1}^{v} [R_{\cdot j} - b(v+1)/2]^2 = \mathbf{T}'\mathbf{T}, \tag{5.11}$$

and then we would reject the null hypothesis if this measure is large. To decide which values of D are to be considered as extreme, we could compare its value with its null expectation. So let us compute the expected value of D, under the null hypothesis. We note that

$$E(D|H_0) = \sum_{j=1}^{v} E[\{R_{\cdot j} - b(v+1)/2\}^2|H_0] = \sum_{j=1}^{v} var[R_{\cdot j}|H_0].$$

Now

$$var(R_{.j}|H_0) = \sum_{i=1}^{b} var(R_{ij}|H_0) = \sum_{i=1}^{b} \frac{v^2 - 1}{12} = b(v^2 - 1)/12.$$

Finally, we have

$$E(D|H_0) = v\, var(R_{.j}|H_0) = vb(v^2 - 1)/12 \equiv v\, var(T_j|H_0), \qquad (5.12)$$

and we could use the ratio $\frac{D}{E(D|H_0)} = \frac{\mathbf{T'T}}{v\, var(T_j|H_0)}$ as a test criterion. A multiple of this ratio is the test statistic of Friedman,

$$Q_F = \frac{(v-1)\mathbf{T'T}}{v\, var(T_j|H_0)} = \frac{12}{bv(v+1)}\mathbf{T'T}.$$

The test is to

$$reject\ H_0\ if\ Q_F > c. \qquad (5.13)$$

The critical value can be determined from the permutation distribution of Q_F. A table of these critical values is available in Hollander and Wolfe (1999, Table A.22). For large b the null distribution of Q_F can be approximated by the chi-square distribution with $(v-1)$ degrees of freedom. So an approximate α-level test is to

$$reject\ H_0\ if\ Q_F > \chi^2_{v-1}(1 - \alpha), \qquad (5.14)$$

and the approximate P-value is

$$P\text{-value} \approx P\big(\chi^2_{v-1} > Q_F(obs)\big), \qquad (5.15)$$

where $Q_F(obs)$ is the observed value of the statistic Q_F.

<u>Remark 5.2.</u> The statistic Q_F can be shown to be equal to $\mathbf{T'\Sigma^- T}$, where $\mathbf{\Sigma^-}$ is a symmetric generalized inverse of $\mathbf{\Sigma}$, the null covariance matrix of $\mathbf{T}$, which is approximately a multivariate normal random vector. This is the reason behind the chi-square approximation for the distribution of Q_F. Further details are given in Appendix A5 and Section 5.4.

<u>Remark 5.3.</u> When there are ties in some blocks, the expression for $E(D|H_0)$ is

$$E(D|H_0) = \sum_{i=1}^{b} \frac{1}{v} \sum_{l=1}^{v} \left(R_{il} - \frac{v+1}{2}\right)^2,$$

where R_{il} are the midranks assigned to the observations in the ith block, and then Q_F becomes

$$Q_F = \frac{v(v-1)\mathbf{T'T}}{\sum_{i=1}^{b}\sum_{l=1}^{v}(R_{il} - \frac{v+1}{2})^2}.$$

Remark 5.4. Iman and Davenport (1980) investigated the adequacy of the chi-square approximation to the null distribution of Friedman's test statistic Q_F. They suggested the use of a related statistic

$$Q_F^* = \frac{(b-1)Q_F}{[b(v-1) - Q_F]},$$
(5.16)

and an F distribution with $(v-1)$ and $(b-1)(v-1)$ degrees of freedom as an approximation to the null distribution of the statistic Q_F^*. As before, large values of the statistic lead to the rejection of the null hypothesis. So an approximate P-value is

$$P\text{-value} \approx Pr[F(v-1, (b-1)(v-1)) > Q_F^*(obs)],$$

where $Q_F^*(obs)$ is the observed value of the statistic Q_F^*.

Remark 5.5. The statistic Q_F^* and the corresponding P-value can be obtained from a two-way analysis of variance computer program with the midranks as a response. If we want the statistic Q_F, it can be obtained as

$$Q_F = b(v-1)\frac{Q_F^*}{b-1+Q_F^*},$$

and the corresponding P-value can be found from equation (5.15).

Remark 5.6. This Friedman's procedure could be applied under somewhat less stringent assumptions. Instead of assuming that the response variables Y_{ij} are independent, we can assume that the vectors $(Y_{i1}, Y_{i2}, \ldots, Y_{iv})$ are independent with densities $f_i(y_1 - \theta_1, y_2 - \theta_2, \ldots, y_v - \theta_v)$, where the function $f_i(z_1, z_2, \ldots, z_v)$ is symmetric in its arguments. Under these assumptions the null distribution of the ranks of the responses in each block is the same as in the earlier discussion. Thus the earlier analysis is also applicable under the new assumptions.

Remark 5.7. When the response is binary, using midranks one can show that the statistic Q_F reduces to the statistic Q_C of (5.7). Thus Friedman's test (5.13) is the same as Cochran's test (5.8), so Cochran's Q_C test can also be used when the responses in a block are not necessarily independent but satisfy the symmetry condition mentioned in Remark 5.6. For a proof of the equality of statistics Q_C and Q_F, see Lehmann (1998, p. 269).

Remark 5.8. In the case of the comparison of two treatments, the statistic Q_F reduces to the sign test statistic. In particular, in this case

$$Q_F = 4b\left(\frac{A}{b} - \frac{1}{2}\right)^2,$$
(5.17)

Table 5.2 *Creatinine levels as determined by four labs*

Labs	Subjects									
	1	2	3	4	5	6	7	8	9	10
A	2.8	1.7	3.1	3.5	2.7	3.2	2.5	1.9	2.7	3.1
	(3)	(2)	(3)	(2)	(2)	(4)	(1)	(1)	(1)	(3)
B	2.6	1.8	3.0	3.6	2.9	3.0	2.6	2.0	2.9	3.8
	(1)	(3)	(2)	(3)	(4)	(2)	(2)	(2)	(3)	(4)
C	2.7	1.9	3.2	3.4	2.8	2.9	2.9	2.2	3.0	3.0
	(2)	(4)	(4)	(1)	(3)	(1)	(1)	(4)	(4)	(2)
D	2.9	1.6	2.8	3.7	2.6	3.1	2.8	2.1	2.8	2.9
	(4)	(1)	(1)	(4)	(1)	(3)	(3)	(3)	(2)	(1)

where A is the number of blocks in which treatment 1 received rank 1 and treatment 2 received rank 2.

Example 5.2. Creatinine levels for each of ten patients with renal problems were determined by four laboratories. The results appear in Table 5.2.

We are interested in testing the hypothesis that the labs are consistent in determining the creatinine levels.

The measurements for each person are ranked and these ranks are given in parentheses under the observations. From computer program output (see Appendix B5) the F-statistic for testing the equality of treatment (lab) effects is found to be $Q_F^* = 0.57$, with the approximate P-value of 0.6367. This is the statistic proposed by Conover and Iman. The statistic of Friedman is

$$Q_F = (10)(4-1)\frac{0.57}{10-1+0.57} = 1.8,$$

with an approximate P-value of 0.6149. Thus we do not reject the null hypothesis that the labs are consistent in determining the levels of creatinine.

5.4 Rank tests for RCB designs

In the case of comparison of several treatments using independent samples, ranks of the observations were used to score the observations and the test criterion is based on the scores. A similar idea can also be applied here, using the ranks within blocks. For the derivation of the test criterion in general, see Woolson and Lachenbruch (1981). Recall that the scores are chosen in relation to the particular population distributions. The popular scores are: Wilcoxon, Mood's median, van der Waerden, and Savage. Now we describe the general method, using the generic notation $a(.)$ for the scores.

We assume that the response variables Y_{ij} are independent and the pdf of Y_{ij} is

$$f_{ij}(y) = f_i(y - \tau_j), \tag{5.18}$$

where τ_j are the treatment effects and f_i are densities depending on the block index i. We are interested in testing the null hypothesis that the treatment effects are equal, that is,

$$H_0 : \tau_1 = \tau_2 = \cdots = \tau_v. \tag{5.19}$$

Friedman's test is a member of the class of tests to be discussed.

As in the case of Friedman's procedure, we rank the observations within blocks and the rank of Y_{ij} will be denoted by R_{ij}. The score attached to the observation Y_{ij} is a known function of the rank, namely, $a(R_{ij})$. The average of the scores within a block is

$$\bar{a} = (1/v) \sum_{j=1}^{v} a(R_{ij}).$$

We compute the sum of the scores received by the jth treatment, giving

$$A_j = \sum_{i=1}^{b} a(R_{ij}), \ j = 1, 2, \cdots, v.$$

The general form of the test statistic is

$$Q = \frac{v-1}{v \ var_0(A_j)} \sum_{j=1}^{v} [A_j - E_0(A_j)]^2,$$

where E_0 and var_0 are the mean and the variance under the null hypothesis. We recall that

$$E_0(A_j) = b\bar{a}; \quad var_0(A_j) = b \left[\frac{1}{v} \sum_{l=1}^{v} (a(l) - \bar{a})^2 \right].$$

Using these expressions, we see that

$$Q = \frac{v-1}{b \sum_{l=1}^{v} [a(l) - \bar{a}]^2} \sum_{j=1}^{v} [A_j - b\bar{a}]^2. \tag{5.20}$$

Large values of Q lead to the rejection of the null hypothesis. This statistic is approximately a chi-square variable with $(v-1)$ degrees of freedom, for large b. This discussion appears in Hájek and Šidák (1967, p. 118). The proof about the asymptotic chi-square distribution is given in Hájek and Šidák (1967, p. 173).

Remark 5.9. The denominator of the test statistic Q of (5.20) needs modification when there are ties in some blocks. Further, the test statistic Q of (5.20) can be retrieved, from the F-statistic, for testing the equality of treatment effects in a two-way ANOVA model with scores as data, using the relation

$$Q = b(v-1) \frac{F}{b-1+F}.$$

When the scores are ranks, the test based on Q is Friedman's test. A procedure based on the median scores was proposed by Brown and Mood (1951). We can use other scores such as the van der Warden scores. The median scores procedure will be described first, followed by Downton's procedure, which uses the exponential ordered scores.

5.4.1 Median procedures

Here median scores are attached to the observations. These scores are obtained by comparing the observations in a block with the median observation of the block. Let $m(R_{ij})$ be the score received by the observation Y_{ij}. Observations that are above the median receive a score of 1, whereas observations that are below the median receive a score of -1. These scores can be related to the ranks calculated for Friedman's procedure. They are given by

$$m(r) = \begin{cases} -1, & r < (v+1)/2 \\ 0, & r = (v+1)/2 \\ 1, & r > (v+1)/2. \end{cases}$$

These are the optimal scores used for constructing the linear rank statistics discussed in Chapter 2. Now sums of these scores received by each treatment

$$M_j = \sum_{i=1}^{b} m(R_{ij}), \quad j = 1, 2, \ldots, v$$

are calculated. The average of these scores is

$$\bar{m} = 0,$$

and

$$var_0(M_j) = b(1/v) \sum_{l=1}^{v} m^2(l) = b(2a/v),$$

where a is $\frac{v-1}{2}$ or $\frac{v}{2}$ depending on whether v is odd or even. Now the test statistic simplifies as

$$Q_{ME} = \frac{(v-1)}{2ba} \sum_{j=1}^{v} M_j^2. \tag{5.21}$$

This quantity is compared to a percentile of the chi-square distribution with $(v-1)$ degrees of freedom. Large values of this statistic lead to the rejection of the null hypothesis. So an approximate P-value is

$$P\text{-value} \approx Pr\left[\chi^2_{v-1} > Q_{ME}(obs)\right],$$

where $Q_{ME}(obs)$ is the observed value of the statistic Q_{ME}.

Table 5.3 *Median scores for the data of Table 5.2*

Labs	Subjects									
	1	2	3	4	5	6	7	8	9	10
A	1	−1	1	−1	−1	1	−1	−1	−1	1
B	−1	1	−1	1	1	−1	−1	−1	1	1
C	−1	1	1	−1	1	−1	1	1	1	−1
D	1	−1	−1	1	−1	1	1	1	−1	−1

Remark 5.10. Brown and Mood (1951) defined the median scores as

$$m_{BM}(r) = \begin{cases} 0, & r \leq (v+1)/2 \\ 1, & r > (v+1)/2. \end{cases}$$

With these scores we get the test proposed by Brown and Mood. Another set of median scores considered by Hájek and Šidák (1967) are defined as

$$m_{HS}(r) = \begin{cases} 0, & r < (v+1)/2 \\ 1/2, & r = (v+1)/2 \\ 1, & r > (v+1)/2. \end{cases}$$

The details of the median(m) scores procedure will be illustrated for the data of the previous example.

Example 5.2 (cont'd.). We compute the median scores $m(r)$ for the data of Table 5.2 and these are given in Table 5.3.

Using the scores given in Table 5.3 as data and a computer procedure for a two-way ANOVA, the F-statistic for testing the equality of treatment (lab) effects is found to be $F = 0.18$. Now the statistic Q_{ME} of (5.21) turns out to be

$$Q_{ME} = (10)(4-1)\frac{0.18}{10-1+0.18} = 0.6.$$

The corresponding approximate P-value is 0.8964. Thus we do not reject the null hypothesis that the labs are consistent in measuring the creatinine levels.

5.4.2 Downton's procedure

A test with Savage scores has been proposed by Downton (1976), which is appropriate under the following PH model. The model specifies the cdf of Y_{ij} as

$$F_{ij}(y) = 1 - (1 - F_i(y))^{\tau_j}. \tag{5.22}$$

The null hypothesis is (5.19), namely

$$H_0 : \tau_1 = \tau_2 = \cdots = \tau_v.$$

Downton derived the test using the partial likelihood method, proposed by Cox (1972). The scores are

$$s(r) = \sum_{l=1}^{r} \frac{1}{v - l + 1} - 1 \equiv \hat{\Lambda}(r) - 1, \quad for \ r = 1, 2, \ldots, v, \qquad (5.23)$$

so that the sum of the scores received by treatment j is

$$S_j = \sum_{i=1}^{b} s(R_{ij}), \quad j = 1, 2, \ldots, v.$$

Here the mean of S_j is zero and the variance of S_j is

$$var_0(S_j) = b(1/v) \left[\sum_{r=1}^{v} s^2(r) \right];$$

thus the statistic Q simplifies to

$$Q_S = \frac{(v-1)}{b \sum_{r=1}^{v} s^2(r)} \sum_j S_j^2. \qquad (5.24)$$

The approximate P-value is

$$P\text{-value} \approx Pr\left[\chi_{v-1}^2 > Q_S(obs)\right],$$

where $Q_S(obs)$ is the observed value of the statistic Q_S. We will illustrate the calculations for the data of Example 5.2.

Example 5.2 (cont'd.). The Savage scores for the data of Table 5.2 are given in Table 5.4. Using a computer program, the F-statistic is found to be $F = 0.8$ and the statistic Q_S is

$$Q_S = (10)(4 - 1)\frac{0.8}{10 - 1 + 0.8} = 2.46.$$

The corresponding approximate P-value is 0.48. Thus we do not reject the null hypothesis that the labs are consistent in measuring the creatinine levels.

Table 5.4 *Savage scores for the data of Table 5.2*

| Labs | Subjects | | | | | | | | | |
	1	2	3	4	5	6	7	8	9	10
A	.08	−.43	.08	−.43	−.43	1.08	−.75	−.75	−.75	.08
B	−.75	.08	−.43	.08	1.08	−.43	−.43	−.43	.08	1.08
C	−.43	1.08	1.08	−.75	.08	−.75	1.08	1.08	1.08	−.43
D	1.08	−.75	−.75	1.08	−.75	.08	.08	.08	−.43	−.75

5.5 General block designs with continuous uncensored data

In this section we consider block designs in a general setting with n_{ij} observations on the jth treatment in the ith block, and without any restrictions on block sizes or (treatment) replication numbers. Due to missing observations, even if an RCB design is used to conduct the experiment, at the analysis stage we may end up with unequal block sizes and replication numbers. In some cases the experimenter may not be able to choose blocks of equal size so a general block design may have to be used. The general results given here will reduce to the results of Section 5.3 in the settings considered there.

Using the numbers n_{ij}, we form the $b \times v$ matrix $\mathbf{N} = (n_{ij})$, which is called the *incidence matrix* of the block design. Let k_i be the number of units in the ith block, for $i = 1, 2, \ldots, b$, and r_j will be the number of units receiving the jth treatment for $j = 1, 2, \ldots, v$. Let $\mathbf{k}' = (k_1, k_2, \ldots, k_b)$ and $\mathbf{r}' = (r_1, r_2, \ldots, r_v)$ be the vectors of block sizes and of replication numbers. It is easy to see that

$$\mathbf{N1} = \mathbf{k}, \quad \mathbf{1}'\mathbf{N} = \mathbf{r}', \tag{5.25}$$

where $\mathbf{1}$ is the column vector of ones of appropriate order.

Let $Y_{ijl}, l = 1, 2, \ldots, n_{ij}$, be the responses from the units of the ith block receiving the jth treatment as indicated in Section 5.1. Note that there is no variable Y_{ijl} when $n_{ij} = 0$. We rank the observations within each block separately. The analysis is done using the within block ranks; however, a similar analysis using other scores can be done. The response variables Y_{ijl} are assumed to be independent and the common pdf is

$$f_{ij}(y) = f(y - \beta_i - \tau_j), \tag{5.26}$$

where β_i are the block effects and τ_j are the treatment effects. The null hypothesis of interest is

$$H_0 : \tau_1 = \tau_2 = \cdots = \tau_v. \tag{5.27}$$

Within each block, under the null hypothesis, all the observations have the same continuous distribution and hence the distribution of the ranks of the observations is completely known. As indicated earlier we rank the observations within each block separately. Let R_{ijl} be the rank of Y_{ijl} for $l = 1, 2, \ldots, n_{ij}; j = 1, 2, \ldots, v; i = 1, 2, \ldots, b$. Now

$$E_0(R_{ijl}) = \frac{1}{k_i} \sum_{u=1}^{k_i} u = \frac{(k_i + 1)}{2}; \tag{5.28}$$

and

$$\begin{aligned} var_0(R_{ijl}) &= \frac{1}{k_i} \sum_{u=1}^{k_i} u^2 - \frac{(k_i + 1)^2}{4} \\ &= \frac{k_i^2 - 1}{12} = \frac{k_i - 1}{k_i} \sigma_i^2, \end{aligned}$$

where $\sigma_i^2 = k_i(k_i + 1)/12$. Further,

$$cov_0(R_{ijl}, R_{ij'l'}) = \frac{1}{k_i} \sum_{u \neq w} uw - \frac{(k_i + 1)^2}{4},$$

$$= -\frac{k_i + 1}{12} = -\frac{1}{k_i}\sigma_i^2.$$

Let $R_{ij} = \sum_{l=1}^{n_{ij}} R_{ijl}$. The null expected value of this sum is $n_{ij}(k_i + 1)/2$ and we consider

$$R_{ij}^* = R_{ij} - n_{ij}(k_i + 1)/2 = R_{ij} - E_0(R_{ij}).$$

Now

$$var_0(R_{ij}^*) = var_0\left(\sum_{l=1}^{n_{ij}} R_{ijl}\right) = \left(n_{ij} - \frac{n_{ij}^2}{k_i}\right)\sigma_i^2.$$

Further,

$$cov_0(R_{ij}^*, R_{ij'}^*) = cov_0\left(\sum_{l=1}^{n_{ij}} R_{ijl}, \sum_{l'=1}^{n_{ij'}} R_{ij'l'}\right) = -\frac{n_{ij}n_{ij'}}{k_i}\sigma_i^2.$$

Let $T_j = \sum_i R_{ij}^*$, where the summation is over those blocks in which the jth treatment occurs. The vector of these sums is $\mathbf{T}' = (T_1, T_2, \ldots, T_v)$. We note the following facts:

$$\mathbf{T}'\mathbf{1} = 0, \quad and \quad E_0(\mathbf{T}) = \mathbf{0}. \tag{5.29}$$

Further,

$$var_0(\mathbf{T}) = \mathbf{\Lambda} = (\lambda_{jj'}),$$

where

$$\lambda_{jj} = \sum_i \left(n_{ij} - \frac{n_{ij}^2}{k_i}\right)\sigma_i^2, \tag{5.30}$$

and

$$\lambda_{jj'} = -\sum_i \frac{n_{ij}n_{ij'}}{k_i}\sigma_i^2. \tag{5.31}$$

The covariance matrix $\mathbf{\Lambda}$ is a singular non-negative definite matrix. The test statistic is

$$Q = \mathbf{T}'\mathbf{\Lambda}^-\mathbf{T}, \tag{5.32}$$

where $\mathbf{\Lambda}^-$ is a symmetric g-inverse of $\mathbf{\Lambda}$. This statistic is the one proposed by Bernard and van Elteren (1953) as shown by Brunden and Mohberg (1975). Large values of Q indicate rejection of H_0. From the asymptotic distribution theory, we have that the distribution of Q can be approximated by a chi-square distribution with ν degrees of freedom where ν is the rank of $\mathbf{\Lambda}$. Frequently designs for which $\nu = v - 1$ are used. For such designs, the P-value is

$$P\text{-value} \approx P(\chi^2_{v-1} > Q_{obs}), \qquad (5.33)$$

where Q_{obs} is the observed value of Q. Now we consider some special cases.

Remarks about computing

From the vector $\mathbf{T}$, delete one component, say T_v. The resulting vector can be denoted by $\mathbf{T_1}$. Let $\mathbf{\Lambda_1}$ be the covariance matrix of $\mathbf{T_1}$. The test statistic Q of (5.32) is equal to

$$Q = \mathbf{T'_1}\mathbf{\Lambda_1^{-1}}\mathbf{T_1}. \qquad (5.34)$$

Here we do not have to calculate a g-inverse.

In the case of proportional cell frequencies the expression for Q simplifies and we give the relevant details.

5.5.1 Proportional cell frequencies

Here the cell frequencies n_{ij} satisfy the relations

$$n_{ij} = k_i r_j / n,$$

where n is the total number of observations so that $n = \sum_i k_i = \sum_j r_j$. In this case the incidence matrix can be expressed as

$$\mathbf{N} = \frac{1}{n}\mathbf{kr'}.$$

Further,

$$\lambda_{jj} = \sum_i \left\{ \frac{k_i r_j}{n} - \frac{k_i^2 r_j^2}{n^2 k_i} \right\} \frac{k_i(k_i + 1)}{12}$$

$$= \frac{r_j}{n}\left(1 - \frac{r_j}{n}\right)\sum_i \frac{k_i^2(k_i + 1)}{12},$$

and

$$\lambda_{jj'} = -\sum_i \left(\frac{k_i r_j}{n}\right)\left(\frac{k_i r_{j'}}{n}\right)\left\{\frac{1}{k_i}\frac{k_i(k_i + 1)}{12}\right\}$$

$$= -\left(\frac{r_j}{n}\right)\left(\frac{r_{j'}}{n}\right)\sum_i \frac{k_i^2(k_i + 1)}{12}.$$

Let $p_j = r_j/n$ be the proportion of times the jth treatment is used. Let $\mathbf{p'} = (p_1, p_2, \ldots, p_v)$. The covariance matrix is

$$\mathbf{\Lambda} = \{\mathbf{D}(p_1, p_2, \ldots, p_v) - \mathbf{pp'}\}\sum_i \frac{k_i^2(k_i + 1)}{12},$$

where $D(., \ldots, .)$ is a diagonal matrix of its arguments. It can be shown that

$$\Lambda^- = \mathbf{D}\left(\frac{1}{p_1}, \frac{1}{p_2}, \ldots, \frac{1}{p_v}\right) \cdot \left\{\sum_i \frac{k_i^2(k_i+1)}{12}\right\}^{-1}.$$

Let R_j be the sum of the ranks received by the jth treatment and then $T_j = R_j - p_j \sum_i \frac{k_i(k_i+1)}{2}$. The test statistic becomes

$$Q_{Prp} = \frac{12}{\sum_i k_i^2(k_i+1)} \sum_{j=1}^{v} \frac{T_j^2}{p_j}$$

$$= \frac{12n}{\sum_i k_i^2(k_i+1)} \sum_{j=1}^{v} \frac{T_j^2}{r_j}.$$

Haux, Schumacher, and Weckesser (1984) proposed and studied a class of test statistics for this problem. Our special case procedure based on Q_{Prp} is a member of the class of test statistics proposed by Haux et al. Further details can be obtained from their paper.

Remark 5.11. If we further assume that we have only one block, the test statistic Q_{Prp} reduces to the Kruskal-Wallis statistic considered in Chapter 4.

In the following we consider two broad classes of designs and discuss the test statistics for them.

5.5.2 Equal block sizes

When the block sizes are all equal to k, we have

$$k_1 = k_2 = \cdots = k_b = k,$$

and for $i = 1, 2, \ldots, b$

$$\sigma_i^2 = k(k+1)/12 (= \sigma^2).$$

A matrix known as the C-matrix plays an important role in the parametric analysis of block designs. It is defined as

$$\mathbf{C} = \mathbf{D}(r_1, r_2, \ldots, r_v) - \mathbf{N}'\mathbf{D}\left(\frac{1}{k_1}, \frac{1}{k_2}, \ldots, \frac{1}{k_b}\right)\mathbf{N}, \qquad (5.35)$$

where $\mathbf{D}(., ., \ldots, .)$ is a diagonal matrix. In the present case this C-matrix simplifies to

$$\mathbf{C_1} = \mathbf{D}(r_1, r_2, \ldots, r_v) - \frac{1}{k}\mathbf{N}'\mathbf{N}.$$

The covariance matrix Λ is

$$\Lambda = \mathbf{C_1}\sigma^2, \qquad (5.36)$$

and the test statistic Q becomes $Q_1 = \mathbf{T}'\mathbf{C_1^-}\mathbf{T}/\sigma^2$.

Table 5.5 *Weight loss (in lb) under three plans in three centers*

Center	Plan		
	I	II	III
1	1, 1.5, 0.5	2, 2.5	3
2	0, 0.5, 1	2	2.5
3	1, 2		1.5, 2.5

5.5.3 Unequal block sizes

In the case of unequal block sizes, we suggest the use of a different test statistic

$$Q_C^* = \mathbf{T}'\left(\sigma_*^2 \mathbf{C}\right)^{-}\mathbf{T} = \mathbf{T}'\mathbf{C}^{-}\mathbf{T}/\sigma_*^2, \tag{5.37}$$

where $\sigma_*^2 = \sum_{i=1}^{b}(k_i - 1)\sigma_i^2/(n - b)$, n being $\sum_i k_i = \sum_j r_j$.

Remarks about computing

The test statistic Q_C^* recommended here can be obtained by running the PROC GLM procedure of SAS, using within block ranks as responses, and blocks and treatments as factors. The Type III treatment SS is the quadratic form $\mathbf{T}'\mathbf{C}^{-}\mathbf{T}$ and σ_*^2 is the within blocks mean square ignoring the treatments, that is,

$$\sigma_*^2 = \{Treatment\ SS(Type\ III) + Error\ SS\}/(n - b).$$

This computing procedure can be adopted even when there are tied observations in one or more blocks.

Example 5.3. An experimenter is interested in comparing three plans for weight reduction. Under Plan I, there are no restrictions. Plan II consists of a 30-minute exercise regimen and Plan III consists of a diet restriction of 1500 calories per day. Three health clinics were selected for this purpose. At each clinic nine volunteers were recruited and three volunteers were assigned to each plan. At the end of 2 months, data were collected on the volunteers strictly following the regimen. The weight loss (in lb) data appear in Table 5.5.

In the analysis, the centers are treated as blocks and the plans are the treatments to be compared. Here $b = 3$ and $v = 3$. The incidence matrix $\mathbf{N}$ for this case is

$$\mathbf{N} = \begin{pmatrix} 3 & 2 & 1 \\ 3 & 1 & 1 \\ 2 & 0 & 2 \end{pmatrix},$$

and thus $\mathbf{k}' = (6, 5, 4)$ and $\mathbf{r}' = (8, 3, 4)$. By ranking the observations within each center and simplifying, we get the quantities R_{ij}^*, which appear in Table 5.6.

Now

$$\sigma_1^2 = 3.5,\ \sigma_2^2 = 2.5,\ \sigma_3^2 = 1.67.$$

Table 5.6 R_{ij}^* values for the data of Table 5.5

Center	Plan		
	I	II	III
1	−4.5	2	2.5
2	−3	1	2
3	−1	0	1
T_j (Col. Total)	−8.5	3	5.5

Further,

$$\lambda_{11} = \left(3 - \frac{3 \times 3}{6}\right)(3.5) + \left(3 - \frac{3 \times 3}{5}\right)(2.5)$$

$$+ \left(2 - \frac{2 \times 2}{4}\right)(1.67) = 9.92,$$

and

$$\lambda_{22} = \left(2 - \frac{2 \times 2}{6}\right)(3.5) + \left(1 - \frac{1 \times 1}{5}\right)(2.5) = 6.67,$$

$$\lambda_{12} = -\left\{\frac{(3)(2)}{6}(3.5) + \frac{(3)(1)}{5}(2.5)\right\} = -5.$$

Thus the covariance matrix of (T_1, T_2) is

$$\mathbf{\Lambda_1} = \begin{pmatrix} 9.92 & -5 \\ -5 & 6.67 \end{pmatrix},$$

and the inverse of this matrix is

$$\mathbf{\Lambda_1^{-1}} = \begin{pmatrix} 0.16 & 0.12 \\ 0.12 & 0.24 \end{pmatrix}.$$

Hence the test statistic (5.34) is

$$Q = \mathbf{T_1'}\mathbf{\Lambda_1^{-1}}\mathbf{T_1} = 7.6,$$

and the associated approximate P-value is $P(\chi_2^2 > 7.6) = 0.0224$. Thus we reject the hypothesis of equality of the effects of the three plans.

Alternate method. It has been suggested that a different test statistic Q_C^* be used. A computer program developed for this purpose is given in Appendix B5. We compute the within block ranks and analyze these ranks using a standard linear model for a block design. From the output we have

$$\mathbf{T'C^-T} = \textit{Type III trt SS} = 19.844.$$

Further,

$$\sigma_*^2 = (\textit{Type III trt SS} + \textit{Error SS})/(n - b)$$
$$= (19.844 + 12.656)/12 = 2.708.$$

Hence,

$$Q_C^* = (19.824/2.708) = 7.327.$$

Alternatively, this statistic can be obtained from the F-statistic, using the relationship

$$Q_C^* = \frac{(n - b)(v - 1)F}{(n - b - v + 1) + (v - 1)F}.$$

Since $F + 7.84$, we have

$$Q_C^* = \frac{12(2)(7.84)}{11 + (2)(7.84)} = 7.327.$$

The P-value for this statistic is $P[\chi_2^2 > 7.237] = 0.0258$. Thus we reject the null hypothesis.

Several authors made contributions to this problem with this or a slightly different formulation. Some references are Durbin (1951), Bernard and van Elteren (1953), Bruden and Mohberg (1976), Quade (1979), Mack and Skillings (1980), and Skillings and Mack (1981).

Now we consider some settings where the general results simplify considerably and lead to known methods. In the earlier discussion under H_0, there cannot be ties as the responses are continuous random variables.

5.5.4 GRCB designs

In this case $n_{ij} = d$, so that

$$k_1 = k_2 = \cdots = k_b = dv, \quad r_1 = r_2 = \cdots = r_v = db,$$

and $\mathbf{N} = d\mathbf{J}_{b,v}$, where $\mathbf{J}_{b,v}$ is a $b \times v$ matrix of all ones. Hence,

$$\mathbf{C} = bd\mathbf{I_v} - \frac{bd}{v}\mathbf{J}_{v,v}.$$

Thus $\mathbf{C}^- = \frac{1}{bd}\mathbf{I}_v$. Further $\sigma^2 = dv(dv+1)/12$. Finally, the test statistic simplifies to

$$Q = \frac{12}{bd(dv)(dv+1)} \sum_{j=1}^{v} T_j^2,$$ (5.38)

where $T_j = \sum_{i,l}(R_{ijl} - \frac{dv+1}{2})$.

<u>Remark 5.12.</u> Now by setting $d = 1$, we get the results for RCB designs. The test statistic for this case is the same as the Friedman's statistic Q_F of Section 5.3.

5.5.5 Wilcoxon scores procedure

In Section 5.4 we mentioned that we could use any set of scores and construct a test criterion. In that section we assumed that $n_{ij} = 1$. We can generalize the discussion of Section 5.4 by using any set of scores that depend on the within block ranks. However, we restrict the discussion to the so-called Wilcoxon scores, which are the optimal expected scores for logistic shift alternatives. These are linear functions of the ranks and we recall that the score a_{ijl} assigned to Y_{ijl} is

$$a_{ijl} = \frac{2}{(k_i+1)}\left(R_{ijl} - \frac{k_i+1}{2}\right) \equiv \frac{1}{(k_i+1)}w_{ijl},$$

where R_{ijl} is the rank of Y_{ijl} and w_{ijl} is the w-rank of Y_{ijl}. Now we sum the scores for the jth treatment, which is

$$T_j^* = \sum_i \sum_l a_{ijl} = \sum_i \frac{2}{k_i+1}(R_{ij} - E_0(R_{ij})).$$

It may be noted that T_j^* is a weighted sum of R_{ij}^*. Let $\mathbf{T}^*$ be the vector $(T_1^*, T_2^*, \ldots, T_v^*)'$ and $\mathbf{\Lambda}^*$ be the null covariance matrix of $\mathbf{T}^*$. The test statistic is

$$Q_3 = \mathbf{T}^{*'}(\mathbf{\Lambda}^*)^-\mathbf{T}^*,$$ (5.39)

and the test rejects the null hypothesis for large values of Q_3. The same chi-square distribution can be used to approximate the P-value.

   This statistic can be viewed as a generalization of the one studied by Skillings and Mack (1981) and it also is a generalization of the statistic considered by Prentice (1979) for the problem of m rankings. Further details can be found in their papers.

   We want to specialize this general method to obtain a procedure for the problem of comparing two treatments in several blocks; the resulting procedure depends on a weighted sum of two-sample Wilcoxon rank sum statistic. Details of this special case will be presented now.

5.5.6 Blocked comparison of two treatments

We now specialize the discussion to the case where $v = 2$. In applications, blocks may correspond to various studies or various centers. Now $T_1^* + T_2^* = 0$, and

$$T_2^* = \sum_i \frac{2}{k_i + 1}(R_{i2} - E_0(R_{i2}))$$

$$= \left(2\sum_i \frac{W_i}{k_i + 1}\right) - E_0\left(2\sum_i \frac{W_i}{k_i + 1}\right),$$

where W_i is the Wilcoxon rank sum statistic for treatment 2 in the ith block. Thus we have

$$T_2^* = 2[W_s - E_0(W_s)],$$

where $W_s = \sum_i \frac{W_i}{k_i+1}$, a statistic suggested by van Elteren (1960) and discussed in detail by Lehmann (1998, p. 135). The statistic Q_3 simplifies to

$$Q_3 = \frac{(T_2^*)^2}{var_0(T_2^*)},$$

which is the same as

$$Q_3 = \frac{[W_s - E_0(W_s)]^2}{var_0(W_s)}, \tag{5.40}$$

where

$$E_0(W_s) = \sum_i \frac{n_{i2}}{2}; \quad var_0(W_s) = \sum_i \frac{n_{i1}n_{i2}}{12(k_i + 1)}.$$

We see that the test based on Q_3 is the same as the test suggested by Lehmann. The P-value can be approximated as $P(\chi^2(1) > Q_3(obs))$, where $Q_3(obs)$ is the observed value of Q_3. For implementing a test for one-sided alternatives, we can approximate the null distribution of W_s by a normal distribution. Nelson (1970) prepared tables of critical values of W_s assuming that n_{ij}'s are equal. This equal cell numbers case is a GRCB design discussed earlier and the test suggested here is equivalent to the test proposed earlier. Now an example with equal cell numbers is given.

Example 5.4. Two types of light bulbs A and B were tested in four houses using each type of bulb in three sockets in each house. Failure times (in thousands of hours) are given in Table 5.7. The within house ranks for the failure times are given in parentheses.

Since $W_s = \frac{1}{k+1}\sum_i W_i = \frac{1}{k+1}R_2$, we have $W_s = \frac{36}{7}$. Further,

$$E_0(W_s) = \binom{3}{2}(4) = 6,$$

Table 5.7 *Failure times (artificial data)*

Bulb (j)	House				R_j
	1	2	3	4	
A	4.8(6)	4.9(6)	4.5(4)	3.8(3)	
(1)	4.5(4)	3.8(1)	4.8(6)	4.3(6)	48
	4.6(5)	4.2(3)	4.4(3)	3.5(1)	
B	3.2(1)	3.9(2)	4.1(1)	3.6(2)	
(2)	3.6(3)	4.3(4)	4.2(2)	4.0(4)	36
	3.5(2)	4.4(5)	4.7(5)	4.1(5)	

and

$$var_0(W_s) = \frac{1}{(12)(7)}(4(3)(3)) = \frac{3}{7}.$$

Thus the value of the test statistic is

$$Q_3 = \frac{\left(\frac{36}{7} - 6\right)^2}{\frac{3}{7}} = \frac{12}{7} = 1.71,$$

and hence

$$P\text{-value} \approx P(\chi^2(1) > 1.71) = 0.191.$$

So we do not reject the null hypothesis at $\alpha = 0.05$.

5.5.7 Balanced incomplete block (BIB) designs

A BIB design is an arrangement of v treatments in b blocks of size k such that

1. every treatment occurs at most once in a block (that is, $n_{ij}=1$ or 0);
2. every treatment occurs in r blocks (that is, $r_j = r$, *for all j*);
3. every pair of treatments occur together in λ blocks.

The constants $v, b, r, k,$ and λ are known as the parameters of the design and they satisfy the relations

$$vr = bk, \quad r(k-1) = \lambda(v-1), \quad b \geq v. \tag{5.41}$$

An example with $v = 6, r = 5, k = 3, \lambda = 2$ consists of 10 blocks (that is, $b = 10$) and the treatments that occur in various blocks are as follows.

$$(A, B, D); \ (B, C, E); \ (C, D, A); \ (D, E, B); \ (E, A, C);$$

$$(A, B, F); \ (B, C, F); \ (C, D, F); \ (D, E, F); \ (E, A, F).$$

For this design

$$\mathbf{N'N} = (r - \lambda)\mathbf{I}_v + \lambda\mathbf{J}_{v,v} \quad and \quad \mathbf{C} = \frac{\lambda v}{k}\mathbf{I}_v - \frac{\lambda}{k}\mathbf{J}_{v,v}.$$

So $\mathbf{C}^- = \frac{k}{\lambda v}\mathbf{I}_v$. Let R_j be the sum of the ranks for the jth treatment and $T_j = R_j - \frac{r(k+1)}{2}$. The vector $\mathbf{T}$ is

$$\mathbf{T} = \mathbf{R} - \frac{r(k+1)}{2}\mathbf{1}_v.$$

Further, $\sigma^2 = \frac{k(k+1)}{12}$. Finally, the proposed test statistic will be

$$Q = \mathbf{T'C}^-\mathbf{T}/\sigma^2 = \frac{12}{\lambda v(k+1)}\sum_j \left(R_j - \frac{r(k+1)}{2}\right)^2. \qquad (5.42)$$

This statistic is the same as the one proposed by Skillings and Mack (1981) and it is equivalent to the one proposed by Durbin (1951).

We now give an example where the above design is used.

Example 5.5. Ten volunteers were asked to rate three out of six brands of a consumer product. The experiment used the illustrated BIB design. Here the volunteers are the blocks. The brands are considered as treatments. The ranks are given in Table 5.8. The test statistic is

$$Q = \frac{12}{2(6)(4)}(36) = 9.$$

The approximate P-value is $P(\chi^2(5) > 9) = 0.1091$, and we do not reject the null hypothesis at 5% level.

Analysis when ties are present

When there are tied observations in one or more blocks, as usual we assign midranks to the tied values and the within block ranks are obtained. Using any computer program, compute the F-statistic for testing the equality of treatment effects. The proposed statistic can be obtained using the following

Table 5.8 *Ranks assigned by volunteers*

Brands	\multicolumn{10}{c}{Volunteers}	R_j
	1	2	3	4	5	6	7	8	9	10	
A	1		3		1	1				1	7
B	2	3		3		3	2				13
C		1	2		3		1	1			8
D	3		1	1				2	1		8
E		2		2	2				3	2	11
F						2	3	3	2	3	13

formula:

$$Q = \frac{(n - b)(v - 1)F}{(n - b - v + 1) + (v - 1)F},$$

and then the approximate P-value can be calculated.

5.6 A multiple comparison procedure using Friedman's ranks

Following the rejection of homogeneity hypothesis by Friedman's test, we may want to compare pairs of treatments. Recall that Friedman's test uses the within block ranks. The procedure to be described is based on these within block ranks.

Let R_{ij} be the rank received by the observation under treatment j in the block i. We use the average of the ranks received by the treatments. These are

$$\bar{R}_{.j} = \frac{1}{b} \sum_i R_{ij}.$$

It seems natural to declare that treatments j and j' are different if

$$|\bar{R}_{.j} - \bar{R}_{.j'}| > c.$$

Using the union-intersection principle, the overall test is derived. This test is to

$$\text{reject } H_0 \text{ of } (5.9) \quad \text{if } max|\bar{R}_{.j} - \bar{R}_{.j'}| > c. \tag{5.43}$$

In order to control overall type I error we need to choose c such that

$$P(max|\bar{R}_{.j} - \bar{R}_{.j'}| > c|H_0) \le \alpha.$$

Using the variance and covariance matrix of $\bar{R}_{.j}$, an approximation for c is

$$c \approx \sqrt{\frac{k(k + 1)}{12b}} q_{k,\infty}^{1-\alpha}, \tag{5.44}$$

where $q_{k,\infty}^{1-\alpha}$ is the percentile of the range statistic.

This procedure has been generalized for other block designs by Skillings and Mack (1981).

5.7 Page test for ordered alternatives in RCB designs

In the previous sections we considered only global tests. However, in some cases the relevant alternatives are restricted. Here we discuss the hypothesis testing problem with completely ordered alternatives

$$H_{CO2} : F_{i1} \text{ st} < F_{i2} \text{ st} < \ldots \text{st} < F_{iv} \text{ for all } i. \tag{5.45}$$

In this problem we assume that the data come from an RCB design. One of the early solutions to this problem is the test proposed by Page (1963). This procedure will be described now. Here we rank the observations within each block.

Let R_{ij} be the rank received by the observation under treatment j in the block i. Then we obtain the sum of ranks, $R_{.j}$, received by the observations under treatment j, that is,

$$R_{.j} = \sum_{i=1}^{b} R_{ij}. \tag{5.46}$$

Now the test statistic is

$$L = 1R_{.1} + 2R_{.2} + \cdots + vR_{.v}, \tag{5.47}$$

and the test

rejects H_0 if $L > c$.

It may be noted that under the alternative (5.43), $R_{.j}$ will be larger than $R_{.j'}$, for $j > j'$ and the test statistic L gives a weight to $R_{.j}$ that is larger than the weight given to $R_{.j'}$, when $j > j'$. Page (1963) provided a table of critical values. We discuss the large sample version of the test. This depends on the standardized statistic, Z_L. To define this standardized statistic we need to find the null mean and null variance of L. It can be shown that

$$E_0(L) = bv(v + 1)^2/4, \tag{5.48}$$

and

$$var_0(L) = [bv^2(v^2 - 1)(v + 1)]/144. \tag{5.49}$$

The derivation of the variance is given in Appendix A5. The expected value derivation is Problem 5. Now

$$Z_L = \frac{(L - E_0(L))}{\sqrt{var_0(L)}}.$$

The distribution of this statistic can be approximated by a standard normal distribution, whenever $v > 3$ and $b > 12$. Thus an approximate α-level test

rejects H_0 if $Z_L > z_{1-\alpha}$.

So an approximation to the P-value is

$$P\text{-value} \approx P[N(0, 1) > Z_L(obs)],$$

where $Z_L(obs)$ is the observed value of the statistic Z_L. StatXact 3 can be used to get the exact P-value. A computer program for implementing this test is given in Appendix B5.

Remark 5.12. Suppose that the alternatives are

$$H_{CO1} : F_{i1} \text{ st} > F_{i2} \text{ st} > \ldots \text{st} > F_{iv} \quad \text{for all } i.$$

Then the treatments are renumbered so that treatment j will be treatment $v - j$ and the hypothesis will read H_{CO2}. After the renumbering we can use the above procedure.

Table 5.9 *Pressley strength index of cotton samples (artificial data)*

Block	Level of Potash			
	P_1	P_2	P_3	P_4
1	7.53(1)	8.21(4)	7.93(2)	7.97(3)
2	7.17(1)	7.58(2)	7.63(3)	7.79(4)
3	7.83(2)	7.85(3)	7.82(1)	7.90(4)
4	7.63(1)	7.79(2)	7.90(4)	7.89(3)
5	7.47(1)	7.72(2)	7.85(3)	8.01(4)
6	8.01(4)	7.92(1)	7.95(2)	8.00(3)
7	7.89(1)	7.90(2)	7.96(3)	7.99(4)
8	7.95(3)	7.80(2)	7.75(1)	7.97(4)
9	7.63(1)	7.75(2)	7.93(4)	7.82(3)
10	7.88(4)	7.78(1)	7.80(2)	7.87(3)
11	7.59(1)	7.75(2)	7.80(4)	7.77(3)
12	7.60(1)	7.65(2)	7.72(3)	7.75(4)
13	7.91(4)	7.84(1)	7.85(2)	7.87(3)
14	8.00(4)	7.74(1)	7.89(3)	7.85(2)
15	7.38(1)	7.57(2)	7.75(3)	7.85(4)
Sum of Ranks	30	29	40	51

The procedure is illustrated in the following example.

Example 5.6. Consider an experimental setting (see p. 108 of Cochran and Cox (1957)) to study the effect of the level of potash on the breaking strength of cotton fibers. Four levels of potash ($P_1 > P_2 > P_3 > P_4$) were applied in a randomized block pattern and 15 blocks were used. A sample of cotton was taken from each plot and the Pressley strength index was determined. Artificial data appear in Table 5.9. The experimenter is interested in seeking evidence that a decrease in the level of potash will increase the fiber strength.

The observations in each block are ranked and the ranks are shown in parentheses in Table 5.9. The test statistic is

$$L = 1(30) + 2(29) + 3(40) + 4(51) = 412,$$

and the normalized statistic is

$$Z_L = \frac{412 - [15(4)(4+1)^2/4]}{\sqrt{15(4^2)(4^2-1)(4+1)/144}} = 3.3094.$$

The approximate P-value is

$$P\text{-value} \approx P[N(0,1) > 3.3094] = 0.0005.$$

There is evidence to support the hypothesis that fiber strength increases with the decreasing potash level.

5.8 RCB designs with censored data

The setup here is similar to the setting considered in earlier chapters. The response variables Y_{ijl} are subjected to right censoring. The censoring variables C_{ijl} are independent of the variables Y_{ijl} and their distribution does not depend on j (the treatment indicator) when the null hypothesis is true. We observe X_{ijl}, which is the minimum of Y_{ijl} and C_{ijl}. We also observe the indicator variables δ_{ijl}, which is 1 or 0 depending on whether X_{ijl} is not censored or censored. We recall that the null hypothesis is

$$H_0 : F_{i1} = F_{i2} = \cdots = F_{iv}, \quad for \; i = 1, 2, \ldots, b. \tag{5.50}$$

In the absence of censoring, the cdf of X_{ij} is F_{ij} and we assume that

$$F_{ij}(x) = F[(x - \tau_j - \mu_i)\sigma^{-1}], \tag{5.51}$$

for some continuous distribution function $F(.)$, with the density function $f(.)$. The μ_i and σ are the unknown nuisance parameters for block $i = 1, 2, \ldots, b$. This assumption is the same as (5.26), when $\sigma = 1$. Now the above null hypothesis translates to

$$H_0 : \tau_j = 0, \quad for \; all \; j. \tag{5.52}$$

Under these assumptions Woolson and Lachenbruch (1981) derived a class of rank tests that are locally the most powerful tests for RCB designs for which $n_{ij} = 1$. The permutation distributions were used to define the critical regions. These results are applied to problems in survival analysis. We now discuss these procedures in Subsection 5.8.1, and in Subsection 5.8.2 we specialize the discussion to the case of $v = 2$ with multiple observations on each treatment in each block.

5.8.1 Woolson-Lachenbruch rank tests

We suppress the suffix l since there is only one observation per treatment in each block. Let a_{ij} be a score attached to the observation X_{ij}, obtained by comparing each observation with the other observations in the ith block. The scores are such that the sum of the scores within a block is zero. In other words

$$\sum_j a_{ij} = 0, \quad for \; all \; i = 1, 2, \ldots, b. \tag{5.53}$$

Now for each treatment $j(j = 1, 2, \ldots, v)$, we compute the sum

$$A_j = \sum_i a_{ij}. \tag{5.54}$$

We note that $\sum_j A_j = 0$. The test statistic is

$$Q_{CE} = \frac{(v-1)}{\sum_i \sum_j a_{ij}^2} \sum_{j=1}^v A_j^2. \tag{5.55}$$

Large values of Q_{CE} lead to the rejection of the null hypothesis. The statistic Q_{CE} is asymptotically a chi-square variable with $(v-1)$ degrees of freedom.

It may be noted that this test statistic is a generalization of the statistic Q suggested in Section 5.4 for a complete data case. The test statistic Q_{CE} will reduce to the statistic Q of Section 5.4, when there is no censoring.

In the following we present Prentice-Peto Wilcoxon and logrank scores. For more detailed discussion the interested reader should refer to the paper of Woolson and Lachenbruch (1981).

To facilitate the definition of the scores we introduce some notation. Consider the distinct uncensored observations in each block and order them in ascending order. For the ith block at the tth distinct ordered uncensored observation let d_{it} be the number of uncensored values and let r_{it} be the number at risk. In terms of these quantities the scores are defined as follows.

Prentice-Peto Wilcoxon scores

The scores are

$$a_{it} = \begin{cases} 1 - 2\hat{S}_{it}, & \text{if } X_{it} = Y_{it}, \\ 1 - \hat{S}_{it}, & \text{if } X_{it} = C_{it}. \end{cases}$$

where

$$\hat{S}_{it} = \Pi_{s \leq t} \frac{d_{is}}{r_{is} + d_{is}}. \tag{5.56}$$

Logrank scores

The logrank scores are

$$a_{it} = \begin{cases} \hat{\Lambda}_{it} - 1, & \text{if } X_{it} = Y_{it}, \\ \hat{\Lambda}_{it}, & \text{if } X_{it} = C_{it}. \end{cases}$$

where

$$\hat{\Lambda}_{it} = \sum_{s \leq t} \frac{d_{is}}{r_{is}}. \tag{5.57}$$

Remark 5.13. The Prentice-Peto scores can be viewed as censored data Wilcoxon scores. When there is no censoring these scores simplify to multiples of w-ranks. The logrank scores are a generalization of the Savage scores considered earlier. In the case of no censoring and no ties, these will simplify to Savage scores defined in (5.23). These are also the negative of the logrank scores considered in Chapter 2.

Example 5.7. A cardiologist is interested in comparing the effectiveness of two treatments on restenosis after angioplasty. The treatments are (1) metal stent, and (2) aspirin regimen. On each of 10 days he selected three patients; two were randomly assigned to the two treatments and the third patient was used as a control. The times (in months) to restenosis for the 30 patients were recorded. During the course of the investigation 5 patients died due to

Table 5.10 *Time to restenosis*

Day (Block)	Metal Stent	Aspirin Regimen	Control
1	120	55	9
2	80+	90	60
3	70	75+	80
4	80	80	90+
5	110	100+	95
6	140	120	10
7	80+	90	60
8	130	100	100
9	50	70	40
10	30	90	10

Table 5.11 *Prentice-Peto Wilcoxon and logrank scores for data of Table 5.10*

	Metal Stent		Aspirin		Control	
Day	P-P W Scores	Logrank Scores	P-P W Scores	Logrank Scores	P-P W Scores	Logrank Scores
1	0.5	0.83	0	−0.17	−0.5	−0.66
2	0.25	0.33	0.25	0.33	−0.5	−0.66
3	−0.5	−0.66	0.25	0.33	0.25	0.33
4	0.2	−0.33	−0.2	−0.33	0.4	0.66
5	0.25	0.33	0.25	0.33	−0.5	−0.66
6	0.5	0.83	0	−0.17	−0.5	−0.66
7	0.25	0.33	0.25	0.33	−0.5	−0.66
8	0.4	0.66	−0.2	−0.33	−0.2	−0.33
9	0	−0.17	0.5	0.83	−0.5	−0.66
10	0	−0.17	0.5	0.83	−0.5	−0.66
Total	1.45	1.98	1.6	1.98	−3.05	−3.96

unrelated causes and the observations from these patients are considered as censored observations. The data appear in Table 5.10 (+ means censored).

The Prentice-Peto Wilcoxon scores and the logrank scores appear in Table 5.11. The test statistic with Prentice-Peto Wilcoxon scores is

$$Q_{CE} = \frac{2}{3.98}[1.45^2 + 1.6^2 + (-3.05)^2] = 7.02,$$

and the associated approximate P-value is 0.0298. We conclude that the three methods have different effects on restenosis.

The test statistic with logrank scores is

$$Q_{CE} = \frac{2}{8.68}[1.98^2 + 1.98^2 + (3.96)^2] = 5.42,$$

and the associated approximate P-value is 0.0665. Based on this test we do not reject the null hypothesis that the methods are equal in their effects at the 5% level.

5.8.2 Comparing two treatments in blocks (or strata)

In this section we consider the problem of the comparison of two treatments, allowing multiple observations on each treatment in each block. Some clinical trials are conducted in more than one center and the data are analyzed by considering the centers as blocks. In some cases variables like age, gender, stage of the disease, etc. are used to define the strata (blocks). The discussion with censored data is somewhat analogous to the corresponding discussion of Section 5.4. For each block, we calculate a two-sample statistic using some scores, combine them over the blocks, and develop a test criterion.

Stratified logrank test

Let U_{iLR} be the logrank statistic computed from the scores for treatment 2 with V_{iLR} as the null variance for the ith block. The stratified logrank statistic is

$$U_{LR} = \frac{(\sum_i U_{iLR})^2}{\sum_i V_{iLR}},$$

with approximate $\chi^2(1)$ distribution. Large values of U_{LR} constitute the critical region for the two-sided alternatives. Thus an approximate P-value is $P[\chi^2(1) > U_{LR}(obs)]$, where $U_{LR}(obs)$ is the observed value of U_{LR}.

Stratified Gehan's Wilcoxon test

Let U_{iGE} be the Gehan's Wilcoxon statistic computed from the scores for treatment 2 with V_{iGE} as the null variance for the ith block. Let k_i be the number of observations in the ith block. The stratified Gehan's Wilcoxon statistic is

$$U_{GE} = \frac{[\sum_i \frac{U_{iGE}}{(k_i+1)}]^2}{\sum_i \frac{V_{iGE}}{(k_i+1)^2}},$$

with approximate $\chi^2(1)$ distribution. Large values of U_{GE} constitute the critical region for the two-sided alternatives. Thus an approximate P-value is $P[\chi^2(1) > U_{GE}(obs)]$, where $U_{GE}(obs)$ is the observed value of U_{GE}. The required quantities to calculate U_{LR} and U_{GE} can be obtained from the lifetest procedure of the SAS system.

We illustrate these calculations with an example.

Example 5.8. A randomized clinical was conducted at three centers. The data are the survival times (in months) under two treatments, listed in Table 5.12.

Table 5.12 *Survival times for a study with three centers*

Center 1		Center 2		Center 2	
Trt. A	Trt. B	Trt. A	Trt. B	Trt. A	Trt. B
240	85	250	100+	120	110
200+	100+	50+	120	60	130
100	150	170	150		150+
180+	170		160+		
	160+				

Table 5.13 *Various statistics needed to get stratified test*

Center	Logrank Test		Gehan-Wilcoxon Test			
	U_{iLR}	V_{iLR}	U_{iGE}	$\frac{U_{iGE}}{k_i+1}$	V_{iGE}	$\frac{V_{iGE}}{(k_i+1)^2}$
1	1.1944	0.9344	6.0	0.6	48.0	0.48
2	0.9000	0.4900	4.0	0.5	10.0	0.1563
3	−1.0167	0.6497	−4.0	−0.67	11.0	0.3056
Total	1.0777	2.0741		0.43		0.9419

Using Proc lifetest of SAS (see Appendix B5 for the program used here) we get the required statistics. They are shown in Table 5.13.

The stratified logrank statistic U_{LR} is

$$U_{LR} = \frac{1.0777^2}{2.0741} = 0.56,$$

and the associated approximate P-value is 0.4543. So we do not reject the null hypothesis that the two treatments are equally effective at the 5% level.

The stratified Gehan's Wilcoxon statistic U_{GE} is

$$U_{GE} = \frac{0.43^2}{0.9419} = 0.196,$$

and the associated approximate P-value is 0.6577. We reach the same conclusion as with the logrank test.

Lininger et al. (1979) conducted a study to compare the above two stratified tests. As in the one stratum case, Gehan's Wilcoxon test is slightly less powerful than the logrank test for proportional hazards alternatives. Comparison of the performance of these two tests for the one stratum case is found in Lee, Desu, and Gehan (1975).

Two other methods of giving scores have been proposed in the literature. One uses the aligned observations for giving scores and the other pools all observations over all strata and assigns scores. These methods along with the method of giving scores within each stratum were compared by Podgor and Gastwirth (1994). For further details the interested reader is referred to their paper. Another suggested reference to gain some insight into procedures for analyzing censored data in block designs setting is Groggel, Schaefer, and Skillings (1987).

5.9 Appendix A5: Mathematical supplement

A5.1 Covariance matrix of **T** *of Section 5.3*

Recall that

$$T_j = R_{.j} - E_0(R_{.j}),$$

so that

$$var_0(T_j) = var_0(R_{.j}) = \sum_i (v^2 - 1)/12 = b(v^2 - 1)/12 = \sigma^2.$$

Further,

$$cov_0(T_j, T_{j'}) = -\sum_i (v+1)/12 = -b(v+1)/12 = -\sigma^2/(v-1).$$

Hence the covariance of **T** is

$$\Sigma = \frac{\sigma^2}{v-1}(v\mathbf{I_v} - \mathbf{J_v}),$$

so that

$$\Sigma^- = \frac{v-1}{v\sigma^2}\mathbf{I_v}.$$

Finally, we have

$$\mathbf{T}'\Sigma^-\mathbf{T} = \frac{v-1}{v\sigma^2}\mathbf{T}'\mathbf{T}.$$

A5.2 Derivation of (5.49)

We want to calculate the variance of L, where

$$L = \sum_j jR_{j..}.$$

From A5.1 we have

$$var_0(R_{.j}) = \sigma^2,$$

and

$$cov_0(R_{.j}, R_{.j'}) = -\sigma^2/(v-1),$$

where $\sigma^2 = b(v^2 - 1)/12$. Hence

$$var_0(L) = \sum_j j^2 \sigma^2 + \sum_{j \neq j'} jj'(-\sigma^2/(v-1))$$

$$= \sigma^2 \left[\sum_j j^2 - \frac{1}{(v-1)} \left(\left(\sum_j j \right)^2 - \left(\sum_j j^2 \right) \right) \right]$$

$$= \sigma^2 \left[\frac{v}{(v-1)} \sum_j j^2 - \frac{1}{(v-1)} \left(\sum_j j \right)^2 \right]$$

$$= \frac{\sigma^2}{v-1} \left[\frac{v^2(v+1)(2v+1)}{6} - \frac{v^2(v+1)^2}{4} \right]$$

$$= bv^2(v^2 - 1)(v+1)/144.$$

5.10 Appendix B5: Computer programs

B5.1 Computation of Friedman's statistic

Here we calculate the Friedman's test statistic for Example 5.2. The response variable is creatinine level, labs are treatments, and subjects are blocks. We get the ranks within blocks and do ANOVA on the ranks. The F-statistic for testing the treatment effects is used to get the Friedman's statistic as explained in the text. An alternative method uses the frequency procedure.

```
data rblock;
input blk trt level@@;
lines;
1 1 2.8 1 2 2.6 1 3 2.7 1 4 2.9
2 1 1.7 2 2 1.8 2 3 1.9 2 4 1.6
3 1 3.1 3 2 3.0 3 3 3.2 3 4 2.8
4 1 3.5 4 2 3.6 4 3 3.4 4 4 3.7
5 1 2.7 5 2 2.9 5 3 2.8 5 4 2.6
6 1 3.2 6 2 3.0 6 3 2.9 6 4 3.1
7 1 2.5 7 2 2.6 7 3 2.9 7 4 2.8
8 1 1.9 8 2 2.0 8 3 2.2 8 4 2.1
9 1 2.7 9 2 2.9 9 3 3.0 9 4 2.8
10 1 3.1 10 2 3.8 10 3 3.0 10 4 2.9
;
proc sort;
by blk;
proc rank out=rc;
```

```
by blk;
var level;
ranks rlevel;
proc anova data=rc;
class blk trt;
model rlevel=blk trt;
title 'ANOVA ON WITHIN BLOCK RANKS';
title2 'EXAMPLE 5.2';
run;
/*Alternative method of computing Friedman's statistic.
  The second CMH statistic(the one against row mean
  scores differ) is the one we want.*/
 proc freq data=rblock;
 table blk*trt*level/noprint cmh2 scores=rank;
 title 'FRIEDMAN'S CHI-SQUARE STATISTIC';
 run;
```

Output

ANOVA ON WITHIN BLOCK RANKS
EXAMPLE 5.2
The ANOVA Procedure

Dependent Variable: RLEVEL   Rank for Variable LEVEL

Source	DF	Sum of Squares	Mean Square	F Value
Model	12	3.00000000	0.25000000	0.14
Error	27	47.00000000	1.74074074	
Corrected Total	39	50.00000000		

Source	DF	Anova SS	Mean Square	F Value
BLK	9	0.00000	0.00000000	0.00
TRT	3	3.00000	1.00000000	0.57

Source	Pr > F
BLK	1.0000
TRT	0.6367

FRIEDMANS CHI-SQUARE STATISTIC
The FREQ Procedure

Summary Statistics for TRT by LEVEL
Controlling for BLK

Cochran-Mantel-Haenszel Statistics
(Based on Rank Scores)

Statistic	Alternative Hypothesis	DF	Value	Prob
1	Nonzero Correlation	1	0.1080	0.7424
2	Row Mean Scores Differ	3	1.8000	0.6149

Total Sample Size = 40

B5.2 Analysis of within block ranks for a design with unequal block sizes

This is the rank analysis of weight loss under three plans. The data are from Table 5.5, which corresponds to Example 5.3. We get the ranks within blocks and do block design analysis using the GLM procedure. The response is y.

```
data diet;
input blk trt y@@;
lines;
1 1 1 1 1 1.5 1 1 0.5
1 2 2 1 2 2.5 1 3 3
2 1 0 2 1 0.5 2 1 1 2 2 2 2 3 2.5
3 1 1 3 1 2 3 3 1.5 3 3 2.5
;
proc sort;
by blk;
proc rank data=diet out=five;
by blk;
var y;
ranks ry;
run;
proc print data=five;
run;
proc glm data=five;
class blk trt;
model ry=blk trt;
title 'ANOVA ON WITHIN BLOCK RANKS';
title2 'EXAMPLE 5.3';
run;
```

Output

Obs	blk	trt	y	ry
1	1	1	1.0	2
2	1	1	1.5	3

```
             3      1      1      0.5      1
             4      1      2      2.0      4
             5      1      2      2.5      5
             6      1      3      3.0      6
             7      2      1      0.0      1
             8      2      1      0.5      2
             9      2      1      1.0      3
            10      2      2      2.0      4
            11      2      3      2.5      5
            12      3      1      1.0      1
            13      3      1      2.0      3
            14      3      3      1.5      2
            15      3      3      2.5      4
```

ANOVA ON WITHIN BLOCK RANKS
EXAMPLE 5.3

The GLM Procedure
Class Level Information

Class	Levels	Values
blk	3	1 2 3
trt	3	1 2 3

Number of observations   15

The GLM Procedure

Dependent Variable: ry   Rank for Variable y

Source	DF	Sum of Squares	Mean Square
Model	4	22.27708333	5.56927083
Error	10	12.65625000	1.26562500
Corrected Total	14	34.93333333	

R-Square	Coeff Var	Root MSE	ry Mean
0.637703	36.68478	1.125000	3.066667

Source	DF	Type III SS	Mean Square	F Value
blk	2	2.76041667	1.38020833	1.09
trt	2	19.84375000	9.92187500	7.84

Source	Pr > F
blk	0.3729
trt	0.0090

B5.3 Computation of Page statistic

We calculate the Page test statistic for Example 5.6. The response variable is the breaking strength of cotton fibers. Potash levels are treatments. Within block ranks are obtained, then they are processed as explained in the text.

```
data rblock;
input blk trt stg@@;
lines;
1 1 7.53 1 2 8.21 1 3 7.93 1 4 7.97
2 1 7.17 2 2 7.58 2 3 7.63 2 4 7.79
3 1 7.83 3 2 7.85 3 3 7.82 3 4 7.90
4 1 7.63 4 2 7.79 4 3 7.90 4 4 7.89
5 1 7.47 5 2 7.72 5 3 7.85 5 4 8.01
6 1 8.01 6 2 7.92 6 3 7.95 6 4 8.00
7 1 7.89 7 2 7.90 7 3 7.96 7 4 7.99
8 1 7.95 8 2 7.80 8 3 7.75 8 4 7.97
9 1 7.63 9 2 7.75 9 3 7.93 9 4 7.82
10 1 7.88 10 2 7.78 10 3 7.80 10 4 7.87
11 1 7.59 11 2 7.75 11 3 7.80 11 4 7.77
12 1 7.60 12 2 7.65 12 3 7.72 12 4 7.75
13 1 7.91 13 2 7.84 13 3 7.85 13 4 7.87
14 1 8.00 14 2 7.74 14 3 7.89 14 4 7.85
15 1 7.38 15 2 7.57 15 3 7.75 15 4 7.85
;
proc sort;
by blk;
proc rank out=rc;
by blk;
var stg;
ranks rstg;
proc sort data=rc;
by trt;
run;
proc summary;
```

```
var rstg;
by trt;
output out=stats sum=r;
proc print data=stats;
title 'PAGE TEST FOR EXAMPLE 5.6';
title2 'SUM OF RANKS R(J)';
```

We calculate the statistic, its mean, and its variance.

```
data two;
set stats;
proc iml;
use stats;
read all variables {R} into M1;
W={1 2 3 4 };
P1=W*M1;
print "PAGE STATISTIC" P1;
v=4;
b=15;
v1=v+1;
MU=b*v*v1*v1/4;
print "MEAN OF PAGE STATISTIC." MU;
var=b*(v**3-v)**2/(144*(v-1));
std1=sqrt(var);
STD=round(std1,.0001);
print "ST.DEV OF PAGE STATISTIC" STD;
zp1=(p1-mu)/STD;
ZP=round(zp1,.0001);
print "STANDARDIZED STATISTIC" ZP;
pv1=1-probnorm(ZP);
PV=ROUND(pv1,.0001);
print "APPROX ONE-SIDED P-VALUE" PV;
exit iml;
run;
```

Output

PAGE TEST FOR EXAMPLE 5.6
SUM OF RANKS R(J)

Obs	TRT	_TYPE_	_FREQ_	R
1	1	0	15	30
2	2	0	15	29
3	3	0	15	40
4	4	0	15	51

                          PAGE TEST FOR EXAMPLE 5.6

                                                        P1

                       PAGE STATISTIC   412

                                                        MU

             MEAN OF PAGE STATISTIC.   375

                                                        STD

           ST.DEV OF PAGE STATISTIC   11.1803

                                                        ZP

           STANDARDIZED STATISTIC   3.3094

                                                        PV

           APPROX ONE-SIDED P-VALUE   0.0005

B5.4 Within strata statistics

Here we calculate Wilcoxon and logrank statistics within each stratum for the
data of Example 5.8. They are combined as explained in the text to obtain a
stratified test statistic.

```
data one;
input group trt time censor@@;
liness;
1 1 240 1 1 1 200 0 1 1 100 1 1 1 180 0
1 2 85 1 1 2 100 0 1 2 150 1 1 2 170 1 1 2 160 0
2 1 250 1 2 1 50 0 2 1 170 1 2 2 100 0 2 2 120 1
2 2 150 1 2 2 160 0
3 1 120 1 3 1 60 1 3 2 110 1 3 2 130 1 3 2 150 0
;
proc lifetest notable;
by group;
time time*censor(0);
strata trt;
title 'WITHIN STRATA STATISTICS';
title2 'EXAMPLE 5.8';
run;
```

Output

<pre>
                    WITHIN STRATA STATISTICS
                         EXAMPLE 5.8
                          GROUP=1
                  The LIFETEST Procedure

         Testing Homogeneity of Survival Curves
                 for TIME over Strata
                   Rank Statistics

         TRT   Log-Rank   Wilcoxon

         1   -1.1944   -6.0000
         2   1.1944   6.0000

    Covariance Matrix for the Log-Rank Statistics

         TRT   1   2

         1   0.934414   -.934414
         2   -.934414   0.934414

    Covariance Matrix for the Wilcoxon Statistics

         TRT   1   2

         1   48.0000   -48.0000
         2   -48.0000   48.0000

            Test of Equality over Strata
                                           Pr >
         Test   Chi-Square   DF   Chi-Square

      Log-Rank   1.5268   1   0.2166
      Wilcoxon   0.7500   1   0.3865

                    WITHIN STRATA STATISTICS
                         EXAMPLE 5.8
                          GROUP=2
                  The LIFETEST Procedure
</pre>

Testing Homogeneity of Survival Curves
for TIME over Strata

Rank Statistics

TRT	Log-Rank	Wilcoxon
1	-0.90000	-4.0000
2	0.90000	4.0000

Covariance Matrix for the Log-Rank Statistics

TRT	1	2
1	0.490000	-.490000
2	-.490000	0.490000

Covariance Matrix for the Wilcoxon Statistics

TRT	1	2
1	10.0000	-10.0000
2	-10.0000	10.0000

Test of Equality over Strata

Test	Chi-Square	DF	Pr > Chi-Square
Log-Rank	1.6531	1	0.1985
Wilcoxon	1.6000	1	0.2059

WITHIN STRATA STATISTICS
EXAMPLE 5.8
GROUP=3
The LIFETEST Procedure
Testing Homogeneity of Survival Curves
for TIME over Strata

Rank Statistics

TRT	Log-Rank	Wilcoxon
1	1.0167	4.0000
2	-1.0167	-4.0000

Covariance Matrix for the Log-Rank Statistics

TRT	1	2
1	0.649722	-.649722
2	-.649722	0.649722

Covariance Matrix for the Wilcoxon Statistics

TRT	1	2
1	11.0000	-11.0000
2	-11.0000	11.0000

Test of Equality over Strata

Test	Chi-Square	DF	Pr > Chi-Square
Log-Rank	1.5909	1	0.2072
Wilcoxon	1.4545	1	0.2278

5.11 Problems

1. Derive the following result of Section 5.2:

$$var(B_j|H_0) = \sum_i \Theta_i(1 - \Theta_i).$$

2. Show that $\hat{\mu}$ of (5.6) is an unbiased estimator of $E(S|H_0)$. (Hint: Note that L_i has a binomial distribution with parameters v and Θ_i.)

3. Prove that Cochran's Q statistic is the same as McNemar's statistic for a two-sided alternative.

4. Prove (5.17) of Remark 5.8 (Hint: Note that $R_1 = A + 2(b - A)$ and $R_2 = 2A + (b - A)$.)

5. Prove the result (5.46).

6. Six new graduates participated in a study in which they were asked to rank the importance of factors in selecting their first job. The resulting data are listed in Table 5.14. Test the null hypothesis that the preference pattern is the same for similar graduates.

Table 5.14 *Graduate preferences*

	Graduates					
Factors	1	2	3	4	5	6
Salary	1	1	1	2	1	2
Vacation	2	4	3	1	2	1
Pension Plan	5	2	4	3	4	5
Flex Hours	3	3	2	5	3	4
Parking	4	5	5	4	5	3

Table 5.15 *Tire wear*

	Tire Brand						
Car	A	B	C	D	E	F	G
1	1	2		3			4
2	2	3	1		4		
3		4	2	1		3	
4			2	1	3		4
5	1			3	2	4	
6		2		1	4	3	
7	2		1			3	4

Table 5.16 *Ranking of various brands*

	Volunteers								
Brands	1	2	3	4	5	6	7	8	9
A	1	3	2	2	1	1		2	1
B	2	2		1	3	4	1	1	3
C	4	1	3	4	2	2	2	3	4
D	3	4	1	3	4	3			2

7. Seven brands of tires were tested for wear. Seven cars were selected for the study and four brands were mounted on the four wheels of each car. The assignment of tire brand was random. After driving the cars for 10,000 miles, the tire depths were measured and ranked for each car. Rank 1 means *least wear* and rank 4 means *most wear*. The data appear in Table 5.15.

Test the hypothesis that the seven brands are different in their wear patterns.

8. Nine volunteers are asked to rank four brands of a product, but they are asked not to rank a brand if they are not familiar with it. The rankings are given in Table 5.16. The underlying response is the preference score, a continuous variable.

Test the hypothesis that the four brands are preferred equally.

Table 5.17 *Times to finish a task*

Noise Level	Volunteers				
	1	2	3	4	5
1	8	10	12	10	9
2	10	12	10	12	10
3	11	15	11	11	12

Table 5.18 *Survival times*

Center 1		Center 2	
Trt. A	Trt. B	Trt. A	Trt. B
18	9	8	10
25+	20	20+	18
23	30	4	8
20+	15+	3	14
10	40+	10	12+
8	6	30+	
	10		
	7		

9. It is hypothesized that workers take a longer time to finish a task if they are exposed to a higher noise level. To test this hypothesis five volunteers are recruited and each is given three identical tasks to perform under three noise levels. The order in which each one is exposed to the noise levels is random. Level 1 means low level and level 3 corresponds to highest level. Times (in minutes) to complete the tasks appear in Table 5.17.

Perform the Page test and summarize your conclusions.

10. The survival times (in months) under two treatments A and B for some cancer patients in two centers are given in Table 5.18.

Use a stratified logrank procedure to test the hypothesis that the two treatments are equally effective.

5.12 References

Bernard, A. and van Elteren, Ph. (1953). A generalization of the method of m rankings. *Indag. Math.*, **15**, 358–369.

Brown, G.W. and Mood, A.M. (1951). On median tests for linear hypotheses. *Proc. 2nd Berkeley Symp.*, **1**, 159–166.

Brunden, M.M. and Mohberg, N.R. (1976). The Bernard-van Elteren statistic and nonparametric compuptation. *Comm. Statist.*, **B5**, 155–162.

Cochran, W.G. (1950). A comparison of percentages in matched samples. *Biometrika*, **37**, 256–263.

Cochran, W.G. and Cox, G.M. (1957). *Experimental Designs*, 2nd edition, John Wiley & Sons, New York.

Cox, D.R. (1972). Regression models and life tables (with discussion), *J. Roy. Statist. Soc. B*, **34**, 187–202.

Downton, F. (1976). Nonparametric tests for block experiments. *Biometrika*, **63**, 137–141.

Durbin, J. (1951). Incomplete blocks in ranking experiments. *Brit. J. Statist. Psy.*, **4**, 85–90.

Friedman, M. (1937). The use of ranks to avoid the assumption of normality in the analysis of variance. *J. Amer. Statist. Assoc.*, **32**, 675–701.

Groggel, D.J., Schaefer, R.L., and Skillings, J.H. (1987). Procedures for analyzing block designs with censored data. *Comm. Statist.*, **A16**, 431–444.

Hájek, J. and Šidák, Z. (1967). *Theory of Rank Tests*, Academic Press, New York.

Haux, R., Schumacher, M., and Weckesser, G. (1984). Rank tests for complete block designs. *Biom. J.*, **26**, 567–582.

Hollander, M. and Wolfe, D.A. (1999). *Nonparametric Statistical Methods*, 2nd edition, John Wiley & Sons, New York.

Iman, R.L. and Davenport, J.M. (1980). Approximations of the critical region of the Friedman statistic. *Comm. Statist.*, **A9**, 571–595.

Lee, E.T., Desu, M.M., and Gehan, E.A. (1975). A Monte-Carlo study of the power of some two-sample tests. *Biometrika*, **62**, 425–432.

Lehmann, E.L. (1998). *Nonparametrics: Statistical Methods Based on Ranks, Revised First Edition*, Prentice-Hall, Upper Saddle River, New Jersey.

Lininger, L., Gail, M.H., Green, S.B., and Byar, D.P. (1979). Comparison of four tests for equality of survival curves in the presence of stratification and censoring. *Biometrika*, **66**, 419–428.

Mack, G.A. and Skillings, J.H. (1980). A Friedman-type rank test for main effects in a two factor ANOVA. *J. Amer. Statist. Assoc.*, **75**, 947–951.

Nelson, L.S. (1970). Tables for Wilcoxon's rank sum test in randomized blocks. *J. Qual. Tech.*, **2**, 207–218.

Page, E.B. (1963). Ordered hypotheses for multiple treatments: a significance test for linear ranks. *J. Amer. Statist. Assoc.*, **58**, 216–230.

Patel, K.M. (1975). A generalized Friedman test for randomized block design when observations are subject to arbitrary censorship. *Comm. Statist.*, **A2**, 373–380.

Patil, K.D. (1975). Cochran's Q test: exact distribution. *J. Amer. Statist. Assoc.*, **69**, 164–168.

Podgor, M.J. and Gastwirth, J.L. (1994). A cautionary note on applying scores in stratified data. *Biometrics*, **50**, 1215–1218.

Prentice, M.J. (1979). On the problem of m incomplete rankings. *Biometrika*, **66**, 167–170.

Quade, D. (1979). Using weighted rankings in the analysis of complete blocks with additive block effects. *J. Amer. Statist. Assoc.*, **74**, 680–683.

Skillings, J.H. and Mack, G.A. (1981). On the use of a Friedman-type statistic in balanced and unbalanced block designs. *Technometrics*, **23**, 171–177.

Woolson, R.F. and Lachenbruch, P.A. (1981). Rank tests for censored randomized block designs. *Biometrika*, **68**, 427–435.

CHAPTER 6

Independence, correlation, and regression

6.1 Introduction

In some situations, we suspect some kind of relationship between two variables. A study is undertaken to examine the degree of relationship, which could profitably be used to predict one variable, using the information on the other variable. Studies which are undertaken to establish relationships may be called *correlation studies* or *association studies*. The term *association* is usually used when the variables are attributes, that is, their values belong to one of several categories. After establishing the existence of a relationship, we may want to model this relationship, which can be used for prediction purposes. Studies undertaken for this purpose can be termed *regression studies*. First we consider the problem of testing for association or correlation, and then we discuss the linear regression problem. Other commonly used nonparametric regression procedures based on projection pursuit and the nearest neighbor are beyond the scope of this book. In projection pursuit regression, the dependent variable is estimated by the sum of general smooth functions of linear projections of several predictor variables iteratively; the interested reader is referred to Friedman and Stuetzle (1981). In the simplest case of nearest neighbor regression, the dependent variable is estimated by the mean or median of the responses from k nearest neighbor points with respect to Euclidean distance function of the training data. For further details see Bhattacharya and Mack (1987), Devroye et al. (1994), and Yang (1998). We will only discuss the *logistic* and *PH regression* models.

6.2 Analysis of a bivariate sample

We consider a bivariate sample, where the response variables are either continuous or categorical. For the case of categorical responses, we are interested in testing for independence and also estimate an association measure. This analysis will be discussed in Subsection 6.2.1. When the number of categories is the same for the two responses, obtaining a measure of agreement is of interest, as discussed in Subsection 6.2.2. When the responses are continuous variables, correlation analysis is undertaken, which is the topic considered in Section 6.3.

Table 6.1 *Frequencies and probabilities*

Y Cat.	X-Category						
	1	2	$\cdots$	j	$\cdots$	c	Total
1	$n_{11}(\phi_{11})$	$n_{12}(\phi_{12})$	$\cdots$	$n_{1j}(\phi_{1j})$	$\cdots$	$n_{1c}(\phi_{1c})$	$n_{1.}(\phi_{1.})$
2	$n_{21}(\phi_{12})$	$n_{22}(\phi_{22})$	$\cdots$	$n_{2j}(\phi_{2j})$	$\cdots$	$n_{2c}(\phi_{2c})$	$n_{2.}(\phi_{2.})$
$\vdots$	$\vdots$	$\vdots$	$\ddots$	$\vdots$	$\ddots$	$\vdots$	$\vdots$
i	$n_{i1}(\phi i1)$	$n_{i2}(\phi_{i2})$	$\cdots$	$n_{ij}(\phi_{ij})$	$\cdots$	$n_{ic}(\phi_{ic})$	$n_{i.}(\phi_{i.})$
$\vdots$	$\vdots$	$\vdots$	$\ddots$	$\vdots$	$\ddots$	$\vdots$	$\vdots$
r	$n_{r1}(\phi_{r1})$	$n_{r2}(\phi_{r2})$	$\cdots$	$n_{rj}(\phi_{rj})$	$\cdots$	$n_{rc}(\phi_{rc})$	$n_{r.}(\phi_{r.})$
Total	$n_{.1}(\phi_{.1})$	$n_{.2}(\phi_{.2})$	$\cdots$	$n_{.j}(\phi_{.j})$	$\cdots$	$n_{.c}(\phi_{.c})$	$n(1)$

6.2.1 *Test for independence between categorical responses*

Here we deal with studies in which a sample $\{(X_1, Y_1), \ldots, (X_n, Y_n)\}$ is obtained. The response variables X and Y are categorical variables, where X values fall into r categories and Y values fall into c categories. The data are summarized by the cell frequencies, n_{ij}. These frequencies and the probability model appear in Table 6.1.

The quantities ϕ_{ij} are the probabilities associated with the cells (i, j). It should be noted that we delete from further consideration those rows and columns where the total frequencies are zero. Here the problem is to test for the independence of the two variables, X and Y. This hypothesis of independence is the null hypothesis, which is

$$H_0 : \phi_{ij} = \phi_{i.}\phi_{.j}, \quad \text{for all } (i, j). \tag{6.1}$$

A popular test procedure is the *chi-square procedure* of Pearson (see Appendix A4). To carry out this procedure we need to calculate the expected frequencies under the null hypothesis and use them to calculate a discrepancy measure. The expected frequencies are

$$e_{ij} = n\hat{\phi}_{i.}\hat{\phi}_{.j} = n\left(\frac{n_{i.}}{n}\right)\left(\frac{n_{.j}}{n}\right). \tag{6.2}$$

A discrepancy measure is

$$T_{(r \times c)} = \sum_{i=1}^{r}\sum_{j=1}^{c}\frac{(n_{ij} - e_{ij})^2}{e_{ij}} = n\sum_{i=1}^{r}\sum_{j=1}^{c}\frac{[n_{ij} - (n_{i.}n_{.j})/n]^2}{n_{i.}n_{.j}}. \tag{6.3}$$

The subscript of T denotes the number of levels of the two variables. This is used as a test statistic, whose null distribution can be approximated by a chi-square distribution with $\nu = (r - 1)(c - 1)$ degrees of freedom. The test

$$\text{rejects } H_0 \quad \text{if} \quad T > \chi_\nu^2(1 - \alpha), \tag{6.4}$$

and the associated P-value is

$$P\text{-value} \approx P(\chi^2_\nu > T(obs)),$$

where $T(obs)$ is the observed value of T. The expression for T simplifies considerably in the case of $r = 2$ and $c = 2$. This expression is

$$T_{(2\times2)} = \frac{n(n_{11}n_{22} - n_{12}n_{21})^2}{n_1.n_2.n_{.1}n_{.2}}. \tag{6.5}$$

When the null hypothesis (6.1) of independence is rejected, it is common practice to indicate the degree of association. To this end, we consider the quantity T/n, and reexpress it in terms of estimates of cell probabilities ϕ_{ij}. It is easy to see that

$$\frac{T}{n} = \sum_{i=1}^{r}\sum_{j=1}^{c} \frac{[\frac{n_{ij}}{n} - \frac{n_{i.}}{n}\frac{n_{.j}}{n}]^2}{\frac{n_{i.}}{n}\frac{n_{.j}}{n}} = \sum_{i=1}^{r}\sum_{j=1}^{c} \frac{[\hat{\phi}_{ij} - \hat{\phi}_{i.}\hat{\phi}_{.j}]^2}{\hat{\phi}_{i.}\hat{\phi}_{.j}}.$$

Cramér's V

To describe the *degree of dependence* between two discrete (or categorical) random variables, Cramér (1945) proposed a measure called the *mean square contingency*. It is defined as

$$\phi^2_{(r\times c)} = \sum_{i=1}^{r}\sum_{j=1}^{c} \frac{[\phi_{ij} - \phi_{i.}\phi_{.j}]^2}{\phi_{i.}\phi_{.j}}. \tag{6.6}$$

This measure is zero if and only if the variables are independent. Cramér showed that $\phi^2 \leq q - 1$, where $q = min(r, c)$. Thus $0 \leq \frac{\phi^2}{q-1} \leq 1$, and so $\frac{\phi^2}{q-1}$ can be used as a measure of dependence or association in a 0–1 scale. It is easy to see that $\frac{T}{n(q-1)}$ is the ML estimator of the association measure $\frac{\phi^2}{q-1}$. In other words, $\frac{T}{n(q-1)}$ is a sample estimate of Cramér's measure of association, $\frac{\phi^2}{q-1}$.

Consider two discrete random variables X and Y, each assuming two values (x_1, x_2) and (y_1, y_2). The correlation coefficient between these random variables is

$$\rho = \frac{(\phi_{11}\phi_{22} - \phi_{12}\phi_{21})}{(\phi_{1.}\phi_{2.}\phi_{.1}\phi_{.2})^{1/2}} sn[|x_1 - x_2| \cdot |y_1 - y_2|],$$

where $sn(a) = 1(-1)$ depending on whether $a > 0(< 0)$. This coefficient does not depend on the actual values of the variables and thus this quantity can be looked upon as a measure of dependence between two categorical variables, each with two categories. In fact,

$$\phi^2_{(2\times2)} = \rho^2(X, Y).$$

Table 6.2 *Voting pattern of wife and husband*

Wife Voted	Husband Voted Republican	Others	Total
Republican	60	20	80
Others	50	70	120
Total	110	90	200

We can take

$$\sqrt{(T_{(2\times 2)})/n} = \frac{(n_{11}n_{22} - n_{12}n_{21})}{(n_{1.}n_{2.}n_{.1}n_{.2})^{1/2}}$$

as an estimate of the correlation coefficient between two categorical variables and it can be used as a measure of dependence or association. As a generalization of this idea, a currently used measure of dependence or association is the statistic V, where

$$V = \sqrt{\frac{T}{n(q-1)}}, \tag{6.7}$$

and it is known as *Cramér's V*.

We will consider an example where the test for independence is illustrated.

Example 6.1. In a sample of 200 households, the voting patterns of wife and husband in an election are observed and the results are summarized in Table 6.2. We are interested in testing the null hypothesis that the voting patterns of wife and husband are independent. The test statistic of (6.5) is $T = 21.55$ and the associated P-value $= P(\chi_1^2 \geq 21.55) \approx 0$. Thus we conclude the voting patterns are not independent. An estimate of Cramér's measure of association is $\frac{T}{n(q-1)} = \frac{21.55}{200(2-1)} = 0.1078$. Thus Cramér's $V = \sqrt{0.1078} = 0.3282$.

A computer program for doing these calculations is given in Appendix B6.

6.2.2 A measure of agreement-κ

Consider a setting where each of two judges classifies n items into r distinct categories. The resulting frequencies of this classification are given in Table 6.3, along with the theoretical probabilities in parentheses.

Here $P_0 = \sum_{i=1}^{r} \theta_{ii}$ is the probability of agreement of the two judges and this probability is estimated by $\hat{P}_0 = (\sum_{i=1}^{r} n_{ii})/n$. Further, $P_e = \sum_{i=1}^{r} \theta_{i.}\theta_{.i}$ is the probability of agreement of the two judges if the classifications are made independently by them. This probability is estimated by $\hat{P}_e = (\sum_{i=1}^{r} n_{i.}n_{.i})/n$.

Table 6.3 *Classification frequencies and probabilities*

Judge 1 Classification Category	Judge 2 Classification Category				
	1	2	$\cdots$	r	Total
1	$n_{11}(\theta_{11})$	$n_{12}(\theta_{12})$	$\cdots$	$n_{1r}(\theta_{1r})$	$n_{1.}(\theta_{1.})$
2	$n_{21}(\theta_{12})$	$n_{22}(\theta_{22})$	$\cdots$	$n_{2r}(\theta_{2r})$	$n_{2.}(\theta_{2.})$
$\vdots$	$\vdots$	$\vdots$	$\vdots$	$\vdots$	$\vdots$
r	$n_{r1}(\theta_{r1})$	$n_{r2}(\theta_{r2})$	$\cdots$	$n_{rr}(\theta_{rr})$	$n_{r.}(\theta_{r.})$
Total	$n_{.1}(\theta_{.1})$	$n_{.2}(\theta_{.2})$	$\cdots$	$n_{r.}(\theta_{r.})$	$n(1)$

A measure of agreement, introduced by Cohen (1960), is a function of P_0 and P_e. It is called kappa, denoted by κ,

$$\kappa = \frac{P_0 - P_e}{1 - P_e}. \tag{6.8}$$

It measures the excess of the judges' agreement over that expected by chance when the judges independently classify the items. Clearly, $\frac{-P_e}{1-P_e} \leq \kappa \leq 1$. When the observed agreement is expected by chance, $\kappa = 0$, and when there is perfect agreement, $\kappa = 1$. The stronger the agreement, the higher the value of κ. Generally, we want a confidence interval for κ.

To get a confidence interval we start with a point estimate, namely,

$$\hat{\kappa} = \frac{\hat{P}_0 - \hat{P}_e}{1 - \hat{P}_e}.$$

An estimate of the variance of this estimate is

$$\hat{var}(\hat{\kappa}) = \frac{1}{n(1 - \hat{P}_e)^2} \left[\hat{P}_e + \hat{P}_e^{\,2} - \sum_{i=1}^{r} \hat{\theta}_{i.} \hat{\theta}_{.i} (\hat{\theta}_{i.} + \hat{\theta}_{.i}) \right],$$

where $\hat{\theta}_{i.} = n_{i.}/n$ and $\hat{\theta}_{.i} = n_{.i}/n$ (for the derivation refer to Fleiss et al. (1969)). Using an approximate normal distribution for the estimate $\hat{\kappa}$, we get a $100(1 - \alpha)\%$ confidence interval for κ and it is

$$\hat{\kappa} \pm z_{(1-\alpha/2)} \sqrt{\hat{var}(\hat{\kappa})}. \tag{6.9}$$

Calculation of the kappa coefficient and the confidence interval (6.9) for the data of Example 6.1 is given in the output of the computer program of Appendix B6.1.

6.3 Testing for correlation between continuous variables

There is a suspicion that there exists a relationship between two characteristics of subjects. To learn about this relationship for a sample of subjects, the two characteristics, X and Y, are observed. As a first step, we want to establish a relationship, and later model this relationship. The problem of establishing a relationship is usually called *correlation analysis* and modeling the relationship is usually called *regression analysis*. In this section we investigate correlation analysis. The hypothesis of no relationship is formulated as the hypothesis of independence. When X and Y are continuous variables, the correlation coefficient is used as a measure of dependence. Recall that independence implies zero correlation; however, zero correlation does not imply independence in general. One case where zero correlation is equivalent to independence is when the joint distribution of X and Y is a bivariate normal distribution. Zero correlation implies that one variable cannot be viewed as a linear function of the other. Recall that the correlation coefficient is

$$\rho(X, Y) = \frac{cov(X, Y)}{\sqrt{var(X)var(Y)}}, \tag{6.10}$$

where the covariance is

$$cov(X, Y) = E(XY) - E(X)E(Y).$$

It is of some interest to see why the covariance is a measure of dependence. Let the joint cdf of X and Y be $H(.,.)$. The marginal distributions are defined by the cdf's $F(.)$ and $G(.)$. A necessary and sufficient condition for the independence of the variables X and Y is

$$H(x, y) = F(x)G(y), \quad for \ all \ real \ x \ and \ y. \tag{6.11}$$

It seems natural to examine the difference $\delta(x, y) = H(x, y) - F(x)G(y)$ and obtain a measure by summarizing these differences. Such a measure could serve as a measure of dependence. One such summary measure is

$$D(X, Y) = \int_{-\infty}^{\infty} \int_{-\infty}^{\infty} \delta(x, y) dx \, dy. \tag{6.12}$$

Hoeffding proved that this quantity is equal to $cov(X, Y)$, which is zero when X and Y are independent. For an easily accessible proof of Hoeffding's result, the reader is referred to Shea (1982). Instead of using covariance as a measure of dependence, it has been customary to use the correlation coefficient as a measure of dependence, since the covariance is not a unit-free measure; also, $-1 \le \rho \le 1$.

Now we can test the independence hypothesis by formulating it as the hypothesis of zero correlation. To do so, we estimate the correlation coefficient from a sample and test for significance. The usual estimate is the sample correlation coefficient $r(X, Y)$, which is

$$r(X, Y) = \frac{\sum (X_i - \bar{X})(Y_i - \bar{Y})}{\sqrt{[\sum (X_i - \bar{X})^2][\sum (Y_i - \bar{Y})^2]}}. \tag{6.13}$$

The sampling distribution of r depends on the population distribution and thus we cannot obtain a distribution-free test. If we use a statistic that depends on the ranks it is possible to obtain a distribution-free test. This task is achieved by the procedure proposed by Spearman (1904), which will be described now.

6.3.1 Spearman's rank correlation test

Let us rank X-values and Y-values independently. Let R_i be the rank of X_i and S_i be the rank of Y_i. Instead of the original data (X_i, Y_i), consider the rank pairs (R_i, S_i) and calculate the correlation coefficient for these rank pairs. This correlation coefficient is called *Spearman's rank correlation coefficient*. It is given by

$$r_{SP} = \frac{\sum (R_i - \bar{R})(S_i - \bar{S})}{\sqrt{[\sum (R_i - \bar{R})^2][\sum (S_i - \bar{S})^2]}},$$ (6.14)

where $\bar{R} = \sum R_i/n$ and $\bar{S} = \sum S_i/n$. Under the hypothesis of independence, each rank vector is a permutation of $(1, 2, \ldots, n)$ and thus we get $(n!)^2$ possible configurations for the rank pairs. For each of the configurations we can calculate the value of r_S and develop the sampling distribution of this statistic. Using this sampling distribution we can devise tests of hypotheses.

However, to develop the critical regions, Hotelling and Pabst (1936) introduced the statistic

$$T_{HP} = \sum_{i=1}^{n} [R_i - S_i]^2,$$ (6.15)

and established a relationship between this statistic and the statistic r_{SP}.

We now establish this relationship assuming that there are no ties among the X-values as well as among the Y-values. We rewrite T_{HP} as

$$T_{HP} = \sum_{i=1}^{n} [(R_i - \bar{R}) - (S_i - \bar{S})]^2,$$

and expanding the squares we get

$$T_{HP} = \sum_{i} (R_i - \bar{R})^2 + \sum_{i} (S_i - \bar{S})^2 - 2\sum_{i} [(R_i - \bar{R})(S_i - \bar{S})].$$ (6.16)

Under the assumption of no ties among both X-values and Y-values we have

$$\sum_{i} (R_i - \bar{R})^2 = \sum_{i} (S_i - \bar{S})^2 = \frac{n(n^2 - 1)}{12}.$$ (6.17)

Using (6.17) in (6.14), we get

$$\sum_{i} (R_i - \bar{R})(S_i - \bar{S}) = \frac{n(n^2 - 1)}{12} r_{SP}.$$ (6.18)

Now using (6.17) and (6.18) in (6.16) and solving for r_{SP}, we get the relation

$$r_{SP} = 1 - \frac{6T_{HP}}{n(n^2 - 1)}. \tag{6.19}$$

We note that the minimum value of T_{HP} is 0 and the maximum value is $(n^3 - n)/3$. Thus r_{SP} takes values in $[-1, 1]$ like the usual correlation coefficient. From this relationship, it is clear that r_{SP} will be close to 1 when T_{HP} is very small, and r_{SP} will be close to -1 when T_{HP} is very large.

To interpret the alternatives indicated by the extreme values of T_{HP}, let us formally state the various alternatives to the null hypothesis of independence, which are of interest. These are

$$H_{A1} : Positive\ association,$$

$$H_{A2} : Negative\ association,$$

$$H_{A3} : H_{A1}\ or\ H_{A2}.$$

By positive association, we mean that large values of X are associated with large values of Y. If small values of X are associated with larger values of Y, we say that there is a negative association.

It is easy to see that if there is a positive association, large X-values will be associated with large Y-values, so that T_{HP} is expected to take small values. On the other hand, if there is a negative association, large X-values will be associated with small Y-values so that T_{HP} is expected to be large. So large values of T_{HP} indicate a negative association, whereas small values of T_{HP} indicate a positive association. Thus values of r_{SP} close to 1 indicate a positive association and values of r_{SP} close to -1 indicate a negative association.

It can be shown that

$$E_0(r_{SP}) = 0,$$

and the null distribution of r_{SP} is symmetric about zero. The critical regions are given in Table 6.4, where c's are appropriate percentiles, $t_R(obs)$ is the observed value of t_R defined in (6.20), and t_{n-2} is a random variable with t_{n-2} distribution.

A table of percentiles of r_{SP} appears in Glasser and Winter (1961). One can use an approximation to the null distribution to get the critical values or to calculate the P-values.

Table 6.4 *Spearman rank correlation tests for association*

Alternative	Critical Region	Approx. P-Value
H_{A1}	$r_{SP} > c_1$	$P_1 = P(t_{n-2} > t_R(obs))$,
H_{A2}	$r_{SP} < c_2$	$P_2 = P(t_{n-2} < t_R(obs))$,
H_{A3}	$r_{SP} < c_3\ or\ r_{SP} > c_4$	$2.min(P_1, P_2)$.

Asymptotic distribution of r_{SP}

The null distribution of

$$t_R = r_{SP}\sqrt{\frac{n-2}{1-r_{SP}^2}}, \tag{6.20}$$

can be approximated by the t-distribution with $(n-2)$ degrees of freedom. This approximation is reasonably accurate for $n > 10$.

Remark 6.1. When there are ties we can use midranks to compute the statistics r_{SP} and T_{HP} and we can also relate these two. Now (6.16) is valid and (6.17) is not true. Using (6.14) in (6.16), we get

$$r_{SP} = \frac{\sigma_R^2 + \sigma_S^2 - (T_{HP}/n)}{2\sqrt{\sigma_R^2 \sigma_S^2}}, \tag{6.21}$$

where

$$\sigma_R^2 = \sum_i (R_i - \bar{R})^2/n, \ \sigma_S^2 = \sum_i (S_i - \bar{S})^2/n.$$

For testing purposes we can also use a normal approximation to the null distribution of T_{HP}. For further details refer to Lehmann (1998, p. 301).

6.3.2 Kendall's tau

Kendall proposed another measure of association. If the variables are dependent, a change in one variable should result in a change in the other variable. This intuitive notion can be formalized as follows. Suppose that (X_1, Y_1) and (X_2, Y_2) are two i.i.d. random vectors. This pair is called a *concordant pair* if $(X_1 - X_2)(Y_1 - Y_2)$ is positive and the pair is called a *discordant pair* if this product is negative.

Let us compute P_C, the probability of observing a concordant pair.

$$P_C = P[(X_1 - X_2)(Y_1 - Y_2) > 0]$$
$$= P[X_1 - X_2 > 0, Y_1 - Y_2 > 0] + P[X_1 - X_2 < 0, Y_1 - Y_2 < 0].$$

Thus this probability can be seen to be

$$P_C = 2P[X_1 > X_2, Y_1 > Y_2]. \tag{6.22}$$

Since the random vectors are continuous random vectors, the probability of a discordant pair is

$$P_D = 1 - P_C. \tag{6.23}$$

Recall that the null hypothesis is

$$H_0 : X \text{ and } Y \text{ are independent.} \tag{6.24}$$

Under the null hypothesis

$$P_C(H_0) = P[a \; concordant \; pair | H_0] = 2P[X_1 > X_2]P[Y_1 > Y_2] = (1/2).$$
(6.25)

Arguing in a similar manner we can show that if X and Y are independent

$$P_D(H_0) = P[a \; discordant \; pair | H_0] = (1/2).$$

Thus under the null hypothesis (6.24), $P_C = P_D = (1/2)$. So a possible measure of dependence is the difference between these two probabilities, namely,

$$\tau = P_C - P_D = P_C - [1 - P_C] = 2P_C - 1 = 1 - 2P_D.$$
(6.26)

This measure was suggested by Kendall and it is known as *Kendall's tau*. This quantity can be estimated from the data and using this estimate we can devise a test of independence hypothesis.

It is customary to call a measure of dependence a *measure of association*, especially in those cases where there is no evidence that one variable is a cause and the other one is the effect. In such cases we have reason to believe that a change in one variable is followed by a change in the other variable. This association may be due to the influence of a third variable on both the variables we are studying. For example, the IQ level of an individual influences the performances on a math test and a science test. So the score on a math test has association with the score on a science test. Now let us formulate the problem of testing for association. The null hypothesis of no association is formulated as

$$H_0^* : \tau = 0.$$
(6.27)

Various alternatives mentioned earlier are restated in Table 6.5.

As before, we have a sample of n pairs (X_i, Y_i), and we compare each pair (X_i, Y_i) with the other pairs $(X_j, Y_j)(j > i)$ and determine the number, N_C, of concordant pairs and the number, N_D, of discordant pairs. Altogether we need to make $\binom{n}{2}$ comparisons. A natural estimate of τ is

$$T_K = \binom{n}{2}^{-1} [N_C - N_D].$$
(6.28)

Table 6.5 *Various alternatives of interest*

Alternative	τ-Version
H_{A1}: Positive association	$\tau > 0$
H_{A2}: Negative association	$\tau < 0$
H_{A3}: There is association	$\tau \neq 0$

Table 6.6 *Kendall's tau tests*

Alternative	Critical Region
H_{A1}	$T_K > c_1$,
H_{A2}	$T_K < c_2$,
H_{A3}	$T_K < c_3$ or $T_K > c_4$.

A formula for computing T_K will be useful. The computing formula can be stated in terms of the sgn function, namely,

$$sgn(a) = 1, \; for \; a > 0,$$
$$= 0, \; for \; a = 0,$$
$$= -1, \; for \; a < 0.$$

Now we define

$$A_{ij} = sgn(X_j - X_i), \; B_{ij} = sgn(Y_j - Y_i). \qquad (6.29)$$

The general formula for T_K is

$$T_K = \frac{\sum_{i=1}^{n-1} \sum_{j=i+1}^{n} A_{ij} B_{ij}}{[(\sum_{i=1}^{n-1} \sum_{j=i+1}^{n} A_{ij}^2)(\sum_{i=1}^{n-1} \sum_{j=i+1}^{n} B_{ij}^2)]^{(1/2)}}. \qquad (6.30)$$

This quantity reduces to the expression in (6.28), when there are no ties among X-values or among Y-values.

The critical regions of the various tests are given in Table 6.6. The critical values are the appropriate percentiles of the null distribution of the estimate T_K. A table of these percentiles was prepared by Best (1973).

Asymptotic distribution of T_K

The null distribution of

$$Z = \frac{3T_K \sqrt{n(n-1)}}{\sqrt{2(2n+5)}},$$

is asymptotically the standard normal distribution and this result can be used to construct approximate α-level tests of hypotheses about τ. This is a good approximation when $n > 10$.

Remark 6.2. Nelson (1986) noted that the normal approximation for the distribution of T_K is a better approximation compared to the normal approximation for the distribution of Spearman's correlation coefficient, r_{SP}.

Example 6.2. The changes in stock prices on two consecutive Mondays for 15 randomly selected stocks are given in Table 6.7 (artificial data).

Table 6.7 *Changes in stock prices*

Stocks	First Monday	Second Monday	Stocks	First Monday	Second Monday
1	3.10	2.50	9	−2.10	−1.20
2	−1.40	0.25	10	0.70	0.50
3	1.75	−0.35	11	1.60	0.80
4	−0.50	−0.45	12	−1.25	0.25
5	2.85	2.10	13	2.50	1.50
6	1.60	−0.70	14	−0.75	−0.50
7	−1.30	0.20	15	1.65	0.60
8	1.60	0.30			

Using the computer program given in Appendix B6, we get

$$r_{SP} = 0.685, \; P\text{-value} = 0.0048,$$

$$T_K = 0.5534, \; P\text{-value} = 0.0046.$$

Thus we reject the null hypothesis of independence based on Spearman's rank correlation test as well as Kendall's tau test.

6.4   Linear regression

At times it is reasonable to assume that the regression function of Y, that is, $E(Y|X = x)$, is a linear function of x. This means

$$E(Y|x) = \alpha + \beta x, \tag{6.31}$$

where α and β are unknown constants. The constant β is known as the slope of the regression line and frequently we are interested in making inference about β. First we examine the problem of testing a hypothesis about the value of β.

6.4.1 *Testing a hypothesis about the slope (Theil's test)*

A set of n distinct x-values, $(x_1, \ldots, x_n)$, is chosen. At each x_i value, we obtain an observation Y_i. Thus we have the data pairs (Y_i, x_i). On the basis of these data we want to test the null hypothesis

$$H_0 : \beta = \beta_0, \tag{6.32}$$

where β_0 is a specified value. We first calculate

$$Y_i^* = Y_i - \beta_0 x_i. \tag{6.33}$$

Under the null hypothesis $E(Y_i^*)$ is α, a constant. There is no association between these variables and x_i. This hypothesis of no association can be tested using Kendall's tau statistic. So we calculate Kendall's tau T_{reg} for the pairs (x_i, Y_i^*) and use it as a test criterion. The alternative hypotheses and corresponding critical regions are shown in Table 6.8.

Table 6.8 *Theil's tests for slope*

Alternatives	Critical Regions
$H_{A1} : \beta > \beta_0$	$T_{reg} \geq c_1,$
$H_{A2} : \beta < \beta_0$	$T_{reg} \leq c_2,$
$H_{A3} : \beta \neq \beta_0$	$T_{reg} \leq c_3 \text{ or } T_{reg} \geq c_4.$

A partial motivation for the critical region in the context of the alternative $H_{A1} : \beta > \beta_0$ in Table 6.8 is the following. When $\beta = \beta_1(> \beta_0)$, we have

$$cov(X_i, Y_i - \beta_1 X_i) = 0.$$

Now we express Y_i^* as

$$Y_i^* = Y_i - \beta_1 X_i + (\beta_1 - \beta_0)X_i,$$

and observe that, for $\beta = \beta_1$, we have

$$cov(Y_i^*, X_i) = cov(Y_i - \beta_1 X_i, X_i) + cov((\beta_1 - \beta_0)X_i, X_i)$$
$$= (\beta_1 - \beta_0)var(X_i) > 0.$$

Thus the correlation between Y_i^* and X_i is positive under the null hypothesis. The results of this section are from Theil (1950).

6.4.2 Estimation of the slope

If in fact the regression is a straight line, an estimate of the slope can be obtained by considering the straight line passing through the two points (x_i, Y_i) and (x_j, Y_j). Clearly the slope of the line passing through these two points is

$$b_{ij} = \frac{Y_i - Y_j}{x_i - x_j}, \qquad (6.34)$$

which is an estimate of the slope; thus we can obtain $N = \binom{n}{2}$ estimates of the slope. The median of these estimates is taken as a point estimate. In other words,

$$\hat{\beta} = med(b_{ij}). \qquad (6.35)$$

To obtain a confidence interval we proceed as follows. Order these slope estimates and let the ordered values be denoted by

$$B_{(1)} < B_{(2)} < \cdots < B_{(N)}.$$

A $100(1 - \alpha)\%$ confidence interval is $(B_{(L)}, B_{(U)})$, for suitably chosen integers L and U. These integers are

$$L = \lfloor (N - Nw_{(1-\alpha/2)})/2 \rfloor; \ U = \lfloor (N + Nw_{(1-\alpha/2)})/2 \rfloor + 1, \qquad (6.36)$$

where w is the $(1 - \alpha/2)$ quantile of the null distribution of Kendall's tau statistic. Derivation of this confidence interval is given in Appendix A6.

Table 6.9 *Data for regression analysis*

Age (X)	Time (Y)
46	30
26	20
31	40
56	28
36	25
41	35

From the asymptotic normal distribution of Kendall's tau, for large n, we have

$$w_{(1-\alpha/2)} \approx \frac{z_{(1-\alpha/2)}\sqrt{2(2n+5)}}{3\sqrt{n(n-1)}}.$$

We now illustrate the details through an example.

Example 6.3. In a random sample of six workers, the age (X) and time to finish a task (in min.) are noted and the data are listed in Table 6.9 (artificial data).

We want to find an interval estimate for the slope of the regression line. The $N = \binom{6}{2} = 15$, b_{ij} values have been calculated and the ordered values are

$$B_{(1)} = -3.0, \ B_{(2)} = -1.0, \ B_{(3)} = -0.8, \ B_{(4)} = -0.67, \ B_{(5)} = -0.5,$$
$$B_{(6)} = -0.47, \ B_{(7)} = -0.2, \ B_{(8)} = -0.15, \ B_{(9)} = 0.27, \ B_{(10)} = 0.5,$$
$$B_{(11)} = 0.5, \ B_{(12)} = 0.5, \ B_{(13)} = 1.0, \ B_{(14)} = 2.0, \ B_{(15)} = 4.0.$$

The median of b_{ij} values is -0.15, which is used as a point estimate of β.

We have $w_{0.975} \approx 0.70$. Using this value in equation (6.36), we get $L = 2$ and $U = 13$. Thus an approximate 95% confidence interval for β is $(B_{(2)}, B_{(13)}) = (-1.0, 1.0)$. As this interval contains the value 0, we do not reject the null hypothesis $H_0 : \beta = 0$ against the two-sided alternatives at an approximate 0.05 level.

6.5 Logistic regression

In the previous section, we examined the linear regression problem when the response variable Y and the explanatory variable X are both continuous variables. However, sometimes Y may be a binary response variable like the occurrence or nonoccurrence of a clinical event. As the expectation of Y, given X in this case, should be between 0 and 1, the model we used earlier is not appropriate. The frequently used model is

$$E(Y|x) \equiv \phi(x) = \frac{e^{\alpha+\beta \cdot x}}{1 + e^{\alpha+\beta \cdot x}}. \tag{6.37}$$

This model is known as the *logistic regression model*. Here it is convenient to consider the *logit* of $\phi(x)$, which is a linear function of x. The logit is defined as

$$logit(\phi(x)) = ln\left(\frac{\phi(x)}{1 - \phi(x)}\right) = \alpha + \beta \cdot x. \qquad (6.38)$$

It may be noted that the ratio $\frac{\phi(x)}{1-\phi(x)}$ is the odds in favor of $Y = 1$ for a given x and thus the $logit(\Phi(x))$ is the *log odds*. Before we proceed further, let us look into the interpretation of the parameters α and β.

6.5.1 Interpretation of α and β

It is easy to see that $logit(\phi(0)) = \alpha$, and so α can be interpreted as the baseline log odds (in favor of $Y=1$). Also

$$logit(\phi(1)) - logit(\phi(0)) = ln\left(\frac{\phi(1)}{1 - \phi(1)} \cdot \frac{1 - \phi(0)}{\phi(0)}\right) = \beta.$$

Thus β represents the change in the log odds when x changes to 1 from 0. Further, e^β can be seen to be the odds ratio.

For example, when x is a binary variable representing the presence or absence of a risk factor and Y is the occurrence or nonoccurrence of a clinical event, β describes the effect of the risk factor on the response.

When x is a continuous variable, we have

$$logit(\phi(x + 1)) - logit(\phi(x)) = \beta,$$

and hence e^β is the change in the odds ratio when x increases by one unit. Studies are undertaken to estimate the odds ratios and now we discuss the problem of estimating the parameters α and β.

6.5.2 Estimation of α and β

Let (x_i, y_i), $(i = 1, 2, \cdots, n)$ be a random sample of size n from a population where the logistic model holds. Thus

$$E(Y_i|x_i) \equiv \phi(x_i) = \frac{e^{\alpha + \beta \cdot x_i}}{1 + e^{\alpha + \beta \cdot x_i}}.$$

The likelihood of Y's is a function of α and β. We can use the maximum likelihood method to get the estimates. The likelihood equations are

$$\sum_i [y_i - \phi(x_i)] = 0; \quad \sum_i x_i[y_i - \phi(x_i)] = 0. \qquad (6.39)$$

We need to solve these nonlinear equations for α and β to get the maximum likelihood (ML) estimators. The derivation of these equations is given in Appendix A6.

The asymptotic covariance matrix of the estimators depends on the information matrix $I(\alpha, \beta)$, which is

$$\mathbf{I}(\alpha, \beta) = - \begin{bmatrix} I_{\alpha,\alpha} & I_{\alpha,\beta} \\ I_{\alpha,\beta} & I_{\beta,\beta} \end{bmatrix}$$

These quantities are partial derivatives of the log likelihood function, $l(\alpha, \beta)$, and are defined as

$$I_{p,q} = E\left[\left(\frac{\partial^2}{\partial p \partial q}\right) l(\alpha, \beta)\right], \quad for \ p, q = \alpha, \beta. \tag{6.40}$$

Let

$$\hat{\phi}(x_i) = \hat{\alpha} + \hat{\beta} \cdot x_i, .$$

In the information matrix we replace the unknown parameters (α, β) by their ML estimators to get

$$I(\hat{\alpha}, \hat{\beta}) = \begin{bmatrix} \sum_i \hat{\phi}(x_i)\{1 - \hat{\phi}(x_i)\} & \sum_i x_i \hat{\phi}(x_i)\{1 - \hat{\phi}(x_i)\} \\ \sum_i x_i \hat{\phi}(x_i)\{1 - \hat{\phi}(x_i)\} & \sum_i x_i^2 \hat{\phi}(x_i)\{1 - \hat{\phi}(x_i)\} \end{bmatrix}.$$

An estimate of the asymptoptic covariance matrix of $(\hat{\alpha}, \hat{\beta})$ is $(I(\hat{\alpha}, \hat{\beta}))^{-1}$.

Using the asymptotic normality of $\hat{\beta}$, tests of hypotheses about β can be constructed and confidence intervals for β also can be obtained, which in turn can be used to get confidence intervals for e^{β}. We need to use a computer program for these calculations. The relevant computer program is given in Appendix B6.

We now give an example to illustrate the details of the discussion.

Example 6.4. Over a period of 1 week, patients in the age group of 50 to 55 years who visited the emergency room of a hospital were classified by gender and the clinical event of having an MI. The data are given in Table 6.10.

Using the output of the relevant computer program (see Appendix B6), we have

$$\hat{\beta} = -0.2485;$$

$$Std. \ error(\hat{\beta}) = 0.8135;$$

Table 6.10 *Data on emergency unit visitors*

Gender (X)	MI Status (Y)		Total
	No (0)	Yes (1)	
Male (0)	26	4	30
Female (1)	25	3	28
Total	51	7	58

and

$$odds\ ratio = 0.780.$$

The chi-square statistic for testing $H_0 : \beta = 0$, is $(-0.2485/0.8135)^2 = 0.0933$, and the associated P-value is 0.7601. So we do not reject $H_0 : \beta = 0$.

Let us examine the calculation of the confidence limits for the odds ratio. First we obtain a 95% confidence interval for β as (β_L, β_U) where

$$\beta_L = \hat{\beta} - 1.96 \cdot Std.error(\hat{\beta}) = -1.84230;$$

and

$$\beta_U = \hat{\beta} + 1.96 \cdot Std.error(\hat{\beta}) = 1.3460.$$

Since the odds ratio is e^β, by exponentiating, we get the confidence limits as e^{β_L} and e^{β_U}. Thus a 95% confidence interval for the odds ratio is (0.158, 3.842). This interval is available in the output of the computer program.

Now the null hypothesis $\beta = 0$ is not rejected since the P-value is 0.76, and thus the risks of MI for both sexes cannot be assumed to be different. This conclusion is also indicated because the confidence interval for the odds ratio includes the value 1.

6.5.3 Logistic regression with several explanatory variables

Sometimes the binary response variable Y is influenced by several explanatory variables. For example, MI occurrence or nonoccurrence discussed earlier might be influenced by age, cholesterol level, race, gender, physical activity, etc. and we need to include all these variables in the model for estimating the probability of MI occurrence.

The explanatory variables are of three types: continuous, binary, and categorical with $k(> 2)$ categories. With a continuous explanatory variable, like age or cholesterol level, we associate one independent X-variable. With a binary explanatory variable, like gender, we associate one independent X-variable that takes two values 1 or 0. For example, if the binary X-variable denotes the occurrence or nonoccurrence of a status, usually value $1(0)$ corresponds to the occurrence (nonoccurrence). With a categorical variable, like race, $k - 1$ independent $dummy\ variables$ will be used, where k is the number of categories. To denote five catogories for race, like White, African American, Hispanic, Oriental, and Others, we need four variables as defined in Table 6.11.

In this manner we define the X-variables that need to be included in the model. Suppose that there are altogether p X-variables we want in our model. Let $\mathbf{x_i}$ be a p-dimensional vector of the x-values and Y_i be the response for the ith individual. The model to be used is

$$P(Y_i = 1|\mathbf{x_i}) = \Phi(\mathbf{x_i}) = \frac{e^{\alpha + \beta' \mathbf{x_i}}}{1 + e^{\alpha + \beta' \mathbf{x_i}}}, \tag{6.41}$$

Table 6.11 *Dummy variables for race*

	Dummy Variables			
Race	1	2	3	4
White	0	0	0	0
Afr. American	1	0	0	0
Hispanic	0	1	0	0
Oriental	0	0	1	0
Others	0	0	0	1

Table 6.12 *Data on visitors to an emergency room*

Y	X_1	X_2	X_3	X_4	X_5	X_6	Y	X_1	X_2	X_3	X_4	X_5	X_6
1	200	0	0	0	0	0	0	210	1	0	0	1	0
0	280	1	0	0	0	1	1	250	1	0	0	0	1
1	250	1	0	1	0	0	0	175	0	1	0	0	0
0	175	0	1	0	0	0	0	180	0	0	0	1	0
0	210	1	1	0	0	0	0	190	1	0	0	0	1
1	300	0	0	0	1	0	1	210	0	0	0	0	0
0	165	0	0	0	0	0	1	300	0	1	0	0	0
1	270	0	0	0	1	0	1	220	1	0	0	0	0
1	200	1	0	1	0	0	0	240	0	0	1	0	0
0	185	1	0	0	0	0	1	290	1	0	0	0	1

and a computer program will enable us to make inference on the components of β. We will illustrate these concepts with the help of an example.

Example 6.5. A random sample of 20 patients visiting an emergency room in a hospital were selected and collected data are summarized in Table 6.12. The response variable $Y = 1$ for the occurrence of MI and 0, otherwise. X_1 is cholesterol level, and X_2 is the gender such that $X_2=1(0)$ for female(male). Further, X_3 to X_6 are dummy variables defined above for the race categories as given in Table 6.11.

From the output, we note that for the variable X_1 (cholesterol) the estimate of the regression coefficient is significantly different from zero, since the P-value $(Pr > chi\text{-}square)$ is 0.0313. Thus we conclude that the cholesterol level has a significant effect on the occurrence of MI. A point estimate of the odds ratio for this variable (e^{β_1}) is 1.061. This implies that when other things are equal, each unit increase in cholesterol level increases the odds of having an MI by 1.061.

Let us examine the odds ratio for the Hispanic versus White races. This is given against the variable X_4, which is 0.168. This means that the odds of having an MI for Hispanics are 0.168 times the odds for Whites.

Let us compare the odds for Hispanics with those for African Americans. This comparison can be made by computing

$$odds\ ratio(Hispanics\ vs.\ African\ Americans) \equiv odds(H\ vs.\ AA).$$

It can be seen that

$$odds(H\ vs\ AA) = \frac{odds(H\ vs.\ Whites)}{odds(AA\ vs.\ Whites)}$$
$$= 0.168/0.016 = 10.5.$$

Thus the odds of having an MI for Hispanics are 10.5 times the odds for African Americans.

For additional details we refer to Cox and Snell (1989), and Hosmer and Lemeshow (2000), among others.

Probit model

In (6.37), the quantity $\phi(x)$ can be re-expressed as

$$\phi(x) = \frac{1}{1 + e^{-(\alpha + \beta \cdot x)}},$$

which can be seen to be $F(\alpha + \beta \cdot x)$, where $F(.)$ is the cdf of standard logistic distribution. When $F(.)$ is taken as the cdf of standard normal distribution, the model is called the probit model, which is often used in bioassay problems. Details about this model can be found in Finney (1971), among others.

6.6 Procedures for censored data

As in the case of complete data considered thus far, we now discuss the problem of testing for association between two variables, where either one or both are censored. A procedure due to Oakes (1982) will be described in Subsection 6.6.1. A frequently used proportional hazards model due to Cox (1972) for examining the effect of variables on survival will be discussed in Subsection 6.6.2.

6.6.1 Test for independence

Let $(X_i, Y_i), (i = 1, 2, \cdots, n)$ be a random sample from a bivariate survival distribution and $(C_i, D_i), (i = 1, 2, \cdots, n)$ be the corresponding censoring times, which is a random sample from a bivariate continuous distribution. We assume that the censoring times are independent of the survival times. Let

$$T_i = min(X_i, C_i), \quad and \quad S_i = min(Y_i, D_i).$$

Thus the observable random vectors are (T_i, S_i). We use a "+" on the observed value of T_i or S_i to indicate that it is a censored observation.

As in the case of complete data we need to establish an ordering between T_i and T_j as well as between S_i and S_j. To define concordant or discordant pairs, we define

$$
\begin{aligned}
L_{ij} &= 1, && \text{if } T_i \text{ is decidedly less than } T_j \\
&= -1, && \text{if } T_i \text{ is decidedly greater than } T_j \\
&= 0, && \text{otherwise.}
\end{aligned} \tag{6.42}
$$

We similarly define M_{ij} in relation to the comparison of S_i with S_j. Now the pairs (T_i, S_i) and (T_j, S_j) are definitely concordant or discordant pairs depending on whether $\psi_{ij} = 1$, or -1, where $\psi_{ij} = L_{ij} \cdot M_{ij}$.

The average of ψ_{ij} over all pairs of bivariate vectors for $i < j$ is

$$
T_c = \binom{n}{2}^{-1} \sum_{i=1}^{n-1} \sum_{j=i+1}^{n} \psi_{ij}. \tag{6.43}
$$

This is the test statistic proposed, which is a natural generalization of Kendall's test statistic T_K of (6.28). It is easy to see that the statistic T_c reduces to T_K when there is no censoring.

To construct critical regions for testing the null hypothesis of independence, we need to find the mean and variance of T_c, under the null hypothesis. Oakes (1982) showed that

$$
E_0(T_c) = 0,
$$

and

$$
var_0(T_c) = \frac{2A}{n(n-1)} + \frac{4(n-2)G}{n(n-1)}. \tag{6.44}
$$

The quantities A and G are

$$
A = 4P(T_1 \text{ is decidedly less than } T_2 \text{ and } S_1 \text{ is decidedly less than } S_2);
$$

$$
G = P(T_1 \text{ is decidedly less than } T_2, T_3
$$
$$
\text{and } S_1 \text{ is decidedly less than } S_2, S_3),
$$

where $(T_1, S_1), (T_2, S_2)$, and (T_3, S_3) are i.i.d. random vectors.

We need to estimate the quantities A and G in order to estimate the variance, which is needed to compute the standardized statistic

$$
Z = \frac{T_c}{\sqrt{v\hat{a}r_0(T_c)}}.
$$

Since A is a probability, we just have to find the number of such configurations to get an estimate. In other words, a natural estimate is

$$
\hat{A} = \frac{4(\#(i,j) pairs \text{ with } i < j \text{ such that } L_{ij} = 1 \text{ and } M_{ij} = 1)}{\binom{n}{2}}. \tag{6.45}
$$

Table 6.13 *Tests for association*

Alternative	Critical Region	P-Value
H_{A1}: Positive association	$Z > z_{1-\alpha}$	$P(N(0,1) > Z_{Obs})$
H_{A2}: Negative association	$Z < z_\alpha$	$P(N(0,1) < Z_{Obs})$
$H_{A3} : H_{A1} \cup H_{A2}$	$|Z| > z_{1-\alpha/2}$	$P(N(0,1) > |Z_{Obs}|)$

Table 6.14 *Times to onset of hypertension*

Older (X)	35	40	28	25	15+	40	32+	31	20	30+
Younger (Y)	33	44	30+	28	16	42	40	34+	19	35

To estimate G, we need to introduce the following quantities. For $i < j < k$, let

$$L_{ijk} = 1, \quad \text{if } T_i \text{ is decidedly less than } T_j, T_k,$$
$$= 0, \quad \text{otherwise}; \tag{6.46}$$

$$M_{ijk} = 1, \quad \text{if } S_i \text{ is decidedly less than } S_j, S_k,$$
$$= 0, \quad \text{otherwise}. \tag{6.47}$$

Now a natural estimate of G is

$$\hat{G} = \frac{\#(i,j,k) \text{ triples with } i < j < k \text{ such that } L_{ijk} = 1 \text{ and } M_{ijk} = 1}{\binom{n}{3}}.$$

The critical regions of approximate size α tests and the approximate P-values are given in Table 6.13.

We illustrate these details with an example.

Example 6.6. Ten sets of twins born to parents of whom at least one has hypertension were constantly monitored for the time of onset of hypertension. In the course of the investigation three left the study and two did not get hypertension at the end of the study. The (artificial) data appear in Table 6.14. We need to calculate $L_{ij}(M_{ij})$ and these values are given in the Table 6.15.

Summing the $\psi_{ij} = L_{ij}M_{ij}$ over all the 45 cells and dividing by 45, we get $T_c = (14)/(45) = 0.31$. Further,

$$\hat{A} = 4(6)/45 = 0.53.$$

Noting that $L_{123} = 0$, $L_{134} = 0$ and $L_{468} = 1$, and calculating L_{ijk} and M_{ijk}, we get

$$\hat{G} = 26/\binom{10}{3} = 26/120 = 0.22.$$

Table 6.15 *Values of $L_{ij}(M_{ij})$*

i	1	2	3	4	5	6	7	8	9	10
						j				
1	1(1)	$-1(0)$	$-1(-1)$	0(-1)	1(1)	0(1)	$-1(1)$	$-1(-1)$	0(1)
2		$-1(0)$	$-1(-1)$	0(-1)	0(-1)	0(-1)	$-1(0)$	$-1(-1)$	0(-1)
3			$-1(-1)$	0(-1)	1(0)	1(0)	1(0)	$-1(-1)$	1(0)
4				0(-1)	1(1)	1(1)	1(1)	$-1(-1)$	1(1)
5					0(1)	0(1)	0(1)	0(1)	0(1)
6						0(-1)	$-1(0)$	$-1(-1)$	0(-1)
7							$-1(0)$	$-1(-1)$	0(-1)
8								$-1(-1)$	0(0)
9									$-1(1)$

Thus

$$\hat{var}_0(T_c) = \frac{2(0.53)}{10(9)} + \frac{4(10-2)(0.22)}{10(9)} = 0.9.$$

and

$$Z = \frac{0.31}{\sqrt{0.9}} = 0.33.$$

This statistic is not significant and the null hypothesis of independence is not rejected. It may be noted that the two-sided P-value is 0.744.

A confidence interval for the slope under a linear regression model was proposed by Ireson and Rao (1985) using a test for association.

6.6.2 Proportional hazards (PH) model

Survival time distributions may depend on several prognostic factors. These factors play the role of explanatory variables in logistic regression and other regression models. It may be recalled that the variables could be continuous, binary or categorical with $k(> 2)$ categories. As in Subsection 6.5.4, we introduce the X-variables by defining dummy variables if needed. Let p be the number of X-variables we want to include in our model, and $\mathbf{X}$ be a p-dimensional vector of the X-variables corresponding to the prognostic factors.

The commonly used model in this connection is the *proportional hazards (PH) model* proposed by Cox (1972). It states that the hazard function for an individual with covariate vector $\mathbf{x}$ is

$$h(t|\mathbf{x}) = h_0(t) \cdot e^{\beta'\mathbf{x}}, \tag{6.48}$$

where $h_0(t)$ is a hazard function, which is called the baseline hazard function. The vector β is the set of regression coefficients representing the influence of the covariates. This model on the hazard function implies that the survival function of the individual with covariate vector $\mathbf{x}$ is

$$S(t|\mathbf{x}) = exp\left(-\int_0^t h(u|\mathbf{x})du\right).$$

We can simplify this as

$$S(t|\mathbf{x}) = \left(exp\left[-\int_0^t h_0(u)du\right]\right)^{\gamma(\mathbf{x})}.$$

Finally, we have

$$S(t|\mathbf{x}) = (S_0(t))^{\gamma(\mathbf{x})}. \qquad (6.49)$$

The survival function $S_0(t)$ is known as the *baseline survival function* and $\gamma(\mathbf{x}) = \mathbf{exp}(\beta'\mathbf{x})$.

Once we have a data set we want to fit the model to the data, which consists of estimating β without any specific parametric model for h_0. Cox proposed the use of partial likelihood for this estimation. We now give a few details of this method.

Let $t_{(1)} < t_{(2)} < \cdots, < t_{(k)}$ be the distinct ordered failure times. Further, let $R(t_{(i)})$ be the set of indices of those individuals with survival or censored times, that are at least $t_{(i)}$. This set is called the risk set. Also let d_i be the number of events at $t_{(i)}$ for $i = 1, 2, \ldots, k$ and let $\mathbf{x}_{(i)l}$ be the covariate vector of the lth individual with survival time $t_{(i)}$, $l = 1, \ldots, d_i; i = 1, 2, \ldots, k$. Then the conditional on the risk set $R(t_{(i)})$, the probability that d_i individuals have failed at time $t_{(i)}$ can be approximated by

$$L_i = \frac{exp(\beta' \sum_{l=1}^{d_i} \mathbf{x}_{(i)l})}{(\sum_{m \in R(t_{(i)})} exp(\beta'\mathbf{x}_m))^{d_i}}.$$

The overall conditional likelihood (partial likelihood) is

$$L_C = \Pi L_i = \Pi \frac{exp(\beta' \sum_{l=1}^{d_i} \mathbf{x}_{(i)l})}{(\sum_{m \in R(t_{(i)})} exp(\beta'\mathbf{x}_m))^{d_i}}. \qquad (6.50)$$

Estimates of β's are obtained from this partial likelihood, treating it as a regular likelihood, that is, by maximizing the partial likelihood function. We need a computer program to do these calculations.

Let us look into the interpretation of a single beta, β_i associated with the variable X_i, which corresponds to a dichotomous risk factor taking values 1 or 0 for the presence or absence of the factor. Then the relative risk due to the factor represented by X_i, when the other factors are fixed at specified levels, is

$$\frac{h(t|X_i = 1)}{h(t|X_i = 0)} = e^{\beta_i}.$$

The relative risk associated with the factor X_i is thus estimated as $e^{\hat{\beta}_i}$. We can get a confidence interval for this relative risk from the confidence limits

Table 6.16 *Dummy variables for three centers*

Center	X_2	X_3
1	0	0
2	1	0
3	0	1

Table 6.17 *Survival times in a multicenter study*

Time	X_1	X_2	X_3	Time	X_1	X_2	X_3
210+	1	0	0	50	1	1	0
200	1	0	0	170+	1	1	0
100+	1	0	0	100	0	1	0
180	1	0	0	120+	0	1	0
85+	0	0	0	150+	0	1	0
100	0	0	0	160	0	1	0
150	0	0	0	120+	1	0	1
170+	0	0	0	60+	1	0	1
160	0	0	0	110+	0	0	1
250+	1	1	0	130+	0	0	1
				150	0	0	1

B_{iL} and B_{iU} for β_i. The limits for β_i are

$$B_{iL} = \hat{\beta}_i + z_{\alpha/2} \cdot Std.error(\hat{\beta}_i),$$

and

$$B_{iU} = \hat{\beta}_i + z_{1-\alpha/2} \cdot Std.error(\hat{\beta}_i),$$

thus a confidence interval for the relative risk is $(e^{B_{iL}}, e^{B_{iU}})$.

We illustrate these details using an example and the computer program output. The relevant program is given in Appendix B6.

Example 6.7. Consider the survival data given in Table 6.17 using the dummy variables described in Table 6.16. We want to estimate the relative risk of treatment A versus treatment B after adjusting for the center effects. We introduce the following X-variables:

$$X_1 = 1, \quad \text{for treatment A,}$$
$$= 0, \quad \text{for treatment B.} \qquad (6.51)$$

Using the output, a summary about the estimates and other details are given in Table 6.18.

As the treatment differences are of primary interest, we want to test for the significance of the coefficient β_1. This is not significantly different from zero,

Table 6.18 *Maximum likelihood estimates*

Variable	Parameter Estimate	Standard Error	Pr> ChiSq	Hazard Ratio
trt	−0.41416	0.69507	0.5513	0.661
X_1	−0.12982	0.73663	0.8601	0.878
X_2	1.11399	0.75498	0.1401	3.046

since the P-value ($Pr > ChiSq$) is 0.5513. So the treatment effects are not significantly different. Here center effects are also not significant. It may be noted that the hazard ratio against treatment is the same as the relative risk of treatment A versus treatment B. This relative risk is $e^{\hat{\beta}_1} = e^{-0.41416} = 0.661$.

Now we calculate a 95% confidence limit for β_1, so that we can get a confidence interval for the hazard ratio. It follows that

$$B_{1L} = -0.4142 - 1.96(0.6951) = -1.7765,$$

and

$$B_{1U} = -0.4142 + 1.96(0.6951) = 0.9481.$$

Thus a 95% confidence interval for relative risk (hazard ratio) is

$$(e^{-1.7765}, e^{0.9481}) = (0.1692, 2.5809).$$

Since 1 is in this confidence interval, we do not reject the null hypothesis that the hazard functions under different treatments are equal.

6.7 Appendix A6: Mathematical supplement

A6.1 Confidence interval for the slope

Here we derive a confidence interval for the slope of a linear regression model. The starting point is the acceptance region of Theil's test for the two-sided alternatives H_{A3}, given in Table 6.8. However, here we need to adopt the test given in Table 6.8 for the case where the null hypothesis is $H_0 : \beta = \beta_0$ against the alternative $H_A : \beta \neq \beta_0$. The acceptance region of an α-level test is

$$c_3 < T_{reg} < c_4,$$

which is the same as

$$N \cdot c_3 < N \cdot T_{reg} < N \cdot c_4,$$

where $N = \binom{n}{2}$. This region can be interpreted as

$$N \cdot c_3 < \sum_{i<j} sgn(x_i - x_j)sgn(Y_i^* - Y_j^*) < N \cdot c_4, \tag{A1}$$

Noting that $sgn(Y_i^* - Y_j^*) = sgn(x_i - x_j)sgn(b_{ij} - \beta_0)$, we can simplify the region (A1) as

$$N \cdot c_3 < \sum_{i<j} sgn(b_{ij} - \beta_0) < N \cdot c_4. \qquad (A2)$$

Now we observe that

$$\sum_{i<j} sgn(b_{ij} - \beta_0) = \#(b_{ij} > \beta_0) - \#(b_{ij} < \beta_0)$$

$$= N - 2\#(b_{ij} < \beta_0).$$

Hence the region (A2) is the same as

$$\frac{N - Nc_4}{2} < \#(b_{ij} < \beta_0) < \frac{N + Nc_3}{2}. \qquad (A3)$$

The left-side inequality is equivalent to $\beta_0 > B_L$, where $L = \frac{N - Nc_4}{2}$ and B_L is the Lth ordered value of b_{ij}. The right-side inequality is true if and only if $\beta_0 < B_U$, where $U = \frac{N + Nc_3}{2} + 1$. Finally, if $w_{1-\alpha/2}$ is the $1 - \alpha/2$ quantile of Kendall's tau, we have

$$c_3 = -w_{1-\alpha/2}, \; c_4 = w_{1-\alpha/2}.$$

Thus we get the interval (6.36).

A6.2 Maximum likelihood equations for logistic regression

From the model, it follows that the likelihood of the data is

$$L = \Pi_{i=1}^n [\phi(x_i)^{y_i} [1 - \phi(x_i)]^{1-y_i}.$$

Hence the log likelihood is

$$l(\alpha, \beta) = \sum_{i=1}^n [y_i ln(\phi(x_i)) + (1 - y_i)ln(1 - \phi(x_i)].$$

Now taking the partial derivative with respect to α, we get

$$\frac{\partial l(\alpha, \beta)}{\partial \alpha} = \sum_{i=1}^n \left[\frac{y_i}{\phi(x_i)} - \frac{1 - y_i}{1 - \phi(x_i)} \right] \frac{\partial \phi(x_i)}{\partial \alpha}.$$

We see that

$$\frac{\partial \phi(x_i)}{\partial \alpha} = \phi(x_i) \cdot (1 - .\phi(x_i)).$$

Using this fact the partial derivative can be simplified and it turns out that

$$\frac{\partial l(\alpha, \beta)}{\partial \alpha} = \sum_{i=1}^{n} [y_i - \phi(x_i)].$$

A similar analysis will yield the expression given in (6.39) for the partial derivative $\frac{\partial l(\alpha, \beta)}{\partial \beta}$.

6.8 Appendix B6: Computer programs

B6.1 Test for independence

Here we compare the voting patterns of husband and wife. The response can belong to two categories. We want to test for the independence. We use the data from Example 6.1.

```
data vote;
input Wife Husband count@@;
lines;
1 1 60 1 2 20
2 1 50 2 2 70
;
proc print;
run;
proc freq;
title 'Analysis of Example 6.1 Data ';
weight count;
table Wife*Husband/norow nocol nopercent expected   chisq;
run;
proc freq data=vote;
title 'Kappa Statistic';
weight count;
table Wife*Husband/agree;
run;
```

Output
_____

The SAS System

Wife	Husband	count
1	1	60
1	2	20
2	1	50
2	2	70

```
              Analysis of Example 6.1 Data
                    The FREQ Procedure
                 Table of Wife by Husband


          Wife          Husband

          Frequency|
          Expected |        1|        2|  Total
          ---------+--------+--------+
               1 |     60 |     20 |     80
                 |     44 |     36 |
          ---------+--------+--------+
               2 |     50 |     70 |    120
                 |     66 |     54 |
          ---------+--------+--------+
          Total         110       90      200
```

```
        Statistics for Table of Wife by Husband

        Statistic           DF    Value    Prob
        -----------------------------------------

        Chi-Square           1  21.5488  <.0001
        Likelihood Ratio     1  22.2755  <.0001
        Chi-Square
        Continuity Adj.      1  20.2231  <.0001
        Chi-Squar
        Mantel-Haenszel      1  21.4411  <.0001
        Chi-Square
        Phi Coefficient           0.3282
        Contingency Coefficient   0.3119
        Cramer's V                0.3282
```

```
                    Kappa Statistic
                   The FREQ Procedure
        Statistics for Table of Wife by Husband

                Simple Kappa Coefficient
        ---------------------------------
        Kappa                        0.3137
        ASE                          0.0636
        95% Lower Conf Limit         0.1890
        95% Upper Conf Limit         0.4384
```

B6.2 Spearman's correlation and Kendall's tau

Here we calculate various correlation coefficients. We use the data from Example 6.2. First Monday data are indicated by y1 and second Monday data are indicated by y2.

```
data scores;
input y1 y2 @@;
lines;
3.10 2.50 -1.40 0.25 1.75 -0.35 -0.5 -0.45
2.85 2.10  1.60 -0.7 -1.30 0.2 1.60 0.3
-2.10 -1.20 0.70 0.5  1.60 0.8 -1.25 0.25
2.50 1.5 -0.75 -0.5 1.65 0.6
;
proc print;

proc corr  spearman kendall;
var y1 y2;
title1 'Output for Example 6.2';
Title2 'Spearman''s Rho, Kendall''s Tau';
run;
```

Output

The SAS System

y1	y2
3.10	2.50
-1.40	0.25
1.75	-0.35
-0.50	-0.45
2.85	2.10
1.60	-0.70
-1.30	0.20
1.60	0.30
-2.10	-1.20
0.70	0.50
1.60	0.80
-1.25	0.25
2.50	1.50
-0.75	-0.50
1.65	0.60

Output for Example 6.2
Spearman's Rho, Kendall's Tau
The CORR Procedure

2 Variables: y1   y2

Simple Statistics

Variable	N	Mean	Std Dev	Median
y1	15	0.67000	1.72459	1.60000
y2	15	0.38667	1.02390	0.25000

Simple Statistics

Variable	Minimum	Maximum
y1	-2.10000	3.10000
y2	-1.20000	2.50000

Spearman Correlation Coefficients, N = 15
Prob > |r| under H0: Rho=0

	y1	y2
y1	1.00000	0.68520
		0.0048
y2	0.68520	1.00000
	0.0048	

Kendall Tau b Correlation Coefficients, N = 15
Prob > |r| under H0: Rho=0

	y1	y2
y1	1.00000	0.55342
		0.0046
y2	0.55342	1.00000
	0.0046	

B6.3 Fitting logistic model for Example 6.4 data

We are fitting a logistic model when X is a binary variable. Enter $Y = 1$ values first and use the order option.

```
data heart;
input x y count;
lines;
```

```
0 1 4
1 1 3
1 0 25
0 0 26
;
proc print;

proc logistic order=data;
weight count;
model y=x;
title 'Logistic regression of MI against Gender';
run;
```

Output

The SAS System

Obs	x	y	count
1	0	1	4
2	1	1	3
3	1	0	25
4	0	0	26

Logistic regression of MI against Gender
The LOGISTIC Procedure
Model Information

Data Set	WORK.HEART
Response Variable	y
Number of Response Levels	2
Number of Observations	4
Weight Variable	count
Sum of Weights	58
Link Function	Logit
Optimization Technique	Fisher's scoring

Response Profile

Ordered Value	y	Total Frequency	Total Weight
1	1	2	7.00000
2	0	2	51.00000

Model Fit Statistics

Criterion	Intercept Only	Intercept and Covariates
AIC	44.722	46.628
SC	44.109	45.401
-2 Log L	42.722	42.628

Testing Global Null Hypothesis: BETA=0

Test	Chi-Square	DF	Pr > ChiSq
Likelihood Ratio	0.0940	1	0.7592
Score	0.0936	1	0.7596
Wald	0.0933	1	0.7601

Analysis of Maximum Likelihood Estimates

Parameter	DF	Estimate	Standard Error	Chi-Square	Pr > ChiSq
Intercept	1	-1.8718	0.5371	12.1459	0.0005
x	1	-0.2485	0.8135	0.0933	0.7601

Logistic regression of MI against Gender

The LOGISTIC Procedure
Odds Ratio Estimates

Effect	Point Estimate	95% Wald Confidence Limits	
x	0.780	0.158	3.842

B6.4 Fitting logistic model with several X-variables

We are fitting a logistic model with six X-variables. Here X_1 is a continuous variable, X_2 is a binary variable and the remaining four X-variables are dummy variables. We use the data from Example 6.5.

```
data heart;
input y x1 x2 x3 x4 x5 x6;
lines;
1 200 0 0 0 0 0
0 280 1 0 0 0 1
1 250 1 0 1 0 0
0 175 0 1 0 0 0
0 210 1 1 0 0 0
1 300 0 0 0 1 0
0 165 0 0 0 0 0
1 270 0 0 0 1 0
1 200 1 0 1 0 0
0 185 1 0 0 0 0
0 210 1 0 0 1 0
1 250 1 0 0 0 1
0 175 0 1 0 0 0
0 180 0 0 0 1 0
0 190 1 0 0 0 1
1 210 0 0 0 0 0
1 300 0 1 0 0 0
1 220 1 0 0 0 0
0 240 0 0 1 0 0
1 290 1 0 0 0 1
;
proc logistic descending;
model y= x1 x2 x3 x4 x5 x6;
title 'Logistic regression of MI ';
title2 'Output for Example 6.5';
run;
```

Output

Logistic regression of MI
Output for Example 6.5

The LOGISTIC Procedure

Model Information

Data Set	WORK.HEART
Response Variable	y
Number of Response Levels	2
Number of Observations	20
Link Function	Logit
Optimization Technique	Fisher's scoring

Response Profile

Ordered Value	y	Total Frequency
1	1	10
2	0	10

Model Fit Statistics

Criterion	Intercept Only	Intercept and Covariates
AIC	29.726	29.474
SC	30.722	36.444
-2 Log L	27.726	15.474

Testing Global Null Hypothesis: BETA=0

Test	Chi-Square	DF	Pr > ChiSq
Likelihood Ratio	12.2523	6	0.0566
Score	9.6689	6	0.1393
Wald	5.0711	6	0.5347

The LOGISTIC Procedure

Analysis of Maximum Likelihood Estimates

Parameter	DF	Estimate	Standard Error	Chi-Square	Pr > ChiSq
Intercept	1	-11.4515	5.7638	3.9473	0.0469
x1	1	0.0597	0.0277	4.6383	0.0313
x2	1	0.6530	1.6394	0.1586	0.6904
x3	1	-4.1580	3.1248	1.7706	0.1833
x4	1	-1.7854	2.0593	0.7517	0.3859
x5	1	-3.1432	2.3980	1.7181	0.1899
x6	1	-4.7011	3.1423	2.2382	0.1346

Odds Ratio Estimates

Effect	Point Estimate	95% Wald Confidence Limits	
x1	1.061	1.005	1.121
x2	1.921	0.077	47.755
x3	0.016	<0.001	7.146
x4	0.168	0.003	9.494
x5	0.043	<0.001	4.744
x6	0.009	<0.001	4.296

B6.5 PH regression model

Here we illustrate the use of the PHREG procedure. We use the data from Example 6.7.

```
data block;
 input time X1 X2 X3 censor@@;
 lines;
 210 1 0 0 1 200 1 0 0 0 100 1 0 0 1
 180 1 0 0 0 85 0 0 0 1 100 0 0 0 0
 150 0 0 0 1 170 0 0 0 1 160 0 0 0 0
 250 1 1 0 1 50 1 1 0 0 170 1 1 0 1
 100 0 1 0 0 120 0 1 0 1 150 0 1 0 1
 160 0 1 0 0 120 1 0 1 1 60 1 0 1 1
 110 0 0 1 1 130 0 0 1 1 150 0 0 1 0
 ;
 proc print;
 proc phreg data=block;
 model time*censor(0)=X1 X2 X3;
 output out=pred1 survival =s;
 run;
```

Output

The SAS System

Obs	time	X1	X2	X3	censor
1	210	1	0	0	1
2	200	1	0	0	0
3	100	1	0	0	1
4	180	1	0	0	0

5	85	0	0	0	1
6	100	0	0	0	0
7	150	0	0	0	1
8	170	0	0	0	1
9	160	0	0	0	0
10	250	1	1	0	1
11	50	1	1	0	0
12	170	1	1	0	1
13	100	0	1	0	0
14	120	0	1	0	1
15	150	0	1	0	1
16	160	0	1	0	0
17	120	1	0	1	1
18	60	1	0	1	1
19	110	0	0	1	1
20	130	0	0	1	1
21	150	0	0	1	0

The PHREG Procedure
Model Information

Data Set	WORK.BLOCK
Dependent Variable	time
Censoring Variable	censor
Censoring Value(s)	0
Ties Handling	BRESLOW

Summary of the Number of Event and Censored Values

Total	Event	Censored	Percent Censored
21	13	8	38.10

Model Fit Statistics

Criterion	Without Covariates	With Covariates
-2 LOG L	56.748	53.279
AIC	56.748	59.279
SBC	56.748	60.973

Testing Global Null Hypothesis: BETA=0

Test	Chi-Square	DF	Pr > ChiSq
Likelihood Ratio	3.4696	3	0.3247
Score	4.1784	3	0.2428
Wald	3.7316	3	0.2919

Analysis of Maximum Likelihood Estimates

Variable	DF	Parameter Estimate	Standard Error	Chi-Square	Pr > ChiSq	Hazard Ratio
X1	1	-0.4142	0.6951	0.3550	0.5513	0.661
X2	1	-0.1298	0.7366	0.0311	0.8601	0.878
X3	1	1.1140	0.7549	2.1772	0.1401	3.046

6.9 Problems

1. Verify (6.5) using (6.3).

2. Verify that (6.30) reduces to (6.28), when there are no ties among X-values and among Y-values.

3. In a random sample of 500 families, the level of education of the first child and the highest level of education of both parents are noted. The collected data appear in Table 6.19. The levels are coded 1 for high school, 2 for some college, and 3 for at least a college degree. Test for the independence of the education levels of a parent and that of the first child. If this hypothesis is rejected, compute a measure of association.

4. Two dentists examined 200 dental X-rays to detect a dental problem and the results of their diagnoses are given in Table 6.20. Find a 95% confidence interval for κ, a measure of agreement between the two dentists.

Table 6.19 *Education levels of a parent and first child*

First Child	Parent Education 1	2	3	Total
1	100	10	40	150
2	50	20	30	100
3	150	20	80	250
Total	300	50	150	500

Table 6.20 *Diagnosis by two dentists*

	Dentist 2		
Dentist 1	Present	Absent	Total
Present	28	7	35
Absent	2	163	165
Total	30	170	200

Table 6.21 *Survey results*

X	2	5	1.5	1.0	3	0.5	1.2	0.7
Y	100	200	40	20	80	5	10	2

Table 6.22 *Pain relief, age and gender*

Patient	1	2	3	4	5	6	7	8	9	10
Age (X_1)	30	50	45	60	40	48	51	55	59	62
Gender (X_2)	F	M	F	F	M	F	F	M	M	F
Y	1	0	1	0	0	1	1	1	0	1

5. Two judges ranked 15 candidates. The rank given by Judge 1(Judge 2) to the ith candidate is $R_i(S_i)$. These rank pairs (R_i, S_i) are
(12, 15), (3, 5), (1, 6), (9, 4), (8, 7), (2, 1), (10, 14), (4, 3)
(11, 8), (7, 13), (14, 12), (5, 2), (15, 10), (13, 11), (6, 9).
Is there evidence that the two rankings are positively correlated? Find the Spearman's rank correlation coefficient as well as Kendall's tau coefficient.

6. In a random sample of eight pharmaceutical companies, the annual revenue, X (in billions of dollars), and the annual research and development expenditure, Y (in millions of dollars), have been gathered. The data appear in Table 6.21. Assuming a linear model for the regression of Y on X, namely, $E(Y|X = x) = \alpha + \beta x$, find a point estimate of the slope β. Also find a 95% confidence interval for β, using Thiel's method.

7. In a random sample of ten rheumatoid arthritis patients, the occurrence or nonoccurrence of pain relief in 20 minutes after a particular dose of aspirin has been recorded along with patient age and gender. The data appear in Table 6.22. M stands for male and F stands for female. $Y = 1$ means that relief occurred and $Y = 0$ means that relief did not occur. Using a logistic regression model, estimate the probability that a 45-year-old female will get relief in 20 minutes. Also estimate the odds ratio for females versus males at a given age to get pain relief in 20 minutes. Obtain a 95% confidence interval for this odds ratio.

Table 6.23 *PHREG output*

Variable	Parameter Estimate	Std. Error	Pr> ChiSq	Hazard Ratio
Age	0.671	0.1525	.00001	1.9562
Gender	0.250	0.1464	0.8771	1.284

Table 6.24 *Time (in years) for major repair*

Home		1	2	3	4	5	6	7	8	9	10
Washer		5	7	6	7	8+	5+	7	5	4	5
Dishwasher	4 + 78	5	7	7	8	6	6	5+	4		

8. A survival data set, along with covariates age (=1 if age> 50; =0, otherwise) and gender (=1, if female; =0 if male), has been analyzed using the PHREG program, and a portion of the output appears in Table 6.23. Which of the two variables, age and gender, has significant effect on the survival distribution? Find a 95% confidence interval for the relative risk for age >50 versus age ≤ 50.

9. The odds ratio of X_1 versus X_2 is 1.4, and X_3 versus X_2 is 0.8. Find the odds ratio for X_1 versus X_3.

10. Model A washers and model B dishwashers were installed in ten new homes and the time the washer or dishwasher needed a major repair was noted. During the study period, some homes changed equipment before any repair and we will treat the times for those homes as censored observations. The data are given in Table 6.24. Are the times needed for major repair for the washers and dishwashers independent?

6.10 References

Best, D.J. (1973). Extended tables for Kendall's tau. *Biometrika*, **60**, 429–430.

Bhattacharya, P.K. and Mack, Y.P. (1987). Weak convergence of k-NN density and regression estimates with varying k and applications. *Ann. Statist.*, **15**, 976–994.

Cohen, J. (1960). A coefficient of agreement for nominal scales. *Edu. Psy. Meas.*, **20**, 37–46.

Cox, D.R. (1972). Regression models and life tables. *J. Roy. Statist. Soc.*, **34B**, 187–220.

Cox, D.R. and Snell, E.J. (1989). *Analysis of Binary Data*, 2nd edition, Chapman & Hall, New York.

Cramér, H. (1945). *Mathematical Methods of Statistics*, Princeton University Press, Princeton, New Jersey.

Devroye, L., Gyorfi, L., Krzyzak, A., and Lugosi, G. (1994). On the strong universal consistency of nearest neighbor regression function estimates. *Ann. Statist.*, **22**, 1371–1385.

Finney, D.J. (1971). *Probit Analysis*, 3rd edition, Cambridge University Press, London.

Fleiss, J.L. (1981). *Statistical Methods for Rates and Proportions*, 2nd edition, John Wiley & Sons, New York.

Fleiss, J.L., Cohen, J., and Everett, B.S. (1969). Large sample standard errors of kappa and weighted kappa. *Psychol. Bull.*, **72**, 323–327.

Friedman, J.H. and Stuetzle, W. (1981). Projection pursuit regression. *J. Amer. Statist. Assoc.*, **76**, 817–823.

Glasser, G.J. and Winter, R.F. (1961). Critical values of the coefficient of rank correlation for testing the hypothesis of independence. *Biometrika*, **48**, 444–448 (Appendix).

Hosmer, D.W. and Lemeshow, S. (2000). *Applied Logistic Regression*, 2nd edition, John Wiley & Sons, New York.

Hotelling, H. and Pabst, M.R. (1936). Rank correlation and tests of significance involving no assumption of normality. *Ann. Math. Statist.*, **7**, 29–43.

Ireson, M.J. and Rao, P.V. (1985). Interval estimation of slope with right censored data. *Biometrika*, **72**, 601–608.

Kendall, M. and Gibbons, J.D. (1990). *Rank Correlation Methods*, 5th edition, Edward Arnold, London.

Lehmann, E.L. (1998). *Nonparametric Statistical Methods Based on Ranks, Revised First Edition*, Prentice-Hall, Upper Saddle River, New Jersey.

Nelson, L.S. (1986). Critical values for sums of squared rank differences in Spearman's rank correlation test. *J. Qual. Tech.*, **18**, 194–196.

Oakes, D. (1982). A concordance test for independence in the presence of censoring. *Biometrics*, **38**, 451–455.

Shea, G.A. (1982). Hoeffding's lemma, in *Encyclopedia of Statistical Sciences*, Vol. 3, Edited by S. Kotz and N.L. Johnson, John Wiley & Sons, New York, 648–649.

Spearman, C. (1904). The proof and measurement of association between two things. *Amer. J. Psychol.*, **15**, 72–101.

Theil, H. (1950). A rank-invariant method of linear and polynomial regression analysis I, II, III. *Pro. Kon. Ned. Akod. Wet. A*, **53**, 386–392, 521–525, 1397–1412.

Yang, Y. (1998). Nearest neighbor median estimation of regression function and its derivatives. *Taiwan J. Math.*, **2**, 313–319.

CHAPTER 7

Computer-intensive methods

7.1 Introduction

In this chapter we give a brief description of methods that require extensive computing. Three classes of procedures that are of interest in the context of nonparametric statistics are (1) permutation tests (2) randomization tests and (3) bootstrap techniques. It should be noted that the tools used for the permutation tests and the randomization tests are the same, but the contexts in which they are used are diffferent. When the data are collected from a designed experiment, treatments usually are assigned randomly to the experimental units, and we use randomization tests, which are distribution-free procedures. If the data are random samples and under the null hypothesis, all the relevant permutations of the data are equally likely, and we use permutation tests. The procedures based on ranks, which have been discussed thus far, are a class of permutation tests.

The discussion is restricted to the analysis of continuous responses and is related to the procedures we have studied in the previous chapters. All these methods to be discussed can be called *resampling methods*. As this name implies, the methods use samples drawn from the data at hand. Since we are sampling the data (the sample) at hand, not the populations, the sampling process is aptly called the *resampling process*. We discuss permutation tests and randomization tests in Section 7.2. The final section deals with bootstrap methods.

7.2 Permutation tests and randomization tests

In some instances *permutation tests* are used in place of rank tests. These tests use the data as it is without considering ranks. The main features of a test are

(a) to form a test statistic that distinguishes the alternative hypothesis from the null hypothesis; sometimes we take a statistic that is used in the parametric analysis of the problem;

(b) permute the data in relation to the investigation and get the distribution of the test statistic using all permutations;

(c) determine the *P*-value, which is the probability of observing the test statistic or extreme values that support the alternative hypothesis.

When the basis of the permutations is the randomization used to assign the treatments to the experimental units, the tests are called *randomization tests*.

Instead of permuting the data, that is, reassigning the data to the units without replacement, if the assignment is with replacement, we get the bootstrap samples, which are used for inference. Methods that use bootstrap samples will be discussed in more detail in Section 7.3.

Lehmann (1998) in the epilogue to his book indicated that "the permutation tests are intermediate between rank tests and normal theory tests." Good (2000) presents a detailed account of the permutation tests and his book includes several applications of these tests and an extensive bibliography. Mielke and Berry (2001) used a distance function approach in developing permutation tests. Another approach to construct permutation tests uses the invariance principle and this method was described in Welch (1990). A book fully devoted to the discussion of randomization tests is Edgington (1995). A lucid account of resampling methods is available in Good (1999).

We describe the details of constructing these tests through some examples.

Example 7.1. (A permutation test) Let F and G be the distribution functions of the number of hours per day people watch TV in rural and urban areas. We assume the shift model, namely,

$$G(x) = F(x - \Delta),$$

and want to test the null hypothesis $H_0 : \Delta = 0$ against the alternative $H_A : \Delta \neq 0$. Random samples of sizes $n_1 = 3$ and $n_2 = 4$ were taken from these areas and the following data were collected.

Rural(X): 3, 4, 5
Urban(Y): 1, 1.5, 2, 2.5

A natural test statistic is $T = |\bar{X} - \bar{Y}|$ and large values give evidence against the null hypothesis.

To calculate the P-value we proceed as follows. When X and Y have the same distribution under H_0, any three observations of the combined sample can be considered as an X-sample. Thus there are $\binom{7}{3} = 35$ possible ways of partitioning the combined sample into two groups of sizes three and four. Each way of partitioning will be attached a probability of $(1/35)$. For each of the partitions the value of T is noted and the distribution of T is developed. It is given in Table 7.1.

The observed value of T is 2.125. So the P-value is the $P(T \geq 2.125)$, which according to the permutation distribution is $(1/35)+(1/35)=0.029$. Thus we reject H_0 at $\alpha = 0.05$.

The underlying principle in this procedure is that all partitions are equally likely under H_0. There is no randomization involved in the study. Hence we call this a permutation test. In this case, the two-sample t-test with pooled variance gives a P-value of 0.015 and this leads to rejection of H_0 for $\alpha = 0.05$. The exact P-value for the Wilcoxon signed rank test is found to be 0.0571, which does not lead to the rejection of H_0 for $\alpha = 0.05$.

Table 7.1 *Permutation distribution of* $T = |\bar{X} - \bar{Y}|$

t	Prob.	t	Prob.
0.083	3/35	1.250	2/35
0.208	4/35	1.375	2/35
0.375	4/35	1.542	2/35
0.500	3/35	1.667	1/35
0.667	2/35	1.833	1/35
0.792	3/35	1.958	1/35
0.958	3/35	2.125	1/35
1.083	2/35	2.250	1/35

Table 7.2 *Time (in min.) for pain relief*

Pain	Pairs				
Reliever	1	2	3	4	5
A	30	25	23	30	28
B	24	22	20	31	25
Difference $D = A - B$	6	3	3	-1	3

Example 7.2. (A randomization test) A scientist interested in comparing two pain relievers A and B with respect to the mean time of relief recruited 10 volunteers. These volunteers were grouped into five similar pairs based on the demographics. Pain relievers A and B are randomly assigned to the individuals in each pair. The assignments are done independently for various pairs. The durations (time) of pain relief were recorded and are given in Table 7.2.

We assume that the two cdf's of time of relief F and G are related through a shift model. In other words

$$G(x) = F(x - \Delta),$$

where $F(G)$ is the cdf of time of relief under $A(B)$. It is of interest to test the null hypothesis $H_0 : \Delta = 0$ against the alternative $H_A : \Delta \neq 0$. If $\bar{d}$ is the mean of the differences, a large value of $|\bar{d}|$ will lead to the rejection of the null hypothesis. Equivalently large values of $T = |\sum_i d_i|$ will lead to the rejection of H_0, where D_i is the difference in the times for pain relief for medications A and B in the ith pair.

In pair 1, the present randomization produced the responses 30 and 24 for medications A and B. Under the null hypothesis that A and B have the same effect, we could have observed 24 and 30 as our responses under medications A and B resulting in a difference of -6. Similar comments apply to the other pairs. This argument results in $2^5 = 32$ different permutations of the d values by attaching a "+" or "−" sign to 6, 3, 3, 1, 3. The distribution of T based on these 32 permutations is given in Table 7.3.

Table 7.3 *Randomization distribution of* $T = |\sum_i d_i|$

t	2	4	8	10	14	16
$f(t)$	8/32	8/32	6/32	6/32	2/32	2/32

The P-value of the data is the probability that $T \geq 14$ under the above distribution and it is $2/32 + 2/32 = 0.125$. Thus we do not reject the null hypothesis at the 0.05 level.

It may be noted that Wilcoxon signed rank test gives a P-value of 0.106 for these data.

7.3 Bootstrap methods

Let us consider the problem of finding a 95% confidence interval for the median. We have a data set consisting of five observations. Due to small sample size, one would consider the range of the sample as a confidence interval. What can we say about the coverage probability of this interval? If the sample were a random sample from a continuous distribution, the coverage probability of the range based on a sample of size n is $1 - \frac{1}{2^{n-1}}$, which in the present case is $(15/16)$ and this quantity is less than 0.95. The minimum sample size for the range to be a $100(1 - \alpha)$ confidence interval for the median is $1 - \frac{ln(\alpha)}{ln(2)}$.

One method for getting the confidence interval of interest is to use a bootstrap sampling scheme. This involves taking samples of size 5 from the data set, where the sampling is with replacement. Each of these samples is called a *bootstrap sample*. Using the medians of the bootstrap samples a distribution is obtained, which is called a *bootstrap distribution of the median*. The 2.5% and the 97.5% percentiles of the bootstrap distribution are taken as the end points of an interval, which is a confidence interval.

This is the idea behind the bootstrap method. Essentially, we are taking samples from the population defined by the sample on hand. The bootstrap distribution is viewed as an estimate of the (sampling) distribution of the statistic of interest. To some extent, how many bootstrap samples to take depends on the objective of the study.

Mathematical framework of a bootstrap procedure

We now introduce some notation and describe the various steps in conducting a bootstrap procedure. Let $(X_1, X_2, \ldots, X_n)$ be a random sample from an unknown probability distribution function $F(x)$. We are interested in the parameter θ, which characterizes the distribution function F. For example, θ could be the median of F. On observing $X_1 = x_1, X_2 = x_2, \ldots, X_n = x_n$, we calculate the empirical (sample) distribution function

$$F_n(x) = \frac{\#x_i \leq x}{n}.$$

This function is an estimate of F. Since we cannot sample from F, we sample from $\hat{F}(.) = F_n(.)$. The sample estimate of θ is denoted by $\hat{\theta}$. This $\hat{F}$

distribution attaches a probability of $(1/n)$ to each of the sample values, x_i. A bootstrap sample $y^* = (x_1^*, x_2^*, \ldots, x_n^*)$ is a random sample of size n drawn *with replacement* from the observed sample $(x_1, x_2, \ldots, x_n)$.

Let $y^*(1), y^*(2), \ldots, y^*(B)$ be B bootstrap samples, where B, the number of bootstrap samples, is usually 25, 50, 100, or 200. For each bootstrap sample $y^*(b), (b = 1, 2, \ldots, B)$ we calculate the statistic of interest, $\hat{\theta}^*(b) = \hat{\theta}(y^*(b))$. The distribution defined by the collection of $\hat{\theta}^*(b)$ is called the *bootstrap distribution* of $\hat{\theta}$. We estimate the variance of $\hat{\theta}$ and/or find a confidence interval for θ.

Let $\hat{\theta}^*(.)$ be the average of the above collection. In other words,

$$\hat{\theta}^*(.) = \sum_{b=1}^{B} \hat{\theta}^*(b)/B. \tag{7.1}$$

The standard deviation of the bootstrap distribution is

$$\hat{SE}_B = \left(\frac{\sum_{b=1}^{B} (\hat{\theta}^*(b) - \hat{\theta}^*(.))^2}{B-1} \right)^{1/2}. \tag{7.2}$$

Standard error, $SE_F(\hat{\theta})$, of $\hat{\theta}$

The standard error of $\hat{\theta}$ does depend on F. We can estimate this quantity from the bootstrap distribution. This estimate is

$$\hat{SE}_F(\hat{\theta}) = \hat{SE}_B. \tag{7.3}$$

Bootstrap confidence interval for θ

Using the estimate of the standard error of $\hat{\theta}$, and approximating the sampling distribution of $\hat{\theta}$ by the normal with mean θ and standard deviation equal to the estimate of the standard error, we get a confidence interval. In other words, an approximate $100(1-\alpha)\%$ confidence interval for θ is

$$(\hat{\theta}^*(.) + z_{(\alpha/2)}\hat{SE}_B, \ \hat{\theta}^*(.) + z_{(1-\alpha/2)}\hat{SE}_B). \tag{7.4}$$

Alternative method-percentile method

Here we assume that B, the number of bootstrap samples, is large, 200 to 1000. Since the bootstrap distribution of $\hat{\theta}$ is an estimate of the sampling distribution, the $100(\alpha/2)$ and $100(1 - \alpha/2)$ percentiles of the bootstrap distribution are taken as the endpoints of the confidence interval.

Since the sampling distribution of $\hat{\theta}$ is continuous, whereas the bootstrap sampling distribution of $\hat{\theta}$ is a step function, we recommend some smoothing to determine the percentile points. Let $\hat{\theta}_{(1)}, \hat{\theta}_{(2)}, \ldots, \hat{\theta}_{(m)}$, be the m distinct ordered values of $\hat{\theta}$ among the bootstrap sample values. Also let $\hat{\theta}_{(i)}$ occur in f_i boot strap samples. Then the cumulative distribution function of the bootstrap distribution is

$$F(\hat{\theta}_{(i)}) = \sum_{j=1}^{i} f_j/B. \tag{7.5}$$

Table 7.4 *Bootstrap distribution of the sample median*

$\hat{\theta}^*(b)$	49	68	69	75	82
Frequency	8	43	73	61	15

A smoothed version of this distribution function is obtained by joining the points $(\hat{\theta}_{(i)}, F(\hat{\theta}_{(i)}))$ and $(\hat{\theta}_{(i+1)}, F(\hat{\theta}_{(i+1)}))$ using a straight line for $i = 1, 2, \ldots,$ $m - 1$ and extending the line joining the points $(\hat{\theta}_{(1)}, F(\hat{\theta}_{(1)}))$ and $(\hat{\theta}_{(2)}, F(\hat{\theta}_{(2)}))$ to the left of $\hat{\theta}_{(1)}$. We recommend this smoothed version of the distribution function for calculating the percentile points.

It may be noted that in this method we are using linear interpolation and some extrapolation.

Now we illustrate the calculation of a confidence interval for the median.

Example 7.3. Suppose the random sample we have is

$$68, 75, 49, 82, 69.$$

The sample median is 69. Since we are interested in constructing a confidence interval for the population median, we generate 200 bootstrap samples and find the median for each sample. The distribution of the bootstrap values is given in Table 7.4.

From Table 7.4, we have

$$F(49) = 0.04, \; F(68) = 0.255.$$

So the required 0.025 quantile is clearly less than 49. Hence we determine this quantile θ_L from the following equation:

$$\frac{49 - \theta_L}{68 - 49} = \frac{0.04 - 0.025}{0.255 - 0.04}. \tag{7.6}$$

Solving this equation we get $\theta_L = 47.7$. Further,

$$F(75) = 0.925, \; F(82) = 1.$$

Thus the required 0.975 quantile θ_U is obtained from the following equation:

$$\frac{82 - \theta_U}{82 - 75} = \frac{1.0 - 0.975}{1.0 - 0.925}. \tag{7.7}$$

Solving this equation we get $\theta_U = 79.7$. A 95% confidence interval for the median is $(47.7, 79.7)$.

Alternatively, we estimate the standard error using equation (7.2). Using Table 7.4, we get

$$\hat{SE}_B = 6.07.$$

Using this value from equation (7.4), we get the interval $(47.1, 80.9)$ as a 95% confidence interval for the median.

Carpenter and Bithell (2000) give a comprehensive summary of the various methods for constructing bootstrap confidence intervals.

Two books of interest are Efron and Tibshirani (1993) and Davison and Hinkley (1996). Some other references are DeAngelis and Young (1998), Efron (1982), Efron and Tibshiran (1986), LePage and Billard (1992), Wu (1986), and Young (1994). It may be noted that many of the computer programs for implementing the bootstrap methods are written in S^+.

7.4 References

Carpenter, J. and Bithell, J. (2000). Bootstrap confidence intervals: when, which, what? A practical guide for medical statisticians. *Statist. Med.*, **19**, 1141–1164.

Davison, A.C. and Hinkley, D.V. (1996). *Bootstrap Methods and Their Applications*, Cambridge University Press, New York.

DeAngelis, D. and Young, G.A. (1998). Bootstrap method, in *Encyclopedia of Biostatistics, Vol. 1*, Edited by P. Armitage and T. Colton, John Wiley & Sons, Chichester, U.K., 426–433.

Edgington, E.S. (1995). *Randomization Tests*, 3rd edition, Marcel Dekker, New York.

Efron, B. (1982). *The Jackknife, the Bootstrap and Other Resampling Plans*, SIAM, Philadelphia.

Efron, B. and Tibshirani, R. (1986). Bootstrap methods for standard errors, confidence intervals, and other measures of statistical accuracy (with discussion). *Statist. Sci.*, **1**, 54–96.

Efron, B. and Tibshirani, R. (1993). *An Introduction to the Bootstrap*, Chapman and Hall/CRC, Boca Raton, Florida.

Good, P. (1999). *Resampling Methods: A Practical Guide to Data Analysis*, Birkhäuser, Boston.

Good, P. (2000). *Permutation Tests: A Practical Guide to Resampling Methods for Testing Hypotheses*, Springer-Verlag, New York.

Lehmann, E.L. (1998). *Nonparametrics: Statistical Methods Based on Ranks, Revised First Edition*, Prentice-Hall, Upper Saddle River, New Jersey.

LePage, R. and Billard, L. (Eds.) (1992). *Exploring the Limits of Bootstrap*, John Wiley, New York.

Mielke, P.W., Jr., and Berry, K.J. (2001). *Permutation Methods: A Distance Function Approach*, Springer-Verlag, New York.

Welch, W.J. (1990). Construction of permutation tests. *J. Amer. Statist. Assoc.*, **85**, 693–698.

Wu, C.F.J. (1986). Jackknife, bootstrap, and other resampling methods in regression analysis (with discussion). *Ann. Statist.*, **14**, 1261–1350.

Young, G.A. (1994). Bootstrap: more than a stab in the dark (with discussion). *Statist. Sci.*, **9**, 382–415.

Answers to selected problems

Chapter 1
3. 90% confidence limits (see 1.49) are (0.3516, 0.8058).
8. $X_{(1)}$, which is 2398.
12. $\hat{S}(250) = \hat{S}(246) = 0.1667$, Std. error $= 0.0982$.

Chapter 2
2. $\chi^2 = 21.9263$, P-value $< .0001$.
8. Logrank $\chi^2 = 0.0277$, P-value $= 0.8677$.
   Wilcoxon $\chi^2 = 0.0663$, P-value $= 0.7967$.
10. Logrank $\chi^2 = 3.1227$, P-value $= 0.0772$.

Chapter 3
1. McNemar statistic $\chi^2 = 12.5522$, P-value $= 0.0004$.
5. Sign test statistic $S = 4$, P-value $= 0.6875$.
   Wilcoxon $V_+ = 14$, P-value $= 0.2500$.

Chapter 4
1. $\chi^2 = 33.0068$, P-value $< .0001$.
7. $\chi^2 = 12.3566$, P-value $= .0894$.
8. KW test, $\chi^2 = 10.8255$, P-value $= .0045$.
12. Logrank $\chi^2 = 2.3988$, P-value $= 0.3014$.
    Wilcoxon $\chi^2 = 1.5591$, P-value $= 0.4586$.

Chapter 5
7. See alternate method of Subsection 5.5.3
   F $= 3.54$, $Q_C^* = 12.31$, P-value $= 0.156$.
9. Take sample C as sample 1, sample B as sample 2 and sample A as
   sample 3. Page statistic $= 66$, mean 60, std. $= 3.1623$, $Z_L = 1.8974$,
   P-value $= 0.0289$.

Chapter 6

3. $\chi^2 = 16.3333$, P-value = 0.0026.

5. Spearman corr. = 0.6821, P-value = 0.0051.
   Kendall's tau = 0.4667, P-value = 0.0153.

7.

Parameter	Estimate	Std. Error	P-Value
X_1	2.3040	1.6453	0.1614
X_2	−0.0102	0.0880	0.9082

Author Index

Subject Index

A
association, 311, 313, 314, 318–320, 322, 329, 332, 347
asymptotic relative efficiency, 122, 194

B
BIB design, 287, 288
binary data, 1, 15, 18, 58, 154, 155, 222
binomial distribution, 2, 3, 7, 10, 18, 35, 41, 78, 121, 128, 137, 181, 187, 307
block designs, 267, 278, 281, 289, 297
bootstrap distribution, 354, 355
bootstrap methods, 351, 354, 357
bootstrap sample, 354, 355
bootstrap techniques, 351

C
C-matrix, 281
categorical data, 92, 95, 160, 222, 224, 245, 260
cdf, 2, 3, 6, 7, 11, 14, 17, 23, 27, 28, 30, 31, 35, 37, 63, 78, 79, 84, 90, 96, 109–111, 134, 139, 167, 181, 187, 202, 276, 292, 316, 329, 353
censoring, 30, 31, 115, 117, 119, 196, 198, 244, 292, 293, 329, 330
Chakraborti and Desu test, 240
clinical equivalence, 63, 69, 129, 155, 178, 181–183, 185, 200
Cochran's test, 272
completely ordered alternatives, 232
concordant pairs, 319
confidence band, 17, 27, 230
confidence bounds, 11
confidence interval, 1, 3, 4, 10, 11, 13–15, 17–22, 33, 34, 38, 46, 49, 58–60, 64, 70–73, 77, 80, 82, 84, 85, 96, 97, 105, 106, 130, 139, 142, 162, 170, 182, 183, 185, 186, 188, 193, 204–206, 208, 214, 315, 323, 324, 327, 332–335, 347–349, 354–356
confidence interval for percentiles, 21, 37, 38
confidence interval for shift parameter, 84, 85, 96, 105, 162
control median test, 76, 78, 80–82, 85, 240
correlation, 234, 311, 313, 314, 316–318, 321, 323, 339
Cramér's V, 313, 314
cross-over design, 193

D
delta method, 39, 40, 59, 71, 152, 201
discordant pairs, 319
Downton's procedure, 270, 275, 276
dummy variables, 327, 328, 332, 334, 342

E
efficacy, 14, 65, 123, 125–127, 260
estimation of cdf, 17
exponential distribution, 26, 98, 111–113, 136, 151, 228
exponential ordered scores, 98, 102, 111, 113, 114, 150, 275
extreme value distribution, 110, 113

F
Fisher's exact test, 66, 73, 74, 76, 132, 225
Fligner-Wolfe test, 239